Encyclopedia of Atmospheric and Climate Sciences

Volume I

Encyclopedia of Atmospheric and Climate Sciences

Volume I

Edited by **Mary D'souza**

R CALLISTO
REFERENCE

New York

Published by Callisto Reference,
106 Park Avenue, Suite 200,
New York, NY 10016, USA
www.callistoreference.com

Encyclopedia of Atmospheric and Climate Sciences: Volume I
Edited by Mary D'souza

International Standard Book Number: 978-1-63239-210-7 (Hardback)

Printed in the United States of America.

Contents

Preface

In contemporary times, the issue of climatic changes, environment conservation and sustainable development have become hot topics. With the urgency to preserve the environment, there has been a rise in analysis and studies on atmospheric and climate sciences. This book would serve as a good starting point to acquaint readers about the various concepts revolving under this subject.

Atmospheric and Climate Sciences is a broad term, which encompasses weather and climate phenomena in all their dimensions. This book provides detailed information about the atmosphere, its processes and effects and how atmospheric changes demonstrate average climates and their change over time. Beginning from the molecular to the global changes, this book traces the evolution in climate which develops over millions of years.

Climate Sciences involves analysing the periodicity of weather events, determining trends and changes in long-term average weather patterns. Since climate is affected by both natural and human-based factors, this book looks at both these factors in detail.

This subject, though essentially scientific and rational, relies on strong theoretical concepts. The various researches in this book are a perfect example of both theory and practice. Also, in order to assist potential researchers in this field, detailed statistical models and research methodologies have been provided, which will also help readers in content analysis.

I would like to thank our researchers and writers from all the parts of the world for sharing their valuable research, which will enlighten upcoming researchers and academicians in this field.

Editor

Chemical Composition and Sources of Rainwater Collected at a Semi-Rural Site in Ya'an, Southwestern China

Min Zhao[1,2], Li Li[1*], Zhilin Liu[1], Bin Chen[1], Jianqiu Huang[1], Jinwang Cai[1], Shihuai Deng[1*]

[1]Provincial Key Laboratory of Agricultural Environmental Engineering, College of Resources and Environment, Sichuan Agricultural University, Chengdu, China

[2]Meishan Environmental Monitoring Center, Meishan, China

ABSTRACT

Rain and snow water samples were collected from Sep. 2010 to Jun. 2011 at a semi-rural site in Ya'an, a city located in the rain-belt along the Tibetan Plateau, to characterize the chemical composition and the sources of precipitation. The collected samples were severely acidified with an annual volume-weighted mean (VWM) pH of 4.03 and an annual acid rain frequency of 79%. SO_4^{2-} and NH_4^+ were the most abundant ions, followed by Ca^{2+}, H^+, NO_3^-, Cl^-, K^+, Na^+, F^- and Mg^{2+}. The acidity of samples was predominantly generated by H_2SO_4 and HNO_3, which were neutralized by NH_4^+ and Ca^{2+} as much as 65%. NH_3 played a major role in neutralizing the acid rain. The average ambient concentration of NH_3 was 174.2 μg/m^3 during sampling periods. Different source apportionment methods, including principle component analysis (PCA), enrichment factor (EF), correlation and back-trajectory analysis were used to track the sources of rainwater. The methods suggested that the pollutants in rainwater were from both local and long-distance transport (1:2.2), or they were from anthropogenic actions (86.4%), sea salts (8.1%) and crustal (5.5%) respectively.

Keywords: Acid Rain; Chemical Composition; Ammonia; Source Apportionment; Ya'an

1. Introduction

Acid rain has received worldwide attention during the past decades for its notably negative effects on aquatic and terrestrial ecosystems. Specifically, acid rain acidifies surface waters and soils, leads to widespread loss of fish population, forest decay and crop yield decline, and accelerates rust process of wild architectures [1,2]. Acid rain can also bring direct and indirect harm to human health. The main risk components are acidity and heavy metal elements [3].

In the past decades, many studies were conducted on acid rains in south China [4-6]. South China has been regarded as the third largest acid region in the world following Northeast America and Central Europe [4,7]. In these previous studies, major attentions were paid to urban areas because more anthropogenic pollutants, *i.e.* SO_2 and NO_x, are emitted in the industrialized sites. Although the negative effects of acid rain on rural areas, such as on forests, farmlands and water bodies, are as important as on urban areas, and acid rains in rural areas

are usually obvious and caused by a significant influence of long range transport air pollutants [8], the knowledge of acid rains in extensive non-urban areas is still limited.

Ya'an is a famous rainy city located on the rain-belt along the Tibetan Plateau in the southwestern China. In Ya'an, about half of the days in a whole year are rainy. Ya'an is dominated by light rains and the annual average precipitation amount is around 1800 mm. Two reasons make us study the chemical composition and the sources of rainwater at a semi-rural site in Ya'an. By the first reason, Ya'an has abundant rains and it is close to the Tibetan Plateau (see in **Figure 1(A)**). Due to these unique topographic and climatic conditions, a large amount of pollutants in the long-distance transport air masses in Ya'an are obstructed by high mountains, washed out by rains and deposited in this area. These pollutants, represented by a majority of regional pollutions, can be used to track the origins of air pollutants in the research area. By the second reason, we are in shortage of a study associated with acid rains at rural-urban transit site on the rain-belt along the Tibetan Plateau. As far as we know, the transit regions are usually heavily influenced

---

*Corresponding author.

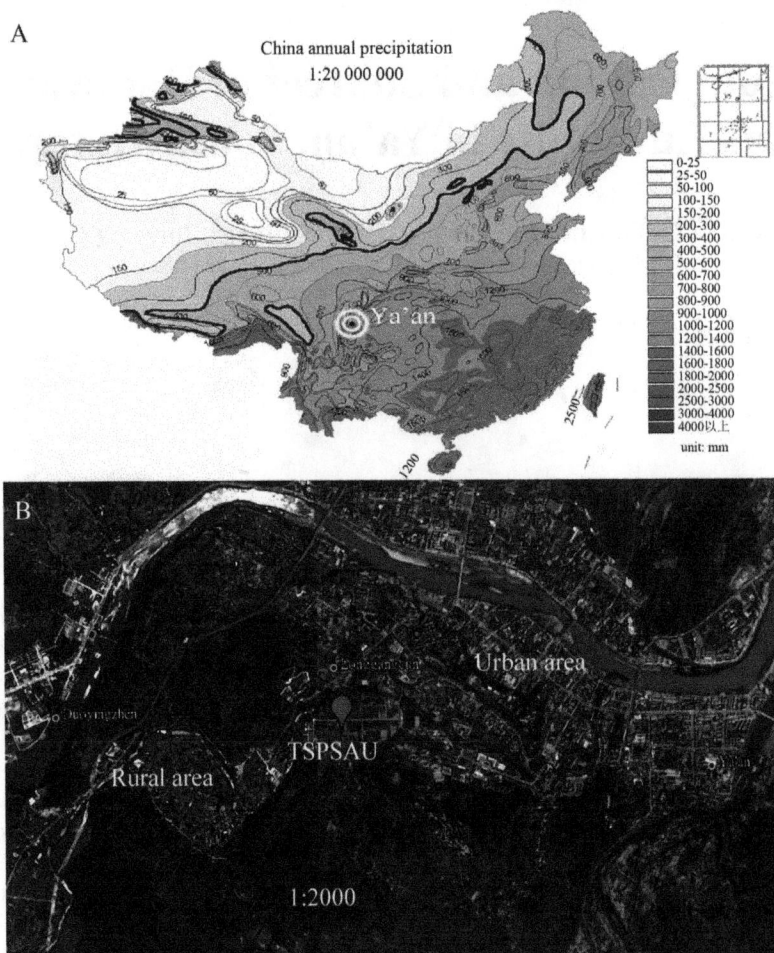

Figure 1. Locations of Ya'an and the sampling site (TSPSAU).

by anthropogenic actions, and in the regions environmental pollution has obvious impacts on extensive ecological lands.

Rain is an effective way to remove particulates and dissolved gases in the atmosphere [9]. The chemical compositions and pH values of rainwater are affected by the scavenging of atmospheric pollutants. These pollutants can be of many origins, for example, SO_2 and NO_x emitted from fuel burning and vehicles transforming into SO_4^{2-} and NO_3^- through photochemical reactions and being washed out by rains; NH_3 coming from agricultural sources, such as livestock breeding, fertilizer, soil emission and biomass burning [10], and undergoing gas-to-particle conversion processes to give rise to NH_4^+. Ca^{2+} mainly originates from daytime convection and vehicle/wind-driven roadside dust [11]. Other ions (e.g. Cl^-, Na^+, Mg^{2+} and K^+) are primarily from natural sources such as soils, forest fires, and sea salts [12]. The acidity of rainwater is a result of the balance between acidic ions and alkaline ones upon their neutralization reactions [13]. To better understand the formation mechanism of the severe acidity of rainwater, it's necessary to gain a deep

insight into the chemical compositions of precipitation.

The objectives of the present study are: 1) to analyze chemical compositions and characterize seasonal variation of precipitation at a semi-rural site in Ya'an; 2) to discuss the formation mechanism of the acidity of the rainwater; and 3) to investigate the possible pollution sources as well as their relative contributions to the rains at the research site.

2. Materials and Methods

2.1. Site Description

The sampling site is situated on the roof of a two-story building (29°58'58" N, 102°58'44" E) about 7 m tall in Teaching and Scientific Park of Sichuan Agricultural University (TSPSAU). The total area of TSPSAU is 33.3 ha. It comprises a dozen of laboratory buildings, a livestock farm (a small number of animals are experimentally fed in this farm), a small wastewater treatment plant and a large area of experimental fields. The sampling site is kept at least 100 m from possible emission source, and no high obstruction stands nearby.

Figure 1 presents the geographic locations of TSPSAU and Ya'an. As we can see from **Figure 1**, the campus of Sichuan Agricultural University, which belongs to urban area, is located in the east of TSPSAU. At the same time, the extensive farmland, a typical rural area, is located in the west of TSPSAU. The collection site is a good representative of transit place between urban and rural area.

2.2. Sampling Method

Rainwater samples of four different seasons were collected in 17-28 Sep. 2010 (autumn), 31 Dec. 2010-14 Jan. 2011 (winter), 1-18 Apr. 2011(spring) and 3-16 Jun. 2011 (summer) respectively. All the samples were manually collected using polystyrene funnels on the event basis 1.5 m above the roof. The funnels were also used as rain gauges to record the precipitation amount. Samples collected from 20:00 on the first day to 20:00 of the next day were mixed as one sample. Before collections, the funnels were cleaned with 6 M hydrochloric acid solution and rinsed several times by Milli-Q water (resistivity: 18.2 MΩ/cm). In order to prevent contaminations from dry deposition, the funnels were opened and covered as quickly as possible at the beginning and the end of each rain event. Meteorological parameters (*i.e.* temperature, relative humidity and wind speed) were measured by a pocket weather meter (4500, Kestrel, USA) three times at 08:00, 14:00 and 20:00 every day. In total, 25 rainwater and 3 snow water samples were collected.

2.3. Chemical Analysis

At the end of each rainfall episode, rainwater was taken back to laboratory immediately. The pH values and conductivities for the rainwater samples were analyzed with a pH meter (pHS-3C+, Fangzhou, China) and a high-precision conductivity meter (DDS-12W, Lida, China) respectively. Subsequently, the samples were filtered through 0.45 μm hydrophilic microporous membrane equipped with a clean syringe. All water filtrates were preserved at 4°C in a refrigerator until for analysis.

Anions (*i.e.*, F^-, Cl^-, NO_3^- and SO_4^{2-}) were measured by ion chromatography (ICS-90, Dionex, USA) equipped with a conductivity detector. Anions were separated on AS11-HC column using 25 mmol/L KOH (EGC II, Dionex) as eluent kept at a flow rate of 1.0 mL/min. Moreover, K^+ and Na^+ were measured by a flame photometer (6400 A, Shanghai Analy. Instru. Co., China), Ca^{2+} and Mg^{2+} were measured by a flame atomic absorption spectrometry (MKⅡM6, Thermo, USA), and NH_4^+ was measured by a flow injection analytical instrument (FIAstar 5000, Foss, Sweden).

Accompanied with the acid rain monitoring, NH_3 in the real-time atmosphere was also collected by a multi-functional sampler (6120, Laoshan elec. Co., China) using 0.005 M sulfuric acid as absorbent solution. The concentrations of NH_3 in the absorbent solutions were analyzed colorimetrically by a visible-light-spectrophotometer. Blank samples were prepared as the same as field samples only no flow rate. More detailed description of the measurement procedure please referred to Li *et al.* 2013 [14].

2.4. QA/QC

The sampling and analytical procedures for rainwater were performed according to the technical specifications established by Chinese Environmental Protection Administration (HJ/T165-2004 [15]). The detection limits (S/N = 3) were 0.056, 0.042, 0.015, 0.046, 0.033, 0.036, 0.056, 0.061 and 0.028 mg/L for F^-, Cl^-, NO_3^-, SO_4^{2-}, K^+, Na^+, Mg^{2+}, Ca^{2+} and NH_4^+, respectively. The relative standard deviations (RSD) for the ionic analysis were better than 5%. Field blanks were taken by pouring ultra-pure water into funnels, and the background values were subtracted from the concentrations of field samples.

Moreover, the data quality for rainwater samples was checked by ionic balance. Data acquired in this study (more than 90%) is within the acceptable range of ion difference given by Rastogi *et al.* 2005 [16], 15% - 30% between cations and anions for rainwater samples with ionic sum higher than 100 μeq/L. The linear regression between Σanions and Σcations is fairly good (see in **Figure 2(A)**, $R^2 = 0.91$).

3. Results and Discussion

3.1. pH Value and Conductivity

Table 1 lists the chemical compositions of the rainwater collected during sampling periods. The frequency distribution of the pH values of the rainwater is presented

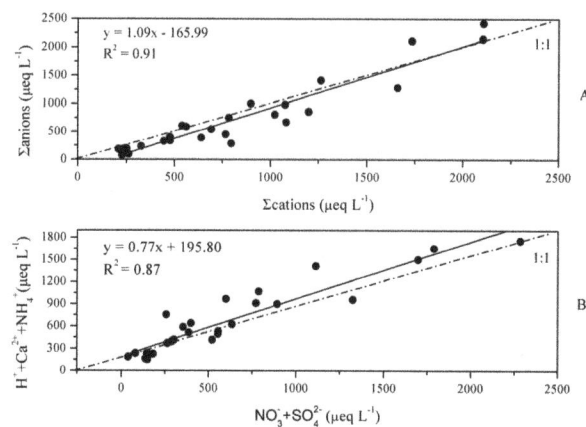

Figure 2. Linear regressions between Σcations and Σanions (A) and between summed concentrations of H^+ + Ca^{2+} + NH_4^+ and NO_3^- + SO_4^{2-} (B).

Table 1. Chemical composition of rainwater collected at TSPSAU.

	Autumn (n = 6)	Winter (n = 6)	Spring (n = 9)	Summer (n = 7)		Annual (n = 28)	
	VWM	VWM	VWM	VWM	VWM	Mean	Min-Max
pH	4.17	3.59	3.77	4.33	4.03	4.61	3.36 - 6.68
H^+ (μeq/L)	68.25	259.46	167.90	46.44	92.75	96.60	0.21 - 436.77
Na^+ (μeq/L)	16.04	71.11	35.63	17.65	24.24	47.48	3.99 - 185.54
K^+ (μeq/L)	31.88	54.16	43.81	13.60	30.05	50.75	5.44 - 254.48
Ca^{2+} (μeq/L)	66.85	302.50	244.78	19.52	98.36	233.75	9.68 - 992.03
Mg^{2+} (μeq/L)	3.35	32.42	36.60	12.21	13.18	37.44	n.d. - 185.85
NH_4^+ (μeq/L)	124.68	427.78	312.96	210.20	203.71	330.06	50.13 - 872.93
F^- (μeq/L)	8.67	40.32	30.27	4.68	13.30	28.19	n.d. - 134.75
Cl^- (μeq/L)	42.59	57.50	52.13	16.82	37.50	69.89	2.48 - 394.57
SO_4^{2-} (μeq/L)	112.00	548.43	382.84	190.34	212.30	419.85	28.81 - 1420.36
NO_3^- (μeq/L)	38.01	363.70	77.67	74.76	84.36	180.54	9.65 - 926.70
$\sum$ions (μeq/L)	512.32	2157.36	1384.59	606.23	809.75	1494.54	302.15 - 4524.22
Conductivity (μS/cm)	83.80	119.95	100.68	53.65	79.59	110.69	30.00 - 274.00
Rainfall (mm)	74.0	14.4	21.2	47.1		156.7	

in **Figure 3**. From **Table 1** and **Figure 3**, it can be seen that the pH values for the individual precipitation are varied from 3.36 to 6.68, with an annual VWM (volume-weight mean) of 4.03. Up to 68% of the pH values were in the range of 3.0 - 4.5, indicating a severe acidification of the rainwater at TSPSAU. On the other hand, 21% samples had the pH values above 5.60, which is the pH value of unpolluted water equilibrated with atmospheric CO_2. Compared with the VWM pH values measured for other areas in China, 5.12 (Beijing [17]); 4.49 (Shanghai [5]); 4.49 (Guangzhou [6]); 4.54 (Jinhua, Zhejiang Province [4]); 4.10 - 4.13 (Tie Shan Ping, Chongqing [8]); 4.44 (Lei Gong Shan, Kaili, Guizhou Province [8]); and for the overseas areas, 4.52 (Tokyo [18]); 4.60 (Newark, New Jersey, USA [19]); 5.20 (São Paulo, Brazil [20]); 6.0 (Kampangsan, Thailand [21]), the VWM pH value for TSPSAU was low. The VMW pH value of the rainwater collected at TSPSAU are even obviously lower than that of Chengdu [22] (5.1, Jan.-Dec., 2008), which is located about 150 km to the northeast of Ya'an. The lacking of alkaline substances like Ca^{2+} in rainwater and the significant anthropogenic influences in the semi-rural atmosphere may be the main causes for the severe acidification of the rainwater collected at TSPSAU.

Electrical conductivity (EC) of rainwater collected at TSPSAU was ranged from 30.00 to 274.00 μS/cm, with a VWM value of 79.59 μS/cm. EC can be looked as a good indicator for the total concentrations of soluble ions. The

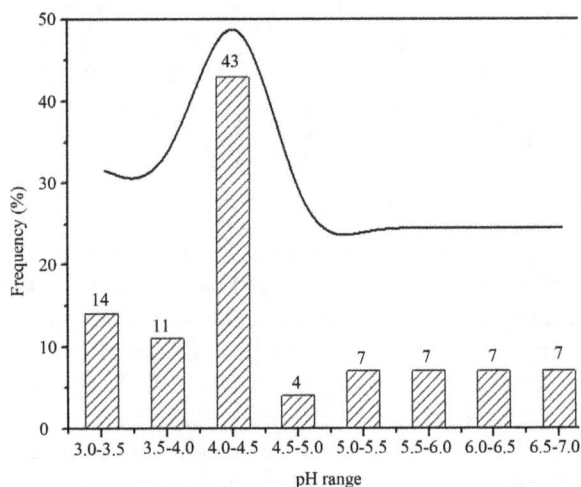

Figure 3. Frequency distribution of pH values for rainwater samples collected at TSPSAU.

VWM conductivity of rainwater collected at TSPSAU was obviously higher than those of coastal sites in the southeastern China, such as Jinhua [4] (20.3 μS/cm) and Shenzhen [23] (25.52 μS/cm), however, comparable to that of Beijing [17] (76.88 μS/cm), and lower than those of inland arid areas, such as Urumqi [24] (91.04 μS/cm) and Jordan [25] (95.00 μS/cm).

3.2. Ionic Composition

Average concentration and standard deviation of $\sum$ions

for rainwater collected at TSPSAU were 1494.54 ± 1193.17 µeq/L, with a VWM value of 809.75 µeq/L during one whole year (see in **Table 1**). Major ions show a general trend: $SO_4^{2-} > NH_4^+ > Ca^{2+} > H^+ > NO_3^- > Cl^- > K^+ > Na^+ > F^- > Mg^{2+}$ in the VWM concentrations. SO_4^{2-}, as the dominant anion with the highest level among all ionic species of a VWM of 212.30 µeq/L, accounted for 26.2% of the total ionic concentrations (see in **Figure 4**). It suggests that acid rain at the research site is of sulfuric-acid type caused mainly by the input of sulfuric acid from anthropogenic sources.

Moreover, NH_4^+ was the predominant cation with a VWM of 203.71 µeq/L during sampling period. The formation of NH_4^+ is tightly connected with gas-to-particle conversion of NH_3 in atmosphere. **Table 2** presents the ambient concentrations of NH_3 monitored during sampling periods. The average concentration of NH_3 was 174.2 µg/m³ at TSPSAU. Biswas et al. 2008 [26] reported that the average ambient concentrations of NH_3 ranged from 21.2 to 81.3 µg/m³ with a mean of 50.1 µg/m³ in Lahore, a large city of Pakistan, during 23 Dec. 2005 to 14 Feb. 2006. The high levels of atmospheric NH_3 in Ya'an were comparable to Lahore. It suggests that NH_3 in the areas around the Tibetan Plateau is often very abundant in the lower troposphere.

As a major crustal element, Ca^{2+} is come from calcareous soils and dusts driven by winds and anthropogenic construction actions, and it accounted for 12.15% of $\sum$ions of rainwater. H^+ accounted for 11.45% of the total ionic concentration, showing an obvious lack of alkaline ions and a severe acidification of rainwater at the research site (see in **Figure 4**).

3.3. Seasonal Variation

Figure 5 shows the temporal trends of water soluble ions in rainwater samples collected during one-year long term observation. From **Figure 5**, it can be seen that the major ions in the autumn rainwater include SO_4^{2-}, NO_3^-, Cl^-, NH_4^+, Ca^{2+}, K^+ and H^+. The concentrations of Cl^- and K^+ were correspondingly higher than those of NO_3^- and NH_4^+ on 27 Sep. 2010, suggesting an obvious influence from biomass burning. Moreover, the dominating ions in the winter and spring samples were SO_4^{2-}, NO_3^-, Ca^{2+} and NH_4^+. The concentrations of Ca^{2+}, an indicator element of soil dust, were higher in the winter and spring than in the summer and autumn, indicating soil dust is heavier in the former seasons than in the latter. Furthermore, the important ions in the summer rainwater were SO_4^{2-}, NO_3^- and NH_4^+, implying a strong photo-oxidation processing during this time. All rainwater chemical species showed the highest concentrations in winter compared to the other three seasons possibly due to the short rainfall amount in this season only except for Mg^{2+} (see in **Table 1**).

3.4. Correlation Analysis

Table 3 presents a Pearson correlation analysis result for

Table 2. Ambient concentrations of NH₃ monitored at TSPSAU (24 hr intervals).

	Autumn 17-28, Sep., 2010		Winter 31 Dec. 2010-14, Jan., 2011		Spring 1-18, Apr., 2011		Summer 3-16, Jun., 2011		All days
	Ave.	Min-Max	Ave.	Min - Max	Ave.	Min-Max	Ave.	Min-Max	Ave.
NH₃ (ppbv)[a]	96.7	46.7 - 148.0	26.5	3.6 - 68.3	207.2	71.6 - 410.9	605.0	354.1 - 963.3	229.0
NH₃ (µg/m³)[a,b]	73.5	35.5 - 112.6	20.2	2.7 - 52.0	157.6	54.4 - 312.5	460.3	269.4 - 732.8	174.2

[a]The ambient concentrations of NH₃ has been subtracted by field blank values. [b]Calculated under the standard state of 0°C, 101.3 kpa.

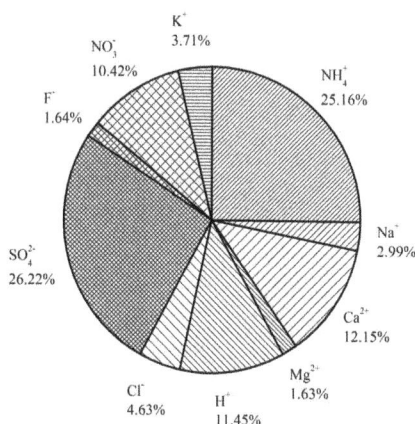

Figure 4. % contribution of different ion species to $\sum$ions for rainwater samples collected in Ya'an (based on VWM values).

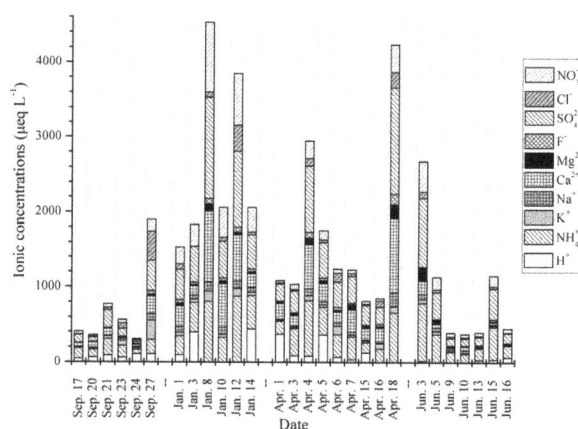

Figure 5. Temporal trends of ionic concentrations for the rainwater samples.

Table 3. Pearson correlation matrix for the ionic species measured in rainwater.

Species	H^+	NH_4^+	K^+	Na^+	Ca^{2+}	Mg^{2+}	F^-	Cl^-	SO_4^{2-}	NO_3^-
NH_4^+	−0.072									
K^+	−0.099	0.328								
Na^+	−0.132	0.709**	0.550**							
Ca^{2+}	−0.163	0.725**	0.428*	0.875**						
Mg^{2+}	−0.220	0.730**	0.244	0.736**	0.664**					
F^-	−0.030	0.630**	0.512**	0.888**	0.879**	0.579**				
Cl^-	−0.199	0.412*	0.831**	0.597**	0.458*	0.290	0.574**			
SO_4^{2-}	−0.110	0.920**	0.443*	0.873**	0.891***	0.825**	0.815**	0.476*		
NO_3^-	−0.089	0.835**	0.436*	0.732**	0.774**	0.536**	0.582**	0.406*	0.851**	

*: Correlation is significant at $P < 0.05$ (two-tailed); **: Correlation is significant at $P < 0.01$ (two-tailed); Underlined: insignificant correlations.

the major ions of the rainwater samples, which was performed using SPSS 19.0. From **Table 3**, it can be seen that the correlations between the major ions are all very good, with the exception of the correlations between K^+ and Mg^{2+}, Cl^- and Mg^{2+}, and NH_4^+ and K^+. The highly significant correlation of K^+ and Cl^- ($R^2 = 0.831$) implied a common origin of them (*i.e.* biomass burning) and possible combining form of KCl. NH_4^+, NO_3^- and SO_4^{2-} showed good relationships between each other ($p < 0.01$), suggesting that they came from one anthropogenic source and shared a common photochemical formation process as of $(NH_4)_2SO_4$ and NH_4NO_3 in the ambient atmosphere. Moreover, good correlations between the acidic ions (NO_3^-, SO_4^{2-} and F^-) and the cations (Ca^{2+}, Mg^{2+} and Na^+) demonstrated that the existence of combining substances, $CaSO_4$, $MgSO_4$, Na_2SO_4, $NaNO_3$, etc., which were formed via neutralization reactions under atmospheric deposition processes.

3.5. Acid neutralization Indicators

Fractional acidity (FA) is an indicator often used to measure acid neutralization capacity of rainwater. FA is expressed in the following form:

$$FA = \frac{\left[H^+\right]}{\left[SO_4^{2-}\right]+\left[NO_3^-\right]} \tag{1}$$

The FA values at TSPSAU were varied from 0.0002 to 2.85, with an average of 0.35, indicating that about 65% of the acidity was neutralized by alkaline constituents. This result was higher than that of Jinhua [4] (0.24), of Guangzhou [27] (0.15), and of Pune [28] (0.013), but lower than that of Newark, New Jersey, USA [19] (0.47) and of Singapore [29] (0.64). FA values were in the middle level in Ya'an when compared to other locations all over the world.

Similarly, the neutralization ability of different ions in rainwater can be estimated by neutralization factors (NF).

NF is expressed as in Equation (2):

$$NF_x = \frac{[X]}{\left[SO_4^{2-}\right]+\left[NO_3^-\right]} \tag{2}$$

where, X denotes the equivalent concentrations of alkaline components in rainwater. Average NF values for NH_4^+, Ca^{2+}, Mg^{2+}, K^+ and Na^+ were 0.71, 0.40, 0.05, 0.13 and 0.10 at TSPSAU, respectively. NH_4^+ was the more important neutralization substance than Ca^{2+} in this study. It is different from many studies on other areas around the world (e.g. Shanghai [5]; Guangzhou [6]; Pune, India [28]), in those areas, the most important neutralization substance is Ca^{2+}. It is agreement with the facts that the obvious NH_3 emission and a small amount of soil dust in the air of Ya'an in continuous rainy and humid climate.

A good linear regression between the concentrations of the main anions ($[SO_4^{2-}] + [NO_3^-]$) and the main cations ($[H^+] + [Ca^{2+}] + [NH_4^+]$) is shown in **Figure 2(B)** ($R^2 = 0.873$). It can be inferred that the acidity of rainwater was primarily affected by SO_4^{2-}, NH_4^+, NO_3^- and Ca^{2+}. In order to assess the balance between acidity and alkalinity of rainwater, ratios between the neutralizing potential (NP) and the acidifying potential (AP) is evaluated in the following Equation (3):

$$\frac{NP}{AP} = \frac{\left[Ca^{2+}\right]+\left[NH_4^+\right]}{\left[SO_4^{2-}\right]+\left[NO_3^-\right]} \tag{3}$$

The NP/AP values were varied from 0.67 to 2.15 with an average of 1.10 in this study. It was close to the values reported for Beijing [30] (1.2) and Shenzhen [23] (1.24). The high NP/AP value indicates strong neutralization effects, alkaline inorganic ions offsetting the acidity caused by high loadings of H_2SO_4 and HNO_3 in rainwater. It is worthy noting that the average pH value of rainwater collected at TSPSAU (VWM pH values: 4.03) is much lower than that of Beijing [30] (6.0), despite the fact that the NP/AP values for the two different sites are similar. The fact is tentatively explained by that the dominant cations in Beijing and Ya'an are different: Ca^{2+} is the dominant cation in Beijing while NH_4^+ in Ya'an. Acidity of rainwater is largely determined by the neutralization reactions between Ca^{2+}/NH_4^+ and SO_4^{2-}/NO_3^-. The pH values of rainwater comprising mainly calcium salts are higher than those of rainwater comprising mainly ammonium salts due to that calcium hydroxide is a stronger base than NH_3.

3.6. Source Analysis

3.6.1. Trajectory Analysis

To explore the origins of pollutants in rainwater, 48-h air mass backward trajectories were computed for each rain event using NOAA HYSPLIT4 Model and Reanalysis meteorological data from 1500 meter above ground level at the starting point. Based on the acquired backward trajectories, the air masses in the rainy days in this campaign can be divided into 3 clusters (see in **Figure 6**).

Air masses in Cluster **1** (SW) contributed 87.5% of the total rainfall during one-year observation. This cluster of air masses, originated from Tibet, Yunnan in China, and Burma and the Bay of Bengal, accounted for 71.9% of $\sum$ions during sampling periods (see in **Table 4**). Air masses in Cluster **2** (NW), accounting for 8.4% of the rainfall, were transferred from Qinghai, Gansu and Xinjiang. This cluster of air masses, characterized by very high loadings of ionic species, was severely polluted due to low rainfall amount and heavy anthropogenic emission

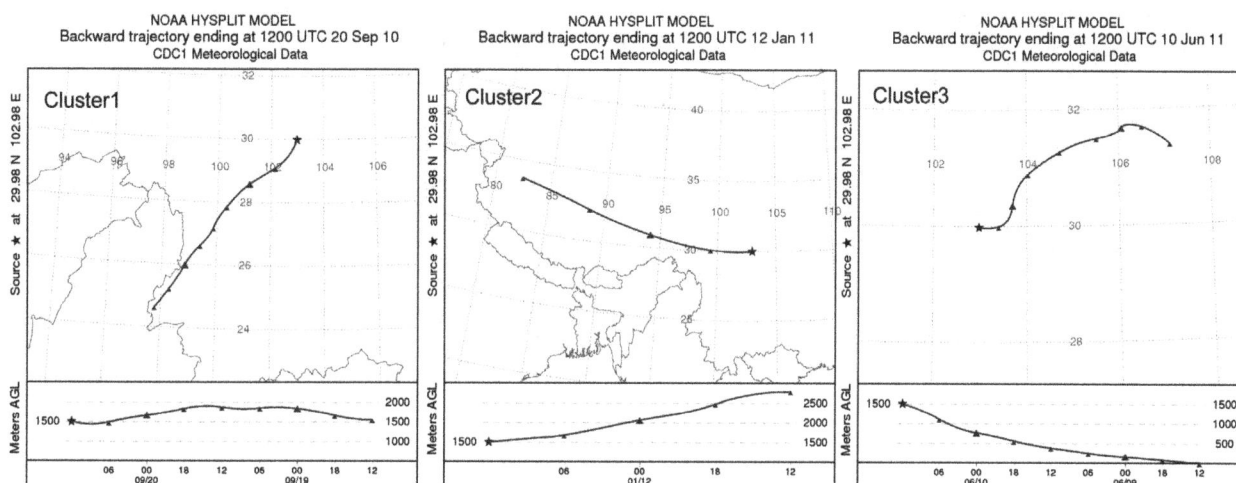

Figure 6. Representative wind directions drawn by HYSPLIT Trajectory Model.

Table 4. The VWM ionic concentrations of rainwater collected under different clusters of air masses and the ionic mass of rainwater collected under different clusters of air masses contributable to those of the total collected rainwater (based on back trajectory analysis).

Cluster	H^+	NH_4^+	K^+	Na^+	Ca^{2+}	Mg^{2+}	F^-	Cl^-	SO_4^{2-}	NO_3^-	$\sum$ions
The VWM ionic concentrations of rainwater collected under different clusters of air masses ($\mu eq \cdot L^{-1}$) [a]											
1	75.24	179.05	27.15	20.44	82.70	11.47	10.80	35.67	177.20	58.52	678.23
2	283.04	466.67	59.08	77.58	330.00	35.36	43.98	62.73	598.29	396.76	2353.49
3	75.60	237.07	36.99	13.70	82.34	13.53	7.73	32.36	257.03	65.78	822.13
Ionic mass of rainwater collected under different clusters of air masses contributable to those of the total collected rainwater (%) [b,c]											
1	70.96	76.17	78.55	71.59	69.90	73.97	70.14	82.53	71.80	58.64	71.91
2	25.71	19.12	16.46	26.17	26.86	21.96	27.51	13.98	23.34	38.28	24.03
3	3.33	4.71	5.00	2.24	3.25	4.07	2.34	3.50	4.86	3.08	4.07

[a]If I_i refers to the ionic concentration of rainwater and R_i refers to the rainfall amount for a rain event, the VWM ionic concentrations of the rainwater collected under a cluster of air masses can be calculated according to the following equation: $C_i = \sum(I_i \times R_i)/\sum R_i$. [b]Total rainfall amounts for the rain events of cluster 1, 2 and 3 are 137.1, 13.2 and 6.4 mm, respectively. [c]If C_i refers to the VWM ionic concentrations of rainwater and R_{ci} refers to the total rainfall amount for a cluster of rain events, the ionic mass of rainwater collected under different clusters of air masses contributable to those of total collected rainwater can be calculated according to the following equation: $\eta = C_i \times R_{ci}/\sum(C_i \times R_{ci})$

along the passing areas in wintertime. Air masses in Cluster **2** accounted for 24.0% of $\sum$ions during sampling periods. Moreover, air masses in Cluster **3** (NE-E-SE) contributing 4.1% of the total sampled rainfall, was primarily from a series of central cities in Sichuan basin, such as Chongqing, Chengdu, Leshan, Yibin and Dazhou. Air masses in Cluster **3** accounted for 4.1% of total ions. Air masses in Cluster **3** were slightly heavier polluted than those of cluster 1. It is agreement with the fact that there are denser population and more obvious anthropogenic influences in Sichuan Basin along path **3** than **1**.

3.6.2. Principal Component Analysis

A varimax rotated principal component analysis was conducted using SPSS 19.0 to reveal the association among the major ions and further identify the source apportionment of chemical species measured in rainwater (the result is shown in **Table 5**). Three factors with high eigenvalues more than 1 were extracted, which accounted in total for 86.0% of the variance. The first factor (PC1) explains 52.3% of the total variance, characterized by high loading of NH_4^+, K^+, Na^+, Ca^{2+}, Mg^{2+}, F^-, Cl^-, SO_4^{2-} and NO_3^-. It demonstrates a well mixed source influenced by crustal, anthropogenic and marine origins, possibly transported over a long distance. The second factor (PC2) is shown by high loading of K^+ and Cl^-, accounting for 23.1% of the total variance and implying a local biomass burning source. The third factor (PC3) is highly loaded by H^+ with a variation of 10.5%.

PC3 is explained to be closely related to the neutralization process of the acidity of rains, which is affected by the NH_3 emission and the gas-to-particle conversion of NH_3 obviously.

3.6.3. Enrichment Factor Analysis

Enrichment factors (EF) have been popularly used in source identification of rainwater elements. In EF method, Na is regarded as the reference element, and Na is assumed to be from only one marine origin. The enrichment factors relative to seawater values ($EF_{seawater}$) for target ions can be calculated according to Equation (4).

$$EF = \frac{\left[\dfrac{X}{Na^+}\right]_{rainwater}}{\left[\dfrac{X}{Na^+}\right]_{seawater}} \qquad (4)$$

In Equation (4), X represents the different target ions. $[X/Na^+]_{seawater}$ refers to the molar concentration ratios between the target ions and Na in seawater, which have already been measured (referred in Keene *et al.*, 1986 [31]). In addition, $[X/Na^+]_{rainwater}$ refers to the molar concentration ratios between the target ions and Na in rainwater samples. Because that NH_4^+, NO_3^- and F^- are not significant components in sea salts, anthropogenic sources are responsible for the most of them, the $EF_{seawater}$ values are calculated only for K^+, Ca^{2+}, Mg^{2+}, SO_4^{2-} and Cl^-.

Table 6 presents the $EF_{seawater}$ values calculated for the

Table 5. Factor analysis on the chemical species of rainwater.

	Rotated component matrix		
	PC1	PC2	PC3
H^+	−0.063	−0.086	**0.985**
NH_4^+	**0.897**	0.131	−0.007
K^+	0.211	**0.922**	−0.014
Na^+	**0.830**	0.442	−0.044
Ca^{2+}	**0.873**	0.300	−0.057
Mg^{2+}	**0.836**	0.018	−0.234
F^-	**0.751**	0.468	0.082
Cl^-	0.257	**0.905**	−0.125
SO_4^{2-}	**0.960**	0.242	−0.031
NO_3^-	**0.820**	0.239	0.016
Eigenvalues	5.229	2.314	1.053
Variance %	52.29	23.14	10.53
Cumulative %	52.29	75.43	85.97

Bold data indicate that the loading for the variables is higher than 0.50.

target ions: the $EF_{seawater}$ value for Mg^{2+} is 3.41, indicating that marine source contributes a small fraction to the ion; K^+, Ca^{2+} and SO_4^{2-} have strikingly high $EF_{seawater}$ values of 64.5, 128.2 and 93.7 respectively, implying almost no marine contribution to these ions in rainwater samples; the $EF_{seawater}$ value for Cl^- is 1.42, demonstrating that Cl^- involves an important proportion of marine origin in this study, which is ascribed to a long-range transport of sea salts from the Indian Ocean. However, the marine origin of Cl^- may be overestimated using this method, because Cl^- evaporates from rainwater to form gaseous HCl during transport when environmental condition is favorable.

3.6.4. Quantification of Major Ions from Different Sources

Providing that contributions from other natural sources are negligible, the ions in rainwater are absolutely from anthropogenic actions, crustal dusts and long-range transport of sea salts. To further evaluate the relative contribution of different sources to each kind of ion in rainwater, Na^+ is used as marine reference, and the non-sea-salt fraction Mg^{2+} is regarded to be from crustal dusts only. The marine contribution for a given ion (i.e. K^+, Ca^{2+}, Mg^{2+}, Cl^- and SO_4^{2-}) can be calculated according to Equation (5):

$$\%[X]_{marine} = \frac{100}{EF_{seawater}} \quad (5)$$

$\%[X]_{marine}$ represents the contribution proportion of sea salt to the target ion in rainwater. Moreover, the crustal sources of K^+ and Ca^{2+} can be evaluated based on the equivalent ratios of $([K^+]/[Mg^{2+}])_{crust}$ and $([Ca^{2+}]/[Mg^{2+}])_{crust}$ of 0.48 and 1.87 respectively [5]. In addition, providing SO_4^{2-} from crustal source is supplied by gypsum, the equivalent $([SO_4^{2-}]/[Ca^{2+}])_{crust}$ ratio is 0.47 [5]. Following this line of thought, the crustal source of

K^+, Ca^{2+} and SO_4^{2-} can be calculated based on the following relationship equations:

$$\%\left[K^+\right]_{crust} = \%\left[Mg^{2+}\right]_{crust} \times 0.48$$
$$\%\left[Ca^{2+}\right]_{crust} = \%\left[Mg^{2+}\right]_{crust} \times 1.87 \quad (6)$$
$$\%\left[SO_4^{2-}\right]_{crust} = \%\left[Ca^{2+}\right]_{crust} \times 0.47$$

Furthermore, the anthropogenic source contributions of K^+, Ca^{2+} and SO_4^{2-} can be easily calculated according to Equation (7):

$$[X]_{anthropogenic} = [X]_{rainwater} - [X]_{crust} - [X]_{marine} \quad (7)$$

Table 7 shows the quantificational result of the source apportionment of the major ions in rainwater samples collected at TSPSAU. Sea salt, crustal and anthropogenic sources accounted for 8.1%, 5.5% and 86.4% of ionic mass for the rainwater samples respectively. Acid rain pollutions were caused mainly by anthropogenic influences in Ya'an.

4. Conclusion

Rainwater samples were collected at a semi-rural site in Ya'an from Sep. 2010 to Jun. 2011. Conductivity, pH values, and ionic concentrations of the rainwater samples were measured. The VWM concentration of the total ions of rainwater during a whole year was 809.75 $\mu eq \cdot L^{-1}$. The annual VWM pH value was 4.03 and the annual acid rain frequency was 79%. Back-trajectory analysis indicated that the air masses in Ya'an can be divided into three types, prevailing in SW, NW and NE-E-SE respectively. Three principle factors were extracted from primary component analysis. The factors were explained by long-distance transport air masses, local biomass burning and the rain neutralization process respectively. Moreover, EF method was also used to identify the sources of rainwater. It revealed that the pollutants in wet deposition

Table 6. $EF_{seawater}$ values calculated for different ionic species.

	Ca^{2+}/Na^+	K^+/Na^+	Mg^{2+}/Na^+	Cl^-/Na^+	SO_4^{2-}/Na^+
$[X/Na^+]_{seawater}$	0.044	0.022	0.227	1.16	0.121
$[X/Na^+]_{rainwater}$	5.64	1.42	0.77	1.65	11.34
$EF_{seawater}$	128.15	64.46	3.41	1.42	93.73

Table 7. Source apportionment of major ions in rainwater.

	K^+	Ca^{2+}	Mg^{2+}	NH_4^+	Na^+	Cl^-	SO_4^{2-}	NO_3^-	F^-	Σions
Marine source	1.55	0.78	29.33	0	100	70.42	1.07	0	0	8.1
Crustal source	14.88	17.71	70.67	0	0	0	3.86	0	0	5.5
Anthropogenic actions	83.57	81.51	0	100	0	29.58	95.08	100	100	86.4

at the research site were from anthropogenic actions (86.4%), sea salts (8.1%) and crustal (5.5%) respectively.

5. Acknowledgements

This work was jointly supported by the Chinese National Natural Science Foundation (No. 41073101) and a key project of Sichuan Provincial Department of Education (No. 10ZA059). Zhilin Liu is supported by a Student Program of Scientific Interest Cultivation Plan of Sichuan Agricultural University.

REFERENCES

[1] T. Larssen, E. Lydersen, D. Tang, Y. He, J. Gao, H. Liu, L. Duan, H. M. Seip, R. D. Vogt, M. Shao, Y. Wang, H. Shang, X. Zhang, S. Solberg, W. Aas, T. Økland, O. Eilertsen, V. Angell, Q. Liu, D. Zhao, R. Xiang, J. Xiao and J. Luo, "Acid Rain in China," *Environment Science & Technology*, Vol. 40, No. 2, 2006, pp. 418-425.

[2] T. Larssen, H. M. Seip, A. Semb, J Mulder, I. P. Muniz, R. D. Vogt, E. Lydersen, V. Angell, T. Dagang and O. Eilertsen, "Acid Deposition and Its Effects in China: An Overview," *Environmental Science & Policy*, Vol. 2, No. 1, 1999, pp. 9-24.

[3] R. A. Goyer, J. Bachmann, T. W. Clarkson, B. G. Ferris, J. Graham, P. Mushak, D. P. Perl, D. P. Rall, R. Schlesinger, W. Sharpe and J. M. Wood, "Potential Human Health Effects of Acid Rain: Report of a Workshop," *Environmental Health Perspectives*, Vol. 60, No. 5, 1985, pp. 355-368.

[4] M. Zhang, S. Wang, F. Wu, X. Yuan and Y. Zhang, "Chemical Compositions of Wet Precipitation and Anthropogenic Influences at a Developing Urban Site in Southeastern China," *Atmospheric Research*, Vol. 84, No. 4, 2007, pp. 311-322.

[5] K. Huang, G. Zhuang, C. Xu, Y. Wang and A. Tang, "The Chemistry of the Severe Acidic Precipitation in Shanghai, China," *Atmospheric Research*, Vol. 89, No. 1, 2008, pp. 149-160.

[6] D. Y. Huang, Y. G. Xu, P. Peng, H. H. Zhang and J. B. Lan, "Chemical Composition and Seasonal Variation of Acid Deposition in Guangzhou, South China: Comparison with Precipitation in Other Major Chinese Cities," *Environment Pollution*, Vol. 157, No. 1, 2009, pp. 35-41.

[7] W. X. Wang and T. Wang, "On the Origin and the Trend of Acid Precipitation in China," *Water, Air and Soil Pollution*, Vol. 85, No. 4, 1995, pp. 2295-2300.

[8] W. Aas, M. Shao, L. Jin, T. Larssen, D. Zhao, R. J. Xiang, J. H. Zhang, J. S. Xiao and L. Duan, "Air Concentrations and Wet Deposition of Major Inorganic Ions Five Non-Urban Sites in China, 2001-2003," *Atmospheric Environ-*

ment, Vol. 41, No. 8, 2007, pp. 1706-1716.

[9] D. Migliavacca, E. C. Teixeira, F. Wiegand, A. C. M. Machado and J. Sanchez, "Atmospheric Precipitation and Chemical Composition of an Urban Site, Guaíba Hydrographic Basin, Brazil," *Atmospheric Environment*, Vol. 39, No. 10, 2005, pp. 1829-1844.

[10] P. A. Roelle and V. P. Aneja, "Characterization of Ammonia Emissions from Soils in the Upper Coastal Plain, North Carolina," *Atmospheric Environment*, Vol. 36, No. 20, 2002, pp. 1087-1097.

[11] K. Ali, G. A. Momin, S. Tiwari, P. D. Safai, D. M. Chate and P. S.-P. Rao, "Fog and Precipitation Chemistry at Delhi, North India," *Atmospheric Environment*, Vol. 38, No. 25, 2004, pp. 4215-4222.

[12] A. Mihajlidi-Zelić, I. Deršek-Timotić, D. Relić, A. Popović and D. Đorđević, "Contribution of Marine and Continental Aerosols to the Content of Major Ions in the Precipitation of the Central Mediterranean," *Science of Total Environment*, Vol. 370, No. 2, 2006, pp. 441-451.

[13] H. Rodhe, F. Dentener and M. Schulz, "The Global Distribution of Acidifying Wet Deposition," *Environment Science & Technology*, Vol. 36, No. 20, 2002, pp. 4382-4388.

[14] L. Li, D. J. Dai, S. H. Deng, J. L. Feng, M. Zhao, J. Wu, L. Liu, X. H. Yang, S. S. Wu, H. Qi, G. Yang, X. H. Zhang, Y. J. Wang and Y. Z. Zhang, "Concentration, Distribution and Variation of Polar Organic Aerosol Tracers in Ya'an, a Middle-Sized City in Western China," *Atmospheric Research*, Vol. 120, No. 2, 2013, pp. 29-42.

[15] HJ/T165, "Technical Specifications for Acid Deposition Monitoring," State Environmental Protection Administration of China, Beijing, 2004. (in Chinese)

[16] N. Rastogi and M. M. Sarin, "Chemical Characteristics of Individual Rain Events from a Semi-Arid Region in India: Three-Year Study," *Atmospheric Environment*, Vol. 39, No. 18, 2005, pp. 3313-3323.

[17] Z. F. Xu and G. L. Han, "Chemical and Strontium Isotope Characterization of Rainwater in Beijing, China," *Atmospheric Environment*, Vol. 43, No. 12, 2009, pp. 1954-1961.

[18] T. Okuda, T. Iwase, H. Ueda, Y. Suda, S. Tanaka, Y. Dokiya, K. Fushimi and M. Hosoe, "Long-Term Trend of Chemical Constituents in Precipitation in Tokyo Metropolitan Area, Japan, from 1990 to 2002," *Science of Total Environment*, Vol. 339, No. 1, 2005, pp. 127-141.

[19] F. Song and Y. Gao, "Chemical Characteristics of Precipitation at Metropolitan Newark in the US East Coast," *Atmospheric Environment*, Vol. 43, No. 32, 2009, pp. 4903-4913.

[20] M. A. Santos, C. F. Illanes, A. Fornaro and J. J. Pedrotti, "Acid Rain in Downtown Sao Paulo City, Brazil," *Water, Air and Soil Pollution*, Vol. 7, No. 1, 2007, pp. 85-92.

[21] M. Panyakapo and R. Onchang, "A Four-Year Investigation on Wet Deposition in Western Thailand," *Journal of Environmental Sciences*, 2008, Vol. 20, No. 4, 2008, pp. 441-448.

[22] H. Wang and G. L. Han, "Chemical Composition of Rainwater and Anthropogenic Influences in Chengdu, Southwest China," Atmospheric Research, 2011, Vol. 99, No. 2, 2011, pp. 190-196.

[23] Y. Huang, Y. Wang and L. Zhang, "Long-Term Trend of Chemical Composition of Wet Atmospheric Precipitation during 1986-2006 at Shenzhen City, China," *Atmospheric Environment*, Vol. 42, No. 16, 2008, pp. 3740-3750.

[24] M. Xu, A. H. Lu, F. Xu and B. Wang, "Seasonal Chemical Composition Variations of Wet Deposition in Urumchi, Northwestern China," *Atmospheric Environment*, Vol. 42, No. 5, 2008, pp. 1042-1048.

[25] K. Ali Al-Khashman, "Chemical Characteristics of Rainwater Collected at a Western Site of Jordan," *Atmospheric Research*, Vol. 91, No. 1, 2009, pp. 53-61.

[26] K. F. Biswas, B. M. Ghauri and L. Husain, "Gaseous and Aerosol Pollutants during Fog and Clear Episodes in South Asian Urban Atmosphere," *Atmospheric Environment*, Vol. 42, No. 33, 2008, pp. 7775-7785.

[27] Y. Cao, S. Wang, G. Zhang, J. Luo and S. Lu, "Chemical Characteristics of Wet Precipitation at an Urban Site of Guangzhou, South China," *Atmospheric Research*, Vol. 94, No. 3, 2009, pp. 462-469.

[28] P. D. Safai, P. S. P. Rao, G. A. Momin, K. Ali, D. M. Chate and P. S. Praveen, "Chemical Composition of Precipitation during 1984-2002 at Pune, India," *Atmospheric Environment*, Vol. 38, No. 12, 2007, pp. 1705-1714.

[29] R. Balasubramanian, T. Victor and N. Chun, "Chemical and Statistical Analysis of Precipitation in Singapore," *Water, Air and Soil Pollution*, Vol. 130, No. 1, 2001, pp. 451-456.

[30] F. Yang, J. Tan, Z. Shi, Y. Cai, K. He, Y. Ma, F. Duan, T. Okuda, S. Tanaka and G. Chen, "Five-Year Record of Atmospheric Precipitation Chemistry in Urban Beijing, China," *Atmospheric Chemistry and Physics*, Vol. 12, No. 4, 2012, pp. 2025-2035.

[31] W. C. Keene, A. A.-P. Pszenny, J. N. Galloway and M. E. Hawley, "Sea-Salt Corrections and Interpretation of Constituent Ratios in Marine Precipitation," *Journal of Geophysical Research*, Vol. 91, No. D6, 1986, pp. 6647-6658.

2

Study of the Oceanic Heat Budget Components over the Arabian Sea during the Formation and Evolution of Super Cyclone, Gonu

P. R. Jayakrishnan, C. A. Babu

Department of Atmospheric Sciences, Cochin University of Science and Technology, Cochin, India

ABSTRACT

Oceans play a vital role in the global climate system. They absorb the incoming solar energy and redistribute the energy through horizontal and vertical transports. In this context it is important to investigate the variation of heat budget components during the formation of a low-pressure system. In 2007, the monsoon onset was on 28th May. A well-marked low-pressure area was formed in the eastern Arabian Sea after the onset and it further developed into a cyclone. We have analysed the heat budget components during different stages of the cyclone. The data used for the computation of heat budget components is Objectively Analyzed air-sea flux data obtained from WHOI (Woods Hole Oceanographic Institution) project. Its horizontal resolution is $1° \times 1°$. Over the low-pressure area, the latent heat flux was 180 Wm^{-2}. It increased to a maximum value of 210 Wm^{-2} on 1st June 2007, on which the system was intensified into a cyclone (Gonu) with latent heat flux values ranging from 200 to 250 Wm^{-2}. It sharply decreased after the passage of cyclone. The high value of latent heat flux is attributed to the latent heat release due to the cyclone by the formation of clouds. Long wave radiation flux is decreased sharply from 100 Wm^{-2} to 30 Wm^{-2} when the low-pressure system intensified into a cyclone. The decrease in long wave radiation flux is due to the presence of clouds. Net heat flux also decreases sharply to -200 Wm^{-2} on 1st June 2007. After the passage, the flux value increased to normal value (150 Wm^{-2}) within one day. A sharp increase in the sensible heat flux value (20 Wm^{-2}) is observed on 1st June 2007 and it decreased thereafter. Short wave radiation flux decreased from 300 Wm^{-2} to 90 Wm^{-2} during the intensification on 1st June 2007. Over this region, short wave radiation flux sharply increased to higher value soon after the passage of the cyclone.

Keywords: Oceanic Heat Budget; Cyclone; Arabian Sea

1. Introduction

The Indian peninsula splits the north Indian Ocean into two basins, the Arabian Sea in the west and Bay of Bengal in the east. In 2007, an organized convection was formed on 28th May near southeastern Arabian Sea. Monsoon onset was declared on 28th May treating the organized convection as monsoon onset surge. This was converged into a well-marked low-pressure area and deepened further into a super cyclone, Gonu on 2nd June 2007. It is important to evaluate the different characteristics of ocean during the formation of a low-pressure system. Hence an attempt is made to study the different heat budget components during the formation of Gonu. A large number of studies were conducted about the heat budget during the past. Hareesh Kumar and Mathew [1] examined the heat budget of the Arabian Sea and found that the heat storage of the sea is mainly controlled by heat change due to horizontal divergence and vertical motion while the effect of heat change due to lateral advection is much larger than expected. Shenoi et al. [2] studied the differences in heat budget of the near-surface Arabian Sea and the Bay of Bengal during summer monsoon season. The influence of heat, moisture and moist static energy budget, over the Arabian Sea and adjoining area (0°N - 30°N and 30°E - 75°E), during the onset and various epochs of Asian summer monsoon, has been analyzed in detail by Mohanty et al. [3]. Duing and Leetmaa [4] found that upwelling plays a significant role in cooling the surface layers of the Arabian Sea during southwest monsoon season. Molinari et al. [5] analyzed the relative importance of air sea fluxes and horizontal advection in the variability of sea surface temperature (SST) for the entire Arabian Sea. In the recent years there were

Study of the Oceanic Heat Budget Components over the Arabian Sea during the Formation and Evolution
of Super Cyclone, Gonu

13

some attempts to study the heat budget of the upper 200 m of the Arabian Sea either with limited datasets (Hasterrath and Lamb [6], Lamb and Bunker [7]; Hastenrath and Greischar [8]) or limited to specific region (Rao [9]; Varma and Kurup [10]). Hsuing *et al.* [11] estimated the annual cycle of heat storage and meridional heat transport in the world oceans.

Oceanic heat budget is the changes in heat stored in the upper layers of the ocean resulting from local imbalance between input and output of heat through the sea surface. The transfer of heat through the surface is called the heat flux. The symbol Q is used to represent the rate of heat flow (measured in Joules per second per square metre, *i.e.* Wm^{-2}). Etter [12] estimated the heat budget using an empirical formula,

$$Q_S = Q_F + Q_V + Q_A$$

where Q_S is rate of oceanic heat storage; Q_F is the net air sea heat flux; Q_V is heat change due to horizontal divergence and vertical advection and Q_A is the heat change due to horizontal advection. Budyko [13] was the first to conduct a study on the spatial and temporal structure of the oceanic heat balance on a global scale. Hastenrath and Lamb [8,14,15] have used long term $1° \times 1°$ averages of surface marine observations to establish a monthly high resolution climatology of the heat budget over parts of the tropical ocean. Oshima *et al.* [16] estimated the surface heat budget of the Sea of Okhotsk during 1987-2001 and the role of sea ice on it.

Halpern and Reed [17] estimated the heat budgets of the upper ocean under light winds. Hastenrath [6] derived the heat budget estimates for the global tropics from recent calculations of the oceanic heat budget and satellite measurements of net radiation at the top of the atmosphere. The local seasonal variation of heat content in the equatorial Atlantic Ocean is found to be about ten times larger than the seasonal variation of the heat gain from the atmosphere through the surface, and is not confined to the upper mixed layer [18]. This annual cycle of heat content appears to be mainly due to vertical movement of the thermocline associated with the dynamical response of the ocean to the seasonally varying winds. The Indian monsoon is partly driven by this air-sea interaction. Even though the broad scale features of the monsoon circulation are repetitive, the total monsoon problem, comprising the year-to-year fluctuations and vagaries in the onset as well as the time space variations in the monsoon activity, is very complex, and it demands intensive study and research [19]. Pisharoty [20]; Das [21] and Mohanty and Mohankumar [22] showed the importance of the energy fluxes over the Indian seas during different epochs of the summer monsoon activity over India. Mohanty *et al.* [23] emphasized the importance to understand the connection between the air-sea fluxes and monsoon activity rather

than SST anomaly and the monsoon activity. They examined the oceanic heat budget components and their variability over the Indian Ocean in relation to the extreme monsoon activity (flood/drought) over the Indian sub continent using Comprehensive Oceanic and Atmospheric Data Set (COADS) for the period 1950-1979. Jayakrishnan and Babu [24] analysed the different surface marine atmospheric boundary layer parameters associated with the evolution of super cyclone Gonu in the Arabian Sea. They could quantify the variations in association with the passage of system. In this analysis, the variability of heat budget components during the evolution and subsequent intensification of Gonu over the Arabian Sea is studied.

2. Materials and Methods

The data used for the computation of heat budget components is Objectively Analyzed air-sea flux data obtained from WHOI (Woods Hole Oceanographic Institution) project. Its horizontal resolution is $1° \times 1°$. The net heat budget equation used here is [1],

$$Q_F = Q_I - Q_B - Q_E - Q_H$$

where Q_F is the net heat flux, Q_I is the incoming short wave radiation flux, Q_B is the effective outgoing long wave radiation flux, Q_H is the sensible heat flux and Q_E is the latent heat flux. In the above budget equation, Q_I contributes significantly to the oceanic heat gain whereas the other terms contribute towards the heat loss from tropical oceanic surfaces. The heat budget components employed for the study are Latent heat flux, Long wave radiation flux, Short wave radiation flux, Net flux and Sensible heat flux during the period of Gonu (26 May to 7 June, 2007). The evolution and intensification of the low pressure system during different stages are described below. **Figure 1** represents track of the system from 26th May to 7th June 2007 (On the track, black circles represent the location of the system before intensification into a cyclone on a daily basis and blue circles represent that for cyclone, on a 6 hourly basis).

On 28th May, monsoon onset was declared by IMD and the onset surge further intensified into a low pressure system over the location 5°N - 6°N and 74°E - 75°E. The low developed into a depression and again intensified into a deep depression at the location 10°N - 11°N and 74°E -75°E. On 2nd June, it developed into a tropical cyclone with centre 15°N - 16°N and 69°E -70°E. It further intensified into a super cyclone Gonu on 3rd June 2007. It moved towards northwest and intensified during 4th and 5th June. At last the landfall occurred on 6th June 2007 in Oman coast (marked in the Figure) and after the landfall the system dissipated. Contours of heat budget components on a daily basis were analyzed from 26th May to 7th June 2007. Also, the area averaged heat

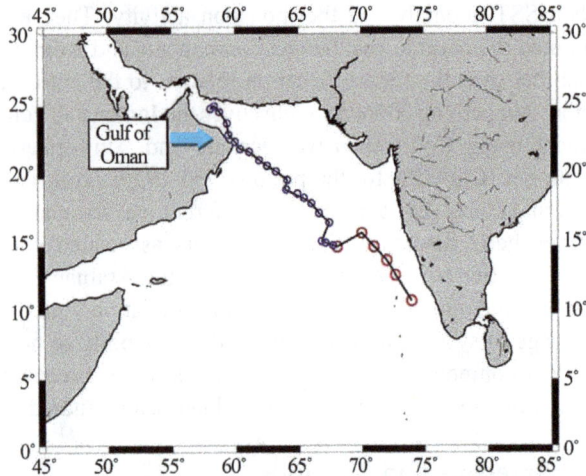

Figure 1. Track of cyclone Gonu from 28 May to 7 June.

budget components at the location of the system when it was deepened into a cyclone were studied. For the analysis of daily variation of heat budget components associated with the low pressure system, a special type of Hovmoller diagram is used in this study. The variation of the parameters at the centre of the low pressure system can be evaluated using this diagram. For this, we first identified the location of centre of the low pressure system on a daily basis during the period of the low pressure system from 26th May to 7th June (13 days). Daily values of heat budget components were extracted at these locations during the 13 days. Then a Hovmoller diagram is plotted with the X-axis as location of points of the low pressure system in the track up to landfall (total 28 points) and the Y-axis as the days starting from 26th May to 7th June. **Table 1** contains the details of the locations in the track as obtained from Unisys weather.

3. Results and Discussions

3.1. Latent Heat Flux

On the onset date *i.e.* 28th May the latent heat flux value was about 200 Wm^{-2} over the oceanic region adjacent to south Indian peninsula (**Figure 2**). On 30th, it increased to 210 Wm^{-2}. On 1st June, it intensified with value of 240 Wm^{-2}. On 3rd and 4th June, the latent heat values increased to 300 Wm^{-2} as the system was intensified into a severe cyclonic storm. Landfall was occurred on 6th June and the flux values decreased abruptly. **Figure 3** gives the Hovmoller diagram for Latent heat flux. The X-axis gives the location of points on the track of the system from the well marked low to the landfall. Y-axis gives the days starting from 26th May to 7th June. From the figure, it is clear that the system was intensified intosevere cyclonic storm on 4th, which is at the middle of the contour and the flux value became 220 Wm^{-2}.

Table 1. Location and track of cyclone as obtained from Unisys weather.

Location					
No	Date	Lat	Lon	Time	Category of cyclone
1	26 May	10N	67E		
2	27 May	6N	72E		
3	28 May	7N	76E		
4	29 May	7N	58E		
5	30 May	11N	74E		
6	31 May	14N	72E		
7	1 June	16N	70E		
8	2 June	15N	68E		
9	2 June	15.10N	67.70E	00Z	TROPICAL STORM
10	2 June	15.30N	67.10E	12Z	TROPICAL STORM
11	2 June	15.40N	66.80E	18Z	TROPICAL STORM
12	3 June	16.80N	67.40E	06Z	TROPICAL STORM
13	3 June	17.50N	66.60E	12Z	CYCLONE-1
14	3 June	18.20N	66.00E	18Z	CYCLONE-2
15	4 June	18.50N	65.50E	00Z	CYCLONE-4
16	4 June	19.20N	64.90E	06Z	CYCLONE-4
17	4 June	19.90N	64.10E	12Z	CYCLONE-5
18	4 June	20.50N	63.20E	18Z	CYCLONE-5
19	5 June	20.90N	62.50E	00Z	CYCLONE-4
20	5 June	21.30N	61.90E	06Z	CYCLONE-4
21	5 June	21.90N	61.10E	12Z	CYCLONE-3
22	5 June	22.10N	60.40E	18Z	CYCLONE-2
23	6 June	22.60N	60.00E	00Z	CYCLONE-1
24	6 June	23.10N	59.50E	06Z	CYCLONE-1
25	6 June	23.90N	59.40E	12Z	CYCLONE-1
26	6 June	24.70N	58.80E	18Z	TROPICAL STORM
27	7 June	25.10N	58.40E	00Z	TROPICAL STORM
28	7 June	24.90N	58.10E	06Z	

Since the system was further intensified, the maximum latent heat flux value was 250 Wm^{-2}, occurred on 5th June. As the system crossed Oman coast on 6th, a decrease in latent heat flux was noticed due to cut off in moisture supply. However, the system was intensified

Study of the Oceanic Heat Budget Components over the Arabian Sea during the Formation and Evolution
of Super Cyclone, Gonu

15

further on 7th June over the Gulf stream (after crossing the Oman coast) and the value became 280 Wm^{-2}, for a short period just before landfall in the Iran coast. The latent heat flux value came back to the normal (100 Wm^{-2}) after dissipation of the system.

3.2. Long Wave Flux

In comparison with the latent heat flux, long wave flux does not show one to one relationship associated with the clouding. It is observed that in response to the intense cloud over the region of system, long wave flux is decreasing. The exact location of the OLR and long wave flux is not matching as in the case of latent heat flux. On 28th, the long wave flux values (**Figure 4**) were high (100 Wm^{-2}). On 1st June, the value came down to 40 Wm^{-2} over the region of the system. The value decreased further as the cyclone intensified on 3rd (20 Wm^{-2}) and 4th (10 Wm^{-2}). After the landfall on 6th June the value started to increase due to the dissipation of Gonu and subsequent revival of monsoon and associated clouding over the Arabian Sea.

Following Waliser and Graham [25]

$$Q_l = (0.94 \times (5.67E - 8) \times SST^4 - (-2800.0 + (SST \times 10.64))$$

where SST is the sea surface temperature. The long wave flux is the outgoing long wave radiation flux which is emitted by the sea surface whereas the OLR is the Outgoing Long wave Radiation on the top of atmosphere. Since the SST is the forcing function for the long wave radiation flux and the response of SST to the passage of cyclone is a slow process, it is difficult to have one to one relationship between the cloud pattern and associated long wave flux. This is the reason why the Hovmoller diagram does not show a similar path corresponding to **Figure 3**.

Figure 5 shows Hovmoller diagram for long wave radiation flux. It is clear that the long wave flux values sharply decreased to 50 Wm^{-2} from 100 Wm^{-2} during the intensification of cyclone. The long wave flux decreased due to the passage of the system (associated with decrease in SST and increase in cloud cover). Long wave flux values were 50 Wm^{-2} during the mature stage of the low pressure system.

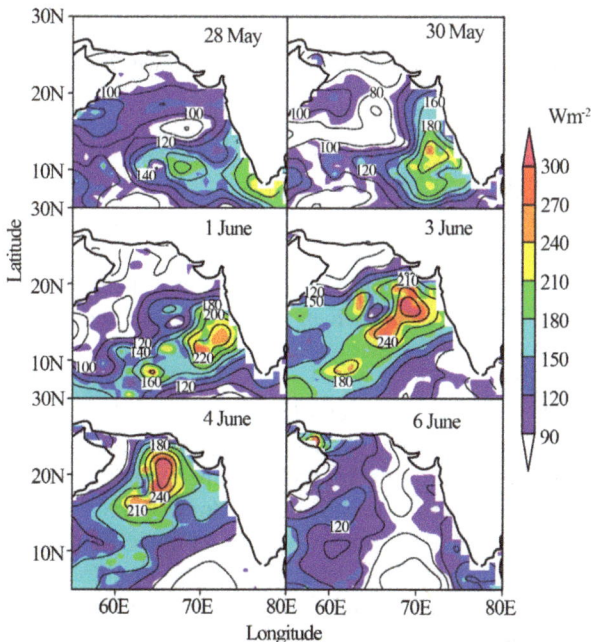

Figure 2. Latent heat flux in the Arabian Sea for various intensities of the cyclone, Gonu during 28 May to 6 June.

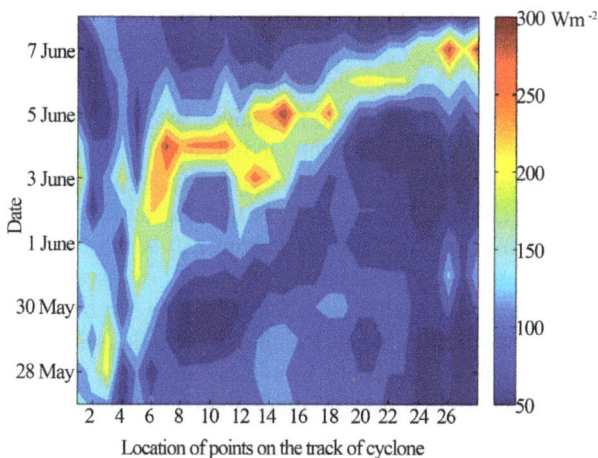

Figure 3. Hovmoller diagram for latent heat flux from 26 May to 7 June 2007.

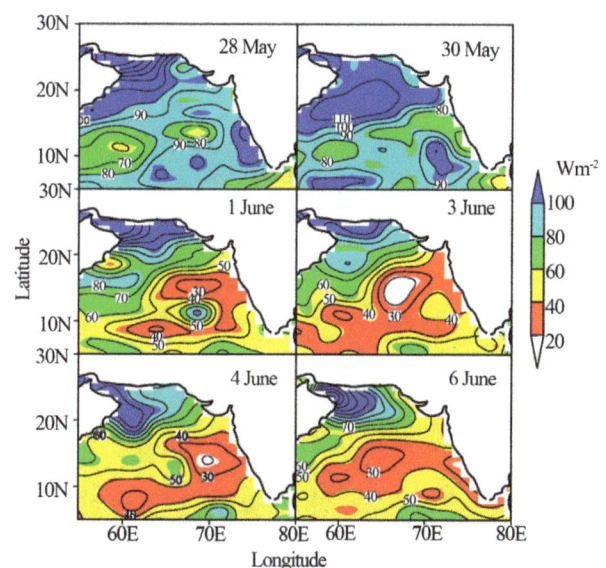

Figure 4. Long wave flux in the Arabian Sea for various intensities of the cyclone, Gonu during 28 May to 6 June.

3.3. Sensible Heat Flux

On 28th the sensible heat flux values (**Figure 6**) increased from 0 Wm^{-2} to 15 Wm^{-2} and on reaching 30th May and 1st June it increased to 20 Wm^{-2}. On 3rd and 4th June it again increased to 30 Wm^{-2} over the cyclone center and after the landfall on 6th June it sharply decreased.

From the Hovmoller diagram (**Figure 7**) it is found that the sensible heat flux values were increased from -10 Wm^{-2} to 30 Wm^{-2}. The maximum values are observed to be on 3rd and 4th June. Sensible heat flux value during the formation of cloud is 10 Wm^{-2}. When there is cloud the value increased to 50 Wm^{-2}.

3.4. Short Wave Radiation Flux

The algorithm to predict net surface insolation from OLR

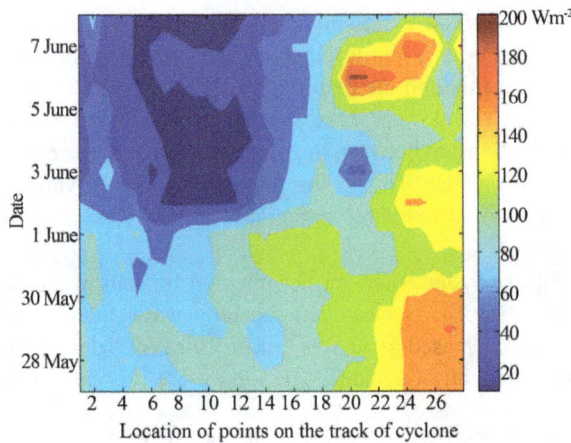

Figure 5. Hovmoller diagram for long wave radiation flux from 26 May to 7 June 2007.

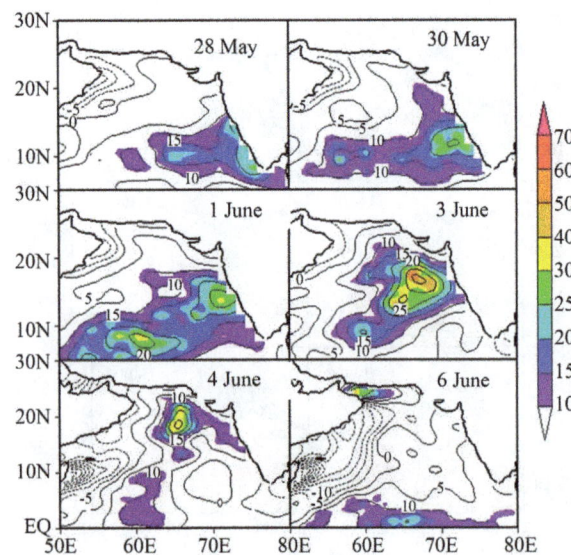

Figure 6. Sensible heat flux in the Arabian Sea for various intensities of the cyclone, Gonu during 28 May to 6 June.

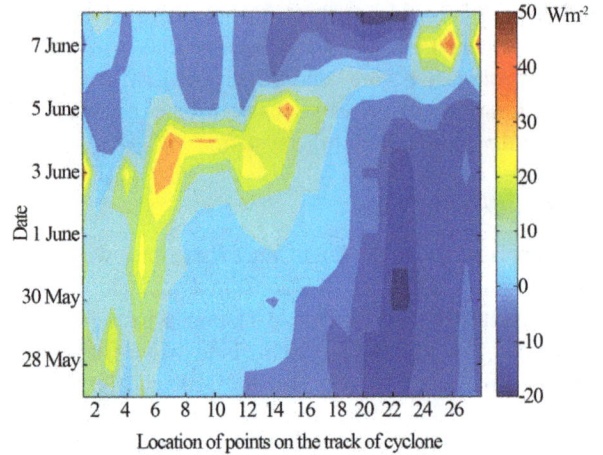

Figure 7. Hovmoller diagram for sensible heat flux from 26 May to 7 June 2007.

is based on linear regression of OLR onto Surface Radiation Budget: Shinoda *et al.* [26]

$$Q_s = 0.93Q_0 - 1.3$$

where Q_s is the net surface insolation in watts per square meter, and Q_0 is the OLR in watts per square meter. **Figure 8** describes variation of short wave radiation flux from the contour charts. Short wave flux decreased to 120 Wm^{-2} over the region of system on May 28th. It was decreased further to 100 Wm^{-2} on 1st June and remained the same value during 3rd and 4th June. After the landfall on 6th it increased to value 260 Wm^{-2}. From the Hovmoller diagram of short wave flux (**Figure 9**) it is observed that short wave radiation flux also decreased from 300 Wm^{-2} to 50 Wm^{-2} during cyclone intensification. After cyclone passage it increased to normal values.

3.5. Net Flux

Over the entire track of the low-pressure system, it is observed that net flux values (**Figure 10**) were negative. On 28th May, the net flux was -150 Wm^{-2}. On 30th, the value decreased further to -250 Wm^{-2} and remained up to 3rd June. At the centre, the values were below -300 Wm^{-2} on 4th June. After the landfall on 6th it increased to positive values. By looking the Hovmoller diagram of Net flux (**Figure 11**) we can note that the net flux values were decreasing from 150 Wm^{-2} to -250 Wm^{-2} over the cyclone area. Large negative values are observed over the centre of the system. It is found that the Net heat flux is mainly varying due to the drastic variations in short wave radiation and latent heat flux. These are the two terms, which are most important and contributing to the net heat flux. Net flux is found to have large negative values (-270 Wm^{-2}) in the wall cloud region of the system. In the absence of low pressure system the net heat flux value is above 60 Wm^{-2}.

Study of the Oceanic Heat Budget Components over the Arabian Sea during the Formation and Evolution of Super Cyclone, Gonu

17

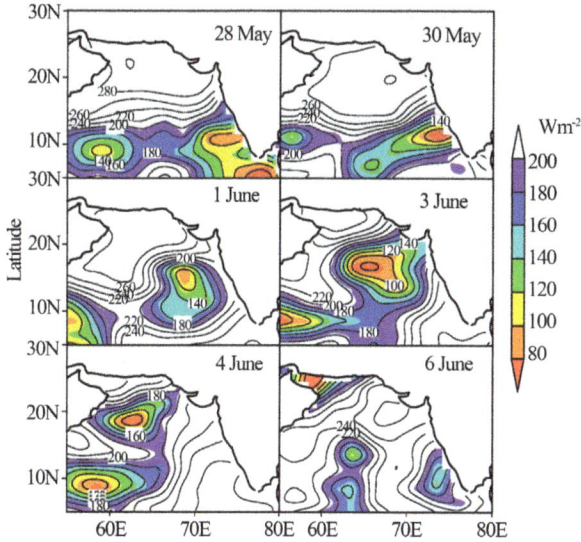

Figure 8. Short wave flux in the Arabian Sea for various intensities of the cyclone, Gonu during 28 May to 6 June.

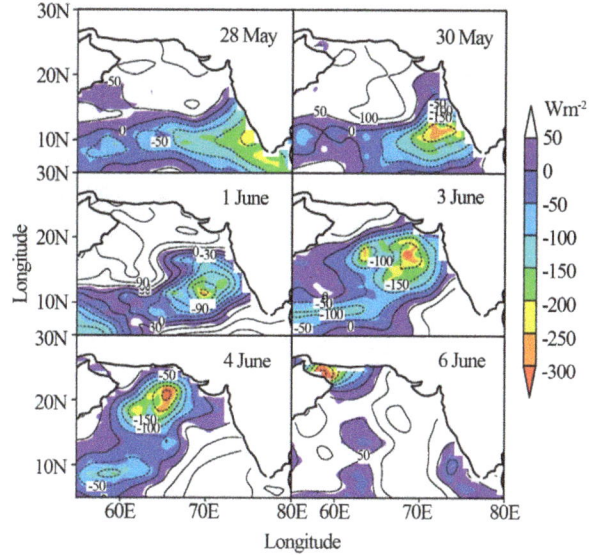

Figure 10. Net flux in the Arabian Sea for various intensities of the cyclone, Gonu during 28 May to 6 June.

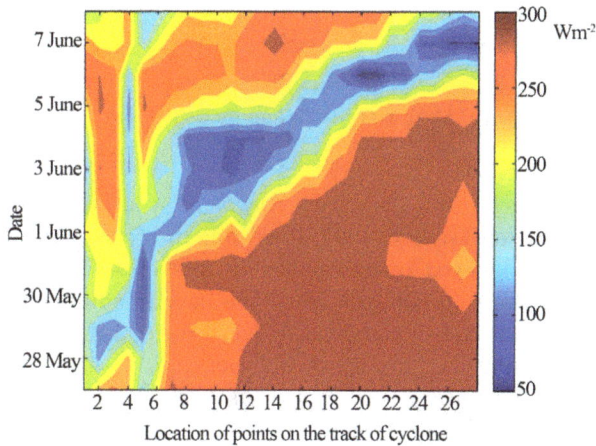

Figure 9. Hovmoller diagram for short wave flux from 26 May to 7 June 2007.

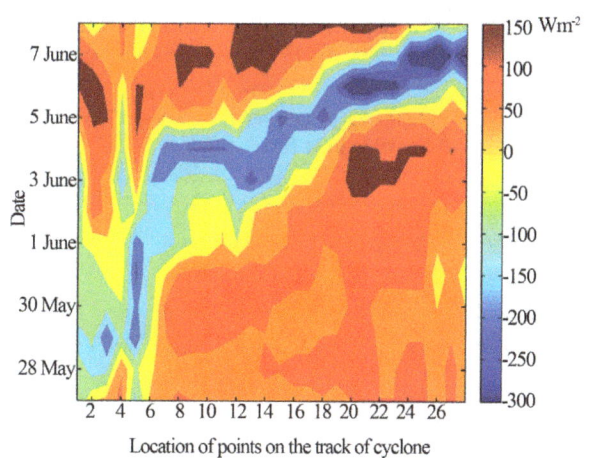

Figure 11. Hovmoller diagram for net radiation flux from 26 May to 7 June 2007.

3.6. Analysis of Fluxes from the Area Averaged Diagrams

Figure 12(a) gives the area averaged latent heat flux over 17°N - 18°N and 66°E - 67°E. The area was chosen because the system was intensified into a cyclonic storm over this location. We can note that the latent heat flux values increased from 120 Wm^{-2} on 26th May to 270 Wm^{-2} on 2nd June. After the passage of cyclone, the values decreased to 60 Wm^{-2}. The variation of long wave flux during the presence of low pressure system is presented in **Figure 12(b)**. The long wave radiation value is found to drop from 100 Wm^{-2} to 20 Wm^{-2} from 26th May to 3rd June. It is found that it remained the same even after the passage of cyclone. The area averaged short wave flux values (**Figure 12(c)**) are found to decrease in the presence of clouds as they block the incom-

ing solar radiation. It is found that the short wave radiation decreased from 300 Wm^{-2} to 90 Wm^{-2} on 3rd June after the intensification of the system. The short wave radiation flux value returns to normal after the landfall of the system. From the area averaged diagram (**Figure 12(d)**) sensible heat flux values are found to be increasing due to the formation of clouds over cyclone area. It is found that the values increased from 5 Wm^{-2} to 30 Wm^{-2} on reaching 3rd June and after the passage of the cyclone it again decreased to normal values. From the area averaged flux pattern at the center of the system, it is found that the average net flux radiation (**Figure 13**) value decreased during the presence of cyclone. The net flux values decreased from 100 Wm^{-2} to –250 Wm^{-2} on 3rd June when cyclone was present. After the passage of cyclone it again increased to normal values.

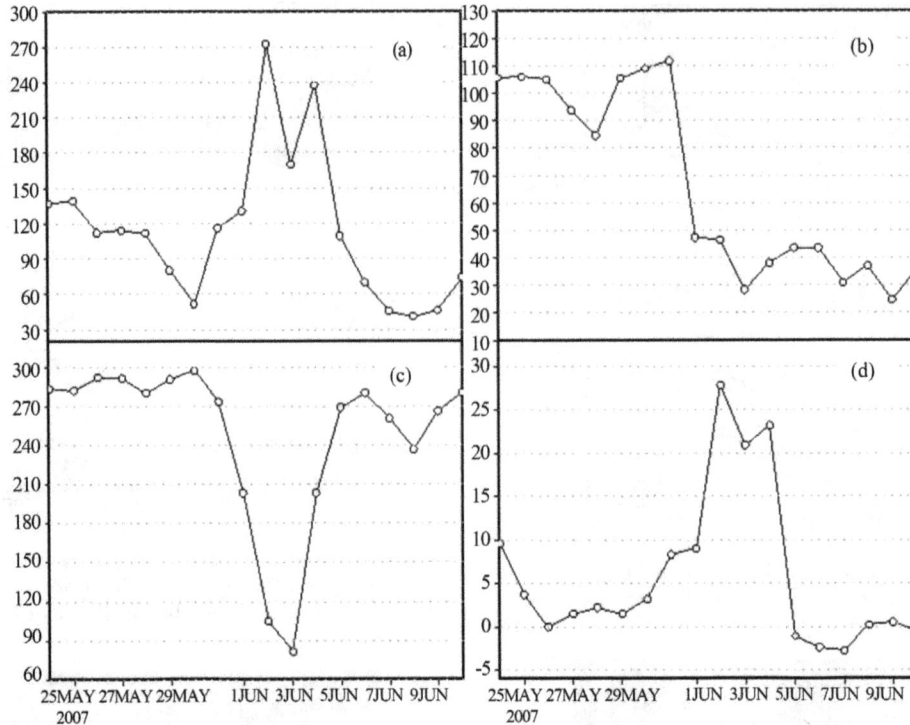

Figure 12. (a) Latent heat flux over the region 17°N - 18°N and 66°E - 67°E; (b) Long wave radiation flux; (c) Short wave radiation flux; (d) Sensible heat flux.

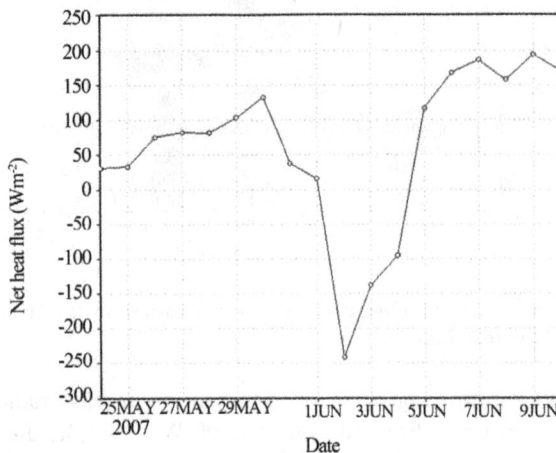

Figure 13. Net heat flux over the region 17°E - 18°E and 66°E - 67°E.

Immediately after the landfall and dissipation of the super cyclone Gonu over the Oman coast, the monsoon cloud bands started to revive over the oceanic region adjacent to south Indian peninsula. The revival of monsoon organized convection helped the formation of intense clouding which was responsible for the low OLR values even after the dissipation of super cyclone Gonu.

4. Conclusion

In 2007, a low pressure system was formed over the Ara-

bian Sea on 29th May and it was developed into a depression on 1st June 2007. It evolved as a cyclone viz. Gonu and moved towards northwest direction and crossed Oman coast on 6th June 2007. We have examined different heat budget components over the Arabian Sea region during the development of Gonu. After analyzing the different heat budget parameters during the formation of Gonu, it is found that latent heat flux increased from 120 Wm^{-2} to a maximum of 210 Wm^{-2} over the cyclone centre. The latent heat flux is increasing due to the intense clouding and associated latent heat release and it is found that the pattern of OLR and latent heat flux has a one to one relationship. For the long wave radiation there is a sharp decrease from 100 Wm^{-2} to 30 Wm^{-2} when the system was intensified into severe cyclone. The long wave radiation is the emitted radiation from the sea surface and it is a function of SST. Since the SST responses are a slow process over the ocean, even though the long wave radiation decreases as the SST decreases, it does not have a similar pattern corresponding to OLR. Net flux decreased sharply to negative values, −200 Wm^{-2} at the centre of the system. The contribution to the net flux is made by the other component fluxes and therefore there is a corresponding decrease observed in the Net flux. There is a sharp increase in the sensible heat flux value to 20 Wm^{-2}. Short wave radiation flux decreases from 300 Wm^{-2} to 90 Wm^{-2} during the formation of cyclone. Short wave radiation flux ix being decreased by

Study of the Oceanic Heat Budget Components over the Arabian Sea during the Formation and Evolution
of Super Cyclone, Gonu

19

the intense cloud cover due to the presence of system and a decrease is observed. From our observations it is concluded that the main meteorological and oceanographic parameters that responsible for the variability in heat budget components are OLR (Cloud cover), SST, and wind magnitude.

5. Acknowledgements

The authors gratefully acknowledge the data received from WHOI centre and the first author is thankful to CSIR (New Delhi) for providing the Senior Research Fellowship. Also the authors are thankful to Dr. P. V. Hareeshkumar, NPOL for his fruitful suggestions in the work. The authors would like to acknowledge CUSAT for providing facilities to carry out research.

REFERENCES

[1] P. V. Hareesh Kumar and B. Mathew, "On the Heat Budget of the Arabian Sea," *Meteorology and Atmospheric Physics*, Vol. 62, No. 3-4, 1997, pp. 215-224.

[2] S. S. C. Shenoi, D. Shankar and S. R. Shetye, "Differences in Heat Budgets of the Near-Surface Arabian Sea and Bay of Bengal: Implications for the Summer Monsoon," *Journal of Geophysical Research*, Vol. 107, No. C6, 2002, p. 3052.

[3] U. C. Mohanty, S. K. Dube and M. P. Singh, "A Study of Heat and Moisture Budget over the Arabian Sea and Their Role in the Onset and Maintenance of Summer Monsoon," *Journal of the Meteorological Society of Japan*, Vol. 61, No. 2, 1983.

[4] W. Duing and A. Leetma, "The Arabian Sea Cooling—A Preliminary heat Budget," *Journal of Physical Oceanography*, Vol. 10, No. 2, 1980, pp. 307-312.

[5] R. L. Molinari, J. C. Swallow and J. F. Festa, "Evolution of Near Surface Thermal Structure in the Western Indian Ocean during FGGE 1979," *Journal of Marine Research*, Vol. 44, No. 4, 1986, pp. 739-762.

[6] S. Hastenrath and P. J. Lamb, "On the Heat Budget of Hydrosphere and Atmosphere in the Indian Ocean," *Journal of Physical Oceanography*, Vol. 10, 1980, pp. 694-707.

[7] P. J. Lamb and A. F. Bunker, "The Annual March of Heat Budget of the Northern and Tropical Arabian Sea," *Journal of Physical Oceanography*, Vol. 12, No. 12, 1982, pp. 1388-1410.

[8] S. Hastenrath and L. L. Greischar, "The Monsoonal Heat Budget of the Hydrosphere-Atmosphere System in the Indian Ocean Sector," *Journal of Geophysical Research*, Vol. 98, No. C4, 1993, pp. 6869-6881.

[9] R. R. Rao, "Seasonal Heat Budget Estimates of the Upper Layers in the Central Arabian Sea," *Mausam*, Vol. 39, 1988, pp. 241-248.

[10] K. K. Varma and R. G. Kurup, "Seasonal Upper Ocean Heat Budget of Northern Part of Somali Current Area," *Indian Journal of Marine Scienes*, Vol. 25, No. 1, 1996, pp. 62-66.

[11] J. Hsuing, R. E. Newell and T. Houghtby, "The Annual Cycle of Oceanic Heat Storage and Oceanic Meridional Heat Transport," *Quarterly Journal of the Royal Meteorological Society*, Vol. 115, No. 485, 1989, pp. 1-28.

[12] E. C. Etter, "Heat and Fresh Water Budget of the Gulf of Mexico," *Journal of Physical Oceanography*, Vol. 13, No. 11, 1983, pp. 2058-2069.

[13] M. I. Budyko, "Atlas of the Heat Balance of the Earth," Idrometcorozdat, Moscoue, 1963, p. 69.

[14] S. Hastenrath and E. J, Lamb, "Climatic Atlas of the Indian Ocean—Part I: Surface Climate and Atmospheric Circulation," The University of Wisconsin Press, Madison, 1979, p. 19.

[15] S. Hastenrath and E. J. Lamb, "Climatic Atlas of the Indian Ocean—Part II: The Oceanic Heat Budget," The University of Wisconsin Press, Madison, 1979, p. 17.

[16] I. Kay, T. Ohshima, Watanabe and S. Nihashi, "Surface Heat Budget of the Sea of Okhotsk during 1987-2001 and the Role of Sea Ice on It," *Journal of the Meteorological Society of Japan*, Vol. 81, No. 4, 2003, pp. 653-677.

[17] D. Halpern and R. K. Reed, "Heat Budget of the Upper Ocean under Light Winds," *Journal of Physical Oceanography*, Vol. 6, No. 6, 1976, pp. 972-975.

[18] J. Merle, "Seasonal Heat Budget in the Equatorial Atlantic Ocean," *Journal of Physical Oceanography*, Vol. 10, 1980, pp. 464-469.

[19] R. Ananthakrishnan, B. Parthasarthy and J. M. Pathan, "Meteorology of Kerala," Contributions to Marine Science, 1979, pp. 123-125.

[20] P. R. Pisharoty, "Evaporation from the Arabian Sea and Indian Southwest Monsoon," *Proceedings of International Indian Ocean Expedition*, Bombay, 22-26 July 1965, pp. 43-54.

[21] P. K. Das, "IMO Monograph on Monsoon," Fifth IMO Lecture Series, World Meteorological Organization, Geneva, 1983, p. 155.

[22] U. C. Mohanty and N. Mohankumar, "A Study of Surface Marine Boundary Layer Fluxes over the Indian Seas during Divergent Epochs Asian Summer Monsoon," *Atmospheric Environment*, Vol. 24A, No. 4, 1990, pp. 823-828.

[23] U. C. Mohanty, K. J. Ramesh, N. Mohankumar and K. V.

J. Potty, "Variability of the Indian Summer Monsoon in Relation to Oceanic Heat Budget over Indian Seas," *Dynamics of Atmospheres and Oceans*, Vol. 21, No. 1, 1994, pp. 1-22.

[24] P. R. Jayakrishnan and C. A. Babu, "Variations of Surface Boundary Layer Parameters Associated with Cyclone Gonu over Arabian Sea Using QuikSCAT Data," *International Journal of Remote Sensing*, Vol. 34, No. 7, 2013, pp. 2417-2431.

[25] D. E. Waliser and N. E. Graham, "Convective Cloud Systems and Warm-Pool Sea Surface Temperatures: Coupled Interactions and Self-Regulation," *Journal of Geophysical Research*, Vol. 98, No. D7, 1993, pp. 881-893.

[26] T. Shinoda, H. H. Hendon and J. Glick, "Intraseasonal Variability of Surface Temperature during the 1998 Summer Monsoon," *Journal of Geophysical Research*, Vol. 28, No. 10, 1998, pp. 2033-2036.

3

Mineral and Sea-Salt Aerosol Fluxes over the Last 340 kyr Reconstructed from the Total Concentration of Al and Na in the Dome Fuji Ice Core

Hironori Sato[1], Toshitaka Suzuki[1], Motohiro Hirabayashi[2], Yoshinori Iizuka[3], Hideaki Motoyama[2], Yoshiyuki Fujii[2]

[1]Department of Earth and Environmental Sciences, Faculty of Science, Yamagata University, Yamagata, Japan
[2]National Institute of Polar Research, Tokyo, Japan
[3]Institute of Low Temperature Science, Hokkaido University, Sapporo, Japan

ABSTRACT

A quantitative analysis of the total concentrations of Al and Na in the Antarctic ice sheet during the past 340 kyr was performed by applying the acid digestion method to the Dome Fuji ice core. Atmospheric fluxes of mineral and sea-salt aerosol to Dome Fuji were calculated from the total concentration. The average fluxes of mineral aerosol to Dome Fuji in the periods of glacial maximum, 18.6 ± 10.1 mg·m^{-2}·yr^{-1}, were larger than the value in the interglacial periods, 3.77 ± 2.20 mg·m^{-2}·yr^{-1}. Conversely, the fluxes of sea-salt have no significant difference between the average value of glacial maximum, 130 ± 55 mg·m^{-2}·yr^{-1}, and that of interglacial, 111 ± 54 mg·m^{-2}·yr^{-1}. The results obtained in this study suggest that the variation of mineral aerosol flux in Dome Fuji, together with climate change, was much larger than that of sea-salt aerosol flux. This result may have occurred because the variety in the intensity of the source and transport during the glacial-interglacial cycle is more significant for mineral aerosol than that for sea-salt aerosol.

Keywords: Ice Core; Antarctica; Dome Fuji; Climate Change; Aerosol Flux; Full Digestion

1. Introduction

Atmospheric aerosols have controlled the energy balance of the Earth through radiative forcing [1,2]. Moreover, their chemical properties have contributed to changes in the climate and environment of the Earth. For example, chemical species in aerosols play an important role in the CO_2 absorption mechanism of the ocean accompanying climate changes. The iron hypothesis, proposed by Martin [3], states that the iron contained in aerosols activated the biological pump in the glacial period. Oba and Pedersen considered that the $CaCO_3$ contained in mineral aerosols was supplied to and dissolved in the ocean to activate the alkalinity pump in the glacial ocean [4]. Changes in the environment of the Earth's surface accompanying a climate change would modify the generation, transport, and the removal processes of mineral and sea-salt aerosols. Modification of aerosol circulation as a result of environmental change further accelerates climate change through the feedback power of atmospheric aerosols. To quantitatively clarify the role of aerosols in past climate changes, analyses of mineral and sea-salt particles in ice cores obtained in the Arctic and Antarctic were performed [5]. Measurement of the metallic components in ice cores is a useful method for quantitative analysis of the particulate matter in these cores. Previous studies include continuous flow analysis (CFA) of the dissolved ionic metals in melted ice cores [6,7], particle induced X-ray emission (PIXE) measurement of metals in the particles collected from ice cores by filtration [8,9], and inductively coupled plasma-sector field mass spectrometry ICP-SFMS measurement of acid leachable iron [10]. It is expected that a large portion of the metallic components contained in snow and ice sheets in Antarctica exist as insoluble particulates. It is well known that nearly all Al and Fe in these mineral particles exist as insoluble aluminosilicates in snow and rain [11,12]. Several recent studies that discuss metallic elements in ice cores also report the significance of insoluble metal in the cores [6,10,13]. Therefore, analysis of the ionic species only is insufficient using ice cores to reconstruct aerosol climate changes.

It is very difficult to collect very small amounts of insoluble particles in an ice core without loss by using a filtration method. In addition, such filtration methods cannot distinguish metallic components originally supplied to the ice sheet as particles that have melted at the time of core melting. Analysis of acid leachable metal is generally recommended, but some minerals continue to release metals into the solution over several weeks [14]. In order to solve such problems, full-digestion analysis of particulate matter in ice cores by using acid and a high-temperature and pressure bomb is effective. While the particulate matter in the ice melt solution and on the filter are inhomogeneous, metallic components in the sample solution prepared by acid full-digestion is completely homogenous. Thus, full-digestion analysis provides accurate and representative data for the total (dissolved + particulate) concentration of metals in ice cores. Few examples of metal analysis in ice cores that apply the full-digestion method have been reported. Gaspari et al. obtained only 16 data of the total concentration of iron in the Dome C ice core by using acid-assisted microwave digestion and reported that the results of acid leachable iron occupied only 30% - 40% of the total iron during the Holocene [10].

In this study, we measured the total concentration of metallic elements by applying a full-digestion analysis to the ice core obtained from Dome Fuji, Antarctica. Such data of total concentration of metals have not been systematically obtained in previous ice core studies. Our results enable quantitative analysis of the role of mineral and sea-salt aerosol in climate change. The Japanese Antarctic Research Expedition conducted deep ice sheet drilling at Dome Fuji (77°19'S, 39°42'E; 3810 ma.s.l) at the peak of the east Dronning Maud Land, Antarcticain 1996 and succeeded in collecting a 2503 m ice core, known as the first Dome Fuji ice core (DF1 core) [15]. Watanabe et al. determined that the profile of oxygen isotope ratio recorded in theDF1 and the Vostok ice core have good agreement over the past three glacial cycles, concluding that the climate change occurring in East Antarctica may be essentially homogeneous [16]. Kawamura et al. performed an analysis of the O_2/N_2 ratio in air trapped in the DF1 core to clarify that the ice core covered the last 340 kyr of climate change and that the climatic record in the core supports the Milankovitch theory [17]. In addition, analysis of the concentration of soluble ion species [18], the volcanic layers [19], the number of microparticles [20], and the chemical composition of salt particles in the DF1 core [21-23] was conducted for clarifying changes in the Earth's climate and environment. In this study, we measured the total concentration of Al and Na in the DF1 core, and we present the record of atmospheric fluxes of mineral and sea-salt aerosols occurring over the past 340 kyr.

2. Experimental

Anice block with a thickness of 7 cm was cut from the entire layer of the DF1 core. To remove contamination on the surface of the ice block, a thickness of approximately 5 mm was sliced from the surface with a ceramic knife. The ice block was then placed in a polyethylene container and rinsed with approximately 1 ml of ultrapure water from a wash bottle. The solution in the container was discarded by decantation. This washing procedure was repeated three times. Next, the ice block was placed at room temperature until 10 ml of the ice melted. The melted water was discarded, and the same washing procedure was repeated again three times. The ice block was then entirely melted within the polyethylene container, and the solution was adjusted to 0.1 N-HNO_3. The solution was entirely evaporated in a Teflon vessel, and the residue was decomposed by the microwave acid digestion method with 0.3 ml of concentrated HNO_3 and 0.2 ml of concentrated HF [24]. The microwave decomposition vessel Type P-25 (San-ai Kagaku) was used in this study. The concentrations of Al and Na in the solution were measured by Inductively Coupled Plasma Mass Spectrometry (Hewlett Packard, HP4500) with a desolvating nebulizer system (CETAC, Aridus). The multi-element solution XSTC-13 (SPEX CertiPrep) was used as a standard for the analysis. In addition, ultrapure water produced by Direct-Q (Millipore) and an ultrapure reagent produced by Kanto Chemical Co. were used in the experiment. All apparatus were soaked in a 1 N-HNO_3 solution for more than three days to remove metal contamination before being used in the experiment. All experiments were performed within a Class 100 laminar flow clean bench.

3. Results and Discussion

Profiles of the concentration of total Al (t-Al) and total Na (t-Na) of the DF1 core for the past 340 kyr are shown in **Figure 1** together with the profile of $\delta^{18}O$ [17]. We can consider that t-Al is a proxy of mineral particles in the ice core; however, we must be aware that t-Na is derived from sea-salt in addition to mineral particles. The concentrations of t-Al in the DF1 core ranged from 1.14 mg·kg^{-1} to 262 mg·kg^{-1}, and the maximum concentration was approximately 230 times the value of the minimum concentration. The concentration peaks of t-Al appeared in the periods of glacial maximum at the past three glacial epochs. In periods other than glacial maximum, particularly the interglacial period, t-Al concentration showed a lower value. This variation qualitatively coincides with the results of the terrigenous metals in the ice cores from Vostok [25] and Dome C [6-10]. It is well known that the arid areas on the continents expanded, and the frequency of dust storms and the activity of meridional

Mineral and Sea-Salt Aerosol Fluxes over the Last 340 kyr Reconstructed from the Total Concentration of Al and Na
in the Dome Fuji Ice Core

23

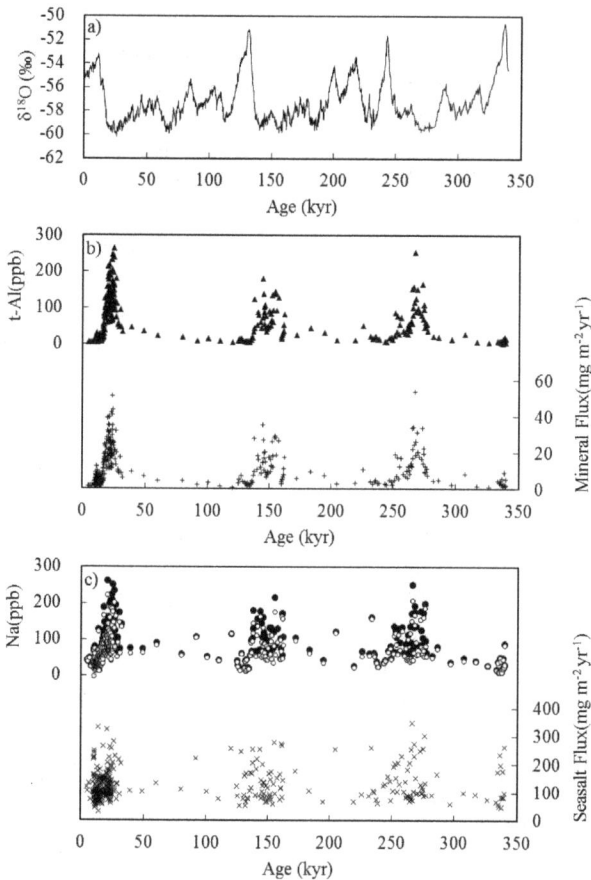

Figure 1. Profiles of (a) $\delta^{18}O$ (Kawamura *et al.*, 2007); (b) t-Al concentration (closed triangles) and mineral flux (plus signs); and (c) t-Na concentration (closed circles), sea-salt-Na concentration (open circles) and sea-salt flux (crosses) of the Dome Fuji ice core during the past 340 kyr.

circulation increased during the Last Glacial Maximum (LGM) [26]. A considerable amount of evidence implies that the atmosphere of the glacial period was generally dustier than that of present conditions [25,27-31]. The record of t-Al concentration in the DF1 core obtained in this study also supports this climate scenario. The maximum concentration of t-Al obtained in this study, 262 mg·kg^{-1}, was twice that of the maximum concentration of Al, approximately 130 mg·kg^{-1}, in the Vostok core obtained through Instrumental Neutron Activation Analysis of a vigorously shake nice melt solution [25]. We suppose that shaking may not produce sufficiently homogenous particulate matter in the solution for effective chemical analysis of the metal. The difference in the maximum concentration of Al may be due to the difference in parameters such as sample depth, coring site, and analytical method. Although we cannot entirely explain which reason is principal, we expected that the difference in the depth of the sample cutting or analytical method may be more significant than spacial variability. The average t-Al concentration and its standard deviation (1 s) during 5 -

15 kyr, roughly corresponding to the Holocene (HOL), was 10.9 ± 6.01 mg·kg^{-1}, and that during 15 - 30 kyr, roughly corresponding to the LGM, was 106 ± 57 mg·kg^{-1}. Therefore, the LGM/HOL ratio of the average concentration of t-Al was approximately 10. The LGM/HOL ratio of mass concentration of micro-particles obtained by the particle counter was reported as 11 in the DF1 [19]. The LGM/HOL ratio of the t-Al concentration agreed with that of the micro-particle concentration, which reflects the extent of LGM-HOL changes in mineral aerosol deposition. Marino *et al.* [9] measured the concentration of Ti contained in the Dome C ice core, in which the stable and insoluble crustal marker was the same as that of Al. The LGM/HOL ratio for Ti concentration calculated from their results was 14, which was also comparable to the results of t-Al concentration analysis obtained in this study. In particular, the results indicate that the mineral aerosol comprising the t-Al concentration in the ice sheet of Dome Fuji increased by approximately ten times and was deposited in LGM in which the cold and dry climate reached its peak. In the Holocene, the warm and wet climate progressed. Glacial-interglacial variations of mineral and sea-salt aerosol fluxes are discussed subsequently.

The range of t-Na concentration in the DF1 was 11.8 - 262 mg·kg^{-1}. Although concentration peaks observed in the periods of glacial maximum were the same as those of t-Al, the ratio of the maximum concentration to minimum concentration was approximately 20, and the extent of concentration variation together with climate change was smaller than that of t-Al. For these reasons, we suppose that the transport of Na through the glacial cycle does not correlate with that of Al because the source of Na to Dome Fuji, generally the sea and sea ice, is significantly closer than that of the source of Al, continental crust. Thus, Na was supplied to Dome Fuji from both the sea and crust, while the most significant source of Al is the crust only. The average concentration of t-Na during 5 - 15 kyr was 41.6 ± 18.1 mg·kg^{-1}, and that during 15 - 30 kyr was 115 ± 43 mg·kg^{-1}; the LGM/HOL ratio was 3. Although no previous studies have measured t-Na concentration by full-digestion of the ice core, the concentration of the dissolved Na in the Holocene and LGM in the Dome C ice core have been reported by Bigler *et al.* [7]. The LGM/HOL ratio of dissolved Na calculated from their result, 5, was somewhat larger than the ratio for t-Na obtained in this study. This ratio difference may indicate that the insoluble fraction of Na in the ice core gained significance in the Holocene. While sea-salt particles are relatively soluble as compared to mineral particles, there is no assurance that an entire sea-salt particle dissolves when the ice core sample melts. In addition, the melted solution of the ice core contained Na, which was originally included as ions in the snow that were dissolved from the particulate matter at the time the ice core melted. Thus, some difficulties persist in the interpretation of the dissolved constituents in the ice cores. Hence, measurements for total concentration of elements by using the full-digestion method provide meaningful data with clear definition for ice core analysis.

The atmospheric fluxes of mineral ($F_{mineral}$) and sea-salt ($F_{seasalt}$) aerosols in Dome Fuji can be estimated from the results of total concentrations of Al and Na. Although Al is a major constituent in the Earth's crust and occupies approximately 8% of the crustal substance, it is a trace element detected in seawater at only 0.8 ppb. Therefore, it is a suitable assumption that t-Al in the aerosol originated entirely from detritus of mineral particles. On the basis of this assumption, $F_{mineral}$ in Dome Fuji is calculated from the following equation:

$$F_{mineral} = [\text{t-Al}] \times 100/8.23 \times R_d, \qquad (1)$$

where [t-Al] is the concentration of t-Al, 8.23 is the mean crustal abundance of Al in percentage [32] and R_d is the snow accumulation rate of the ice sheet in Dome Fuji at a depth of sample, d (Dome Fuji Ice Core Consortium, personal communication).

The flux of sea-salt aerosol, $F_{sea-salt}$ is similarly obtained by the following equation:

$$F_{sea-salt} = [\text{sea-salt-Na}] \times 100/1.06 \times R_d, \qquad (2)$$

where [sea-salt-Na] is the concentration of Na of sea-salt origin, and 1.06 is the mean seawater composition of Na in percentage [33]. While Na is a major element of seawater, it is also a major component of the crust, occupying 2.36% [32]. Hence, t-Na in the atmosphere cannot be assumed as a tracer of sea-salt aerosol at a time or place that included an abundant supply of mineral aerosol [34]. It should be restricted that Na in aerosol is treated as a tracer of sea-salt when the contribution of mineral aerosol can be neglected. Bigler et al. [7] reported that Na of acrustal origin became significant at the high-dust period of a glacial stage in the Dome C ice core. Therefore, [sea-salt-Na] must be used for calculation of $F_{sea-salt}$. Assuming that t-Al in the ice core is of crustal origin, [sea-salt-Na] can be calculated by the following formula:

$$[\text{sea-salt-Na}] = [\text{t-Na}] - (\text{Na/Al})_{crust} \times [\text{t-Al}], \qquad (3)$$

where [t-Na] is the concentration of t-Na in the core, and (Na/Al)$_{crust}$ is the Na/Al ratio in the crust, 0.29 [32]. The results of the calculation of [sea-salt-Na] indicate that the percentage of [sea-salt-Na] to [t-Na] was a maximum of 99.3% and a minimum of 37.5% (**Figure 1**). In particular, roughly 60% of t-Na in the DF1 core was of crustal origin in the maximum case. The average proportion of [sea-salt-Na] to [t-Na] in the Holocene was 91.1% ± 6.7% and that in the LGM was 74.0% ± 12.1%. These results suggest that most of the Na transported to Dome Fuji

during a warm and low-dust period is considered to be of sea-salt origin; however, a significant amount of Na was transported to Dome Fuji from a remote continental crust during a cold and high-dust environment in the LGM. The maximum contributions of crustal Na in the three glacial maximums, recognized in **Figure 1**, were 62.5%, 36.9%, and 53.4%, respectively, in order of recent epochs. In ice core analysis, Na of crustal origin must be considered, particularly in the periods of glacial maximum. Transport of the sea-salt aerosols to inland Antarctica would not significantly vary in the time scale of glacial-interglacial cycle because production of the sea-salt aerosols occurred in the sea and sea ice near the continent. Conversely, deposition of mineral aerosol on the Antarctic ice sheet became significant in glacial maximums at the end of each glacial epoch because 1) the continental shelf exposure increased, accompanied by a decrease in sea level; 2) the arid region on the continents expanded in cold and dry climates; and 3) the frequency of dust storms in the continents and the activity of meridional circulation increased.

The average of atmospheric fluxes of mineral and sea-salt aerosol in Dome Fuji during the glacial maximum and interglacial stage of the last three glacial epochs are presented in **Table 1**. Abbreviations of climatic divisions, IG (interglacial) and GM (glacial maximum), were defined by the range of age mentioned in the table and were numbered in order of younger stages. The divisions IG1, IG2, and IG3 roughly correspond to stages 1, 5e, and 7e, and GM1, GM2, and GM3 roughly correspond to stages 2, 6b, and 8d of the Marine Isotope Stage, respectively. The whole mean of each stage of IG and GM is also shown in the table. The average atmospheric fluxes of mineral aerosol in Dome Fuji at GM periods, 14.0 - 20.2 mg·m^{-2}·yr^{-1}, were approximately one order of

Table 1. Average atmospheric fluxes of mineral and sea-salt in Dome Fuji, Antarctica, during glacial maximum and interglacial stages of last three glacial cycles.

Climatic division	Period Kyr	Mineral flux ± 1 s mg·m^{-2}·yr^{-1}	Sea-salt flux ± 1 s mg·m^{-2}·yr^{-1}
IG1	5 - 14	4.15 ± 2.42	112 ± 54
GM1	15 - 30	19.9 ± 9.9	126 ± 49
IG2	120 - 134	3.22 ± 1.67	115 ± 66
GM2	135 - 149	14.0 ± 8.0	133 ± 55
IG3	235 - 249	3.71 ± 1.03	105 ± 34
GM3	265 - 279	20.2 ± 10.8	155 ± 85
IG$_{mean}$		3.77 ± 2.20	111 ± 54
GM$_{mean}$		18.6 ± 10.1	130 ± 55

IG$_{mean}$ and GM$_{mean}$ represent the whole mean for IG and GM periods, respectively.

magnitude larger than the value at IG periods, 3.22 - 4.15 mg·m^{-2}·yr^{-1}. These results indicate that the generation and transport of mineral aerosols were activated during periods of glacial maximum. The mineral flux in GM2 was somewhat smaller than that in GM1 and GM3, which may have occurred because interstadials were more frequent within the period of GM2 than those in GM1 and GM3 (**Figure 1**). The average fluxes of mineral aerosol in the periods of interglacial and glacial maximum were 3.77 mg·m^{-2}·yr^{-1} and 18.6 mg·m^{-2}·yr^{-1}, respectively. The ratios of the mineral flux in GM and that in IG, the GM/IG ratio, were 4 - 5, and the GM$_{mean}$/IG$_{mean}$ ratio was 4.9. Gaspari *et al.* [10] reported that the LGM/HOL ratio of acid-leachable Fe flux in the Dome C ice core is 36. Moreover, it was reported that the LGM/ HOL ratio of micro-particle flux in the Vostok ice core is ~15 [5]. The ratios obtained in this study were smaller than those in previous studies, which used different methods. We consider that the ratios obtained in this study are highly appropriate for explaining differences in mineral aerosol transport between the periods of glacial maximum and interglacial because our results were obtained by a full-digestion homogenous particulate analysis. As evidence, the LGM/HOL ratios of the atmospheric mineral flux restored from the deep-sea sediment core, 3 - 4 [31,35-37], were comparable to the values obtained in this research. The fluxes of sea-salt show no remarkable differences in IG and GM, and the values of IG$_{mean}$ and GM$_{mean}$ were 111 ± 54 mg·m^{-2}·yr^{-1} and 130 ± 55 mg·m^{-2}·yr^{-1}, respectively. The GM/IG ratios obtained in this study were near unity. This result indicates that although no significant difference was evident in the average sea-salt flux in Dome Fuji between the period of glacial maximum and interglacial, the flux varied widely among data. The results obtained in this study suggest that the variation of mineral flux in Dome Fuji together with climate change was much larger than that of sea-salt flux. Conversely, the variation in sea-salt flux may be affected more by short-term phenomena than by climate change. To evaluate the relationship between climate change and variability of aerosol flux, the coefficients of variation (CV) of the age-moving average of $F_{mineral}$ and $F_{sea-salt}$ in various age intervals were calculated. The CV was obtained by the standard deviation of the age-moving average divided by the average. Age intervals for the moving average were set at 0 - 1 kyr, 1 - 5 kyr, 5 - 15 kyr, and 15 - 25 kyr and subsequently at intervals of 10 kyr until 95 - 105 kyr. For example, 271 datasets were obtained by calculation through 340 kyr at an age interval of 0 - 1 kyr. The average CV obtained by calculation in each age interval is shown in **Figure 2**. The CV of $F_{mineral}$ increased with age interval. The rate of increase is significant in the range of 0 - 1 kyr to 5 - 15 kyr, which may reflect sudden climate changes. In the age interval of 0 - 1 kyr, the CV of $F_{mineral}$ and

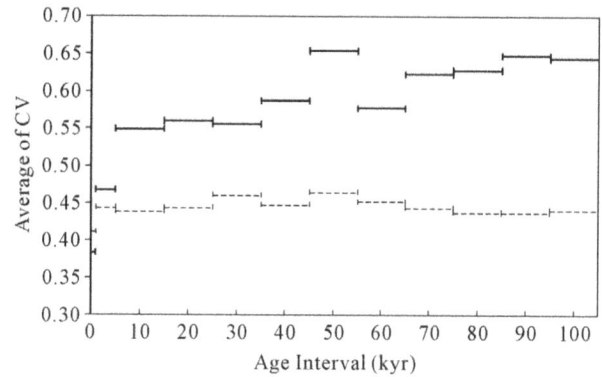

Figure 2. Average of coefficients of variation (CV) of mineral (solid lines) and sea-salt (dashed lines) fluxes to Dome Fuji in various age intervals.

$F_{sea-salt}$ were 0.38 and 0.41, respectively, which suggests that the supply of mineral and sea-salt aerosols to the Antarctic ice sheet have nearly the same variation in the 0 - 1 kyr time interval. The CV of $F_{mineral}$ gradually increased from the age interval of 5 - 15 kyr to 95 - 105 kyr, while that of $F_{sea-salt}$ was nearly constant during the entire age interval. These results support the theory that the supply of mineral aerosol was much more varied than that of sea-salt aerosol, which may be attributed to the following points: 1) The variation in mineral aerosol may be much more sensitive to changes in atmospheric circulation than that of sea-salt aerosol because the source of mineral aerosol, continental crust, is significantly far from the Antarctic ice sheet; 2) changes in the surface environment of the continental crust together with climate changes, e.g., area [Remark 1] and dryness, may be more significant than that of the sea surface; and 3) the change in the sea-salt flux to the Antarctic ice sheet would not change significantly with the glacial cycle because both open sea and sea ice are significant sources for sea-salt aerosol.

4. Conclusion

The profiles of total concentration of Al and Na in the Antarctic ice sheet during the past 340 kyr were clarified by full-digestion analysis of the Dome Fuji ice core. The fluxes of mineral and sea-salt aerosol to Dome Fuji were estimated from the total concentration of Al and Na in the core. The average aerosol fluxes in the periods of interglacial and glacial maximum were 3.77 ± 2.20 mg·m^{-2}·yr^{-1} and 18.6 ± 10.1 mg·m^{-2}·yr^{-1}, respectively, for mineral aerosol and 111 ± 54 mg·m^{-2}·yr^{-1} and 130 ± 55 mg·m^{-2}·yr^{-1}, respectively, for sea-salt aerosol. On the basis of the results obtained in this study, we consider the following scenario for the variation in aerosol flux to the Antarctic ice sheet with the glacial cycle: Transport of sea-salt aerosols generated in the sea and sea ice near Antarctica did not vary significantly in the time scale of the glacial-interglacial cycle. Transport of sea-salt aerosol may be affected by short-term scale phenomena such as meteorological conditions rather than climate change. In the glacial maximum at the end of each glacial epoch, deposition of mineral aerosol on the Antarctic ice sheet reached a maximum because the exposure of the continental shelf increased, accompanied by a decrease in sea level, the arid region on the continents was expanded by the cold and dry climate, and the frequency of dust storms in the continents and the activity of meridional circulation increased. Further investigation including analysis in high time-resolution and examination of the chemical compositions of particulates in the DF1 core will be performed to clarify the role of atmospheric mineral and sea-salt aerosols in climate change.

5. Acknowledgments

The authors are deeply indebted to the Dome Fuji field members of the Japanese Antarctic Research Expedition for collecting the ice core. We would also like to acknowledge the members of the Dome Fuji Ice Core Consortium for fruitful discussions. This study was supported partly by the Grant for Joint Research Program of the National Institute of Polar Research and by the Grant for Joint Research Program of the Institute of Low Temperature Science, Hokkaido University.

REFERENCES

[1]   I. Tegen, A. A. Lacis and I. Fung, "The Influence on Climate Forcing of Mineral Aerosols from Disturbed Soils," *Nature*, Vol. 380, 1996, pp. 419-422.

[2]   S. Solomon, D. Qin, M. Manning, Z. Chen, M. Marquis, K. B. Averyt, M. Tignor and H. L. Miller, "Climate Change 2007: The Physical Science Basis. Contribution of Working Group I to the Fourth Assessment Report of the Intergovernmental Panel on Climate Change," Cambridge University Press, Cambridge, 2007.

[3]   J. H. Martin, "Glacial-Interglacial CO_2 Change: The Iron Hypothesis," *Paleoceanography*, Vol. 5, 1990, pp. 1-13.

[4]   T. Oba and T. F. Pedersen, "Paleoclimatic Significance of Eolian Carbonates Supplied to the Japan Sea during the Last Glacial Maximum," *Paleoceanography*, Vol. 14, 1999, pp. 34-41.

[5]   H. Fischer, M.-L. Siggaard-Andersen, U. Ruth, R. Röthlisberger and E. Wolff, "Glacial/Interglacial Changes in Mineral Dust and Sea-Salt Records in Polar Ice Cores: Sources, Transport, and Deposition," *Reviews of Geophysics*, Vol. 45, 2007, pp. 1-26.

[6]   R. Traversi, C. Barbante, V. Gaspari, I. Fattori, O. Largiuni, L. Magaldi and R. Udisti, "Aluminum and Iron Record for the Last 28 kyr Derived from the Antarctic EDC96 Ice Core Using New CFA Methods," *Annals of Glaciology*, Vol. 32, 2004, pp. 300-306.

[7]   M. Bigler, R. Röthlisberger, F. Lambert, T. F. Stocker and D. Wagenbach, "Aerosol Deposited in East Antarctica over the Last Glacial Cycle: Detailed Apportionment of Continental and Sea-Salt Contributions," *Journal of Geophysical Research*, Vol. 111, 2006, Article ID: D08205.

[8]   G. Ghermandi, R. Cecchi, M. Capotosto and F. Marino, "Elemental Composition Determined by PIXE Analysis of the Insoluble Aerosol Particles in EPICA-Dome C Ice Core Samples Representing the Last 27,000 Years," *Geophysical Research Letters*, Vol. 30, 2003, pp. 21-76.

[9]   F. Marino, V. Maggi, B. Delmonte, G. Ghermandi and J. R. Petit, "Elemental Composition (Si, Fe, Ti) of Atmospheric Dust over the Last 220 kyr from the EPICA Ice Core (Dome C, Antarctica)," *Annals of Glaciology*, Vol. 39, 2004, pp. 110-118.

[10] V. Gaspari, C. Barbante, G. Cozzi, P. Cescon, C. F. Boutron, P. Gabrielli, G. Capodaglio, C. Ferrari, J. R. Petit and B. Delmonte, "Atmospheric Iron Fluxes over the Last Deglaciation: Climatic Implications," *Geophysical Research Letters*, Vol. 33, 2006, Article ID: L03704.

[11] R. Losno, J. L. Colin, N. Lebris, G. Bergametti, T. Jickells and B. Lim, "Aluminum Solubility in Rainwater and Molten Snow," *Journal of Atmospheric Chemistry*, Vol. 17, 1993, pp. 29-43.

[12] S. Kawakubo and M. Iwatsuki, "Speciation of Iron in Rain Water by Size Fractionation, Acid Decomposition, Acid Extraction and Catalytic Determination," *Analytical Sciences*, Vol. 16, 2000, pp. 945-949.

[13] M.-L. Siggaard-Andersen, P. Gabrielli, J. P. Steffensen, T. Strømfeldt, C. Barbante, C. Boutron, H. Fischer and H. Miller, "Soluble and Insoluble Lithium Dust in the EPICA Dome C Ice Core—Implications for Changes of the East Antarctic Dust Provenance during the Recent Glacial-Interglacial Transition," *Earth and Planetary Science Letters*, Vol. 258, 2007, pp. 32-43.

[14] R. H. Rhode, J. A. Baker, M.-A. Millet and N. A. N. Bertler, "Experimental Investigation of the Effects of Mineral Dust on the Reproducibility and Accuracy of Ice Core Trace Element Analyses," *Geochemical Geology*, Vol. 286, 2011, pp. 207-211.

[15] Dome-F. Deep Coring Group, "Deep Ice-Core Drilling at Dome Fuji and Glaciological Studies in East Dronning Maud Land, Antarctica," *Annals of Glaciology*, Vol. 27, 1998, pp. 333-337.

[16] O. Watanabe, J. Jouzel, S. Johnsen, F. Parrenin, H. Shoji and N. Yoshida, "Homogeneous Climate Variability across East Antarctica over the Past Three Glacial Cycles," *Nature*, Vol. 422, 2003, pp. 509-512.

[17] K. Kawamura, F. Parrenin, L. Lisiecki, R. Uemura, F. Vimeux, J. P. Severinghaus, M. A. Hutterli, T. Nakazawa, S. Aoki, J. Jouzel, M. E. Raymo, K. Matsumoto, H. Na-

kata, H. Motoyama, S. Fujita, K. Goto-Azuma, Y. Fujii and O. Watanabe, "Northern Hemisphere Forcing of Climatic Cycles in Antarctica over the Past 360,000 Years," *Nature*, Vol. 448, 2007, pp. 912-916.

[18] Dome-F. Ice Core Research Group, "Preliminary Investigation of Palaeoclimate Signals Recorded in the Ice Core from Dome Fuji Station, East Dronning Maud Land, Antarctica," *Annals of Glaciology*, Vol. 27, 1998, pp. 338-342.

[19] Y. Fujii, M. Kohno, S. Matoba, H. Motoyama and O. Watanabe, "A 320 K-Year Record of Microparticles in the Dome Fuji, Antarctica Ice Core Measured by Laser-Light Scattering," *Memories of National Institute of Polar Research Special Issue*, Vol. 57, 2003, pp. 46-62.

[20] Y. Fujii, M. Kohno, H. Motoyama, S. Matoba, O. Watanabe, S. Fujita, N. Azuma, T. Kikuchi, T. Fukuoka and T. Suzuki, "Tephra Layers in the Dome Fuji (Antarctica) Deep Ice Core," *Annals of Glaciology*, Vol. 29, 1999, pp. 126-130.

[21] H. Ohno, M. Igarashi and T. Hondoh, "Salt Inclusions in Polar Ice Core: Location and Chemical Form of Water-Soluble Impurities," *Earth and Planetary Science Letters*, Vol. 232, 2005, pp. 171-178.

[22] Y. Iizuka, T. Miyake, M. Hirabayashi, T. Suzuki, S. Matoba, H. Motoyama, Y. Fujii and T. Hondoh, "Constituent Elements of Insoluble and Non-Volatile Particles during the Last Glacial Maximum Exhibited in the Dome Fuji (Antarctica) Ice Core," *Annals of Glaciology*, Vol. 55, 2009, pp. 552-562.

[23] Y. Iizuka, R. Uemura, H. Motoyama, T. Suzuki, T. Miyake, M. Hirabayashi and T. Hondoh, "Sulphate-Climate Coupling over the Past 300,000 Years in Inland Antarctica," *Nature*, Vol. 490, 2012, pp. 81-84.

[24] T. Suzuki and M. Sensui, "Application of Microwave Acid Digestion Method to the Decomposition of Rock Samples," *Analytica Chimca Acta*, Vol. 245, 1991, pp. 43-48.

[25] M. De Angelis, N. I. Barkov and V. N. Petrov, "Aerosol Concentrations over the Last Climatic Cycle (160 kyr) from an Antarctic Ice Core," *Nature*, Vol. 325, 1987, pp. 318-321.

[26] CLIMAP Project Members, "The Surface of the Ice-Age Earth," *Science*, Vol. 191, 1976, pp. 1131-1144.

[27] M. Ram, R. I. Gayley and J. R. Petit, "Insoluble Particles in Antarctic Ice: Background Aerosol Size Distribution and Diatom Concentration," *Journal of Geophysical Research*, Vol. 93, 1988, pp. 8378-8382.

[28] P. A. Mayewski, L. D. Meeker, S. Whitlow, M. S. Twickler, M. C. Morrison, P. Bloomfield, G. C. Bond, R. B. Alley, A. J. Gow, P. M. Grootes, D. A. Meese, M. Ram, K. C. Taylor and W. Wumkes, "Changes in Atmospheric Circulation and Ocean Ice Cover over the North Atlantic during the Last 41,000 Years," *Science*, Vol. 263, 1994, pp. 1747-1751.

[29] J. P. Steffensen, "The Size Distribution of Microparticles from Selected Segments of the Greenland Ice Core Project Ice Core Representing Different Climatic Periods," *Journal of Geophysical Research*, Vol. 102, 1997, pp. 26755-26763.

[30] M. C. Reader, I. Fung and N. McFarlane, "The Mineral Dust Aerosol Cycle during the Last Glacial Maximum," *Journal of Geophysical Research*, Vol. 104, 1999, pp. 9381-9398.

[31] T. Irino and R. Tada, "High-Resolution Reconstruction of Variation in Aeolian Dust (Kosa) Deposition at ODP Site 797, the Japan Sea, during the Last 200 ka," *Global and Planetary Change*, Vol. 35, 2003, pp. 143-156.

[32] S. R. Taylor, "The Abundance of Chemical Elements in the Continental Crust: A New Table," *Geochimicaet Cosmochimica Acta*, Vol. 28, 1964, pp. 1273-1285.

[33] W. S. Broecker and T.-H. Peng, "Tracers in the Sea," Eldigio Press, Palisades, New York, 1982.

[34] T. Suzuki and S. Tsunogai, "Origin of Calcium in Aerosols over the Western North Pacific," *Journal of Atmospheric Chemistry*, Vol. 6, 1988, pp. 363-374.

[35] S. A. Hovan, D. K. Rea, N. G. Pisias and N. J. Shackleton, "A Direct Link between the China Loess and Marine $d^{18}O$ Records: Aeolian Flux to the North Pacific," *Nature*, Vol. 340, 1989, pp. 296-298.

[36] S. C. Clemens and W. L. Prell, "Late Pleistocene Variability of Arabian Sea Summer Monsoon Winds and Continental Aridity: Eolian Records from the Lithogenic Component of Deep-Sea Sediments," *Paleoceanography*, Vol. 5, 1990, pp. 109-145.

[37] P. B. deMenocal, W. F. Ruddiman and E. M. Pokras, "Influences of High- and Low-Latitude Processes on African Terrestrial Climate: Pleistocene Eolian Records from Equatorial Atlantic Ocean Drilling Program Site 663," *Paleoceanography*, Vol. 8, 1993, pp. 209-242.

4

Seasonal Variations of the Surface Fluxes and Surface Parameters over the Loess Plateau in China

Wei Li[1], Tetsuya Hiyama[2], Nakako Kobayashi[3]
[1]Graduate School of Environmental Studies, Nagoya University, Nagoya, Japan
[2]Research Institute for Humanity and Nature, Kyoto, Japan
[3]Hydrospheric Atmospheric Research Center, Nagoya University, Nagoya, Japan

ABSTRACT

Turbulent fluxes were measured by an eddy covariance system at three levels over an intricate land surface on the southern part of the Loess Plateau, consisting of heterogeneous flat terrain and a large valley 500 m away from the observation site to the southeast. The surface roughness length, the seasonal variation of bulk transfer coefficient for sensible heat (C_H), and the seasonal variation of surface moisture availability (β) were also analyzed based on the observation. The flux footprint was carefully considered in this study. A relatively dry period of the experimental area existed from June to the first week of July 2004 when the land surface offered turbulent energy to the atmospheric surface layer mainly by sensible heat flux with a maximum value of around 230 Wm^{-2}. A wet duration lasted from the second week of July to the end of September 2004 with very frequent rainfall events in conditions when the winds were mainly from the southeast;latent heat flux was dominant during the wet season and reached a peak value of around 280 Wm^{-2}. The surface parameters of C_H and β were calculated when the mean winds coming from the flat terrain, *i.e.*, from the northwest direction. The values of C_H ranged between 0.004 and 0.006 during the observational year of June 2004 to June 2005. The surface moisture availability β changed with seasons as anticipated with high values during June and July 2004 and lowest values around 0.03 in February 2005. Its peak value of 0.91 occurred in July; the mean value of β during the wet season was 0.29. Furthermore, the relationship between the surface soil water content and β indicated that changes in soil water content contributed much to variations of surface moisture availability β.

Keywords: Loess Plateau; Surface Fluxes; Bulk Transfer Coefficient; Soil Moisture Availability

1. Introduction

Turbulent fluxes in the atmospheric surface layer (ASL) are not only governed by properties of the airflow, they are also closely connected to the structure and function of the underlying surface. Generally, land surfaces are not homogeneous; the vertical structure of the atmospheric boundary layer (ABL) is modified over heterogeneous surfaces compared with that over homogeneous surfaces. As Wyngaard [1] ever pointed out, it is always a difficult challenge to determine accurate and representative surface fluxes over heterogeneous surfaces.

The Loess Plateau in China, as a semi-arid plateau, located within the middle reaches of the Yellow River basin and being affected by the Asian monsoon activities in summer, possesses a unique heterogeneity of topography and consequently induces typical processes of energy budget and fluxes transportations in the ABL. Measuring accurate surface energy fluxes and surface parameters over the Loess Plateau have significant meanings for the ABL evolution study over such a heterogeneous land-scape, the Asian monsoon and climate change studies, and the research of the water cycle systems of the Yellow River Basin and even of the Eastern Asia. The field campaign in this study which was equipped with a flux tower and a wind profile Radar was the first ABL field observation conducted over the Loess Plateau [2].

In literatures, Kimura *et al*. [3,4] ever studied seasonal variations of heat balance including sensible and latent heat fluxes and soil moisture on the Loess Plateau using a three-layer soil model and verified the model by observations over bare soil fields. Liu *et al*. [5] simulated regional evapotranspiration during 20 years by three complementary models over the Yellow River basin within which the Loess Plateau lies. These studies ignored the effects of surface heterogeneity on surface fluxes, evapotranspiration, and surface heat balance. However, many studies have been conducted to estimate the heat transfer and evapotranspiration over complex terrains utilizing synergetic methods. Tamagawa [6] obtained mean bulk transfer coefficient for sensible heat from observed flux data and mast profiles over a desert area. A multi-layer energy budget model was proposed by Kondo and Watanabe [7] which indicated that the bulk transfer coefficients for sensible and latent heat were sensitive to meteorological conditions when using the radiative surface temperature as the mean surface temperature. Kafle and Yamaguchi [8] estimated the spatial distribution of evapotranspiration over an intricate topography using a topography-considered two-source energy balance model based on remote sensing data and meteorological data. Kustas *et al.* [9] also studied the effect of pixel resolution of the remote sensing inputs on model flux estimates for an agricultural region.

Based the ABL experiment in this study, many works have been done to investigate the effects of topography and surface heterogeneity on the ABL properties over the experimental region on the Loess Plateau. Li *et al.* [10] studied the characteristics of turbulent spectra when winds from the flat terrain and from the southeastern valley, and found that for flow from the valley direction the valley did significantly influence the shape of the lateral spectrum indicating the modification of the overlying ABL structure. The effect of the topography on local circulation and cumulus generation over this region was evaluated by Nishikawa *et al.* [11] through numerical simulations of ABL using a cloud-resolving non-hydrostatic model. Takahashi *et al.* [12] investigated the diurnal variation of water vapor mixing between the ABL and the free atmosphere over the Loess Plateau based on the measurements of water vapor and wind by the ground-based microwave radiometer and the wind profile Radar.

In this paper, we will estimate the surface energy fluxes over the experimental site. The main objective is to study the variability of surface fluxes with the surface wetness. The seasonal variations of the bulk transfer coefficient for sensible heat and the surface moisture availability will be presented to investigate the properties of energy transfer and evapotranspiration over the studied region on the Loess Plateau.

2. Study Area

2.1. Experimental Site

The ABL observation in this study was carried out at a field site (35°12'N, and 107°40'E) located in the middle southern area of the Loess Plateau in China. The Loess Plateau has a semi-arid land climate, and lies within 34° to 40°N and 100° to 115°E. The Loess Plateau consists generally of strongly dissected flat terrain and gullies extending long to several tens kilometers with typical depths in the order of 100 m and surface widths in the order of 1 km [10]. Its total area is 0.62 million km^2 (35% tableland; 65% gully slopes). The annual precipitation ranges from 150 to 750 mm, 70% of which falls in summer. Because loess soil is susceptible to erosion, the Loess Plateau landscape includes numerous steep gullies. Approximately 90% of the soil deposited into the Yellow River each year originates from the Loess Plateau [3,4, 13].

The land cover of the studied site is heterogeneous, consisting of wheat, apple trees, and residences; the area is mainly flat, except for a large valley with the nearest edge about 500 m away from the observation tower in the southeast. The valley has the depth of around 100 m, a surface width of about 1 km, and the length around 20 km extending to the southeast (see **Figure 1**).

2.2. Climatology

The monthly precipitation and mean air temperature averaged during 45 years from 1957 to 2001 at the experimental site are shown in **Figure 2**. These long-term statistical data show that large precipitation over the experimental area occurs mainly during the three months from July to September, and that the maximum air temperature occurs in July. The mean annual air temperature at the experimental site is 9.1°C. The dominant wind direction is from the southeast in summer and from the northwest during the other three seasons [10].

The daily precipitation observed during 13 months from June 2004 to June 2005 is shown in **Figure 3**. Due to the high altitude and the semi-arid climate of the Loess Plateau, the experimental region gets into winter from the middle of October and the winter lasts to the end of March; during winter the region has low air temperature, low solar radiation, and low precipitation (see **Figures 2** and **3**).

In **Figure 3**, it is obvious that from June through to the first week of July in 2004, the observed precipitation was

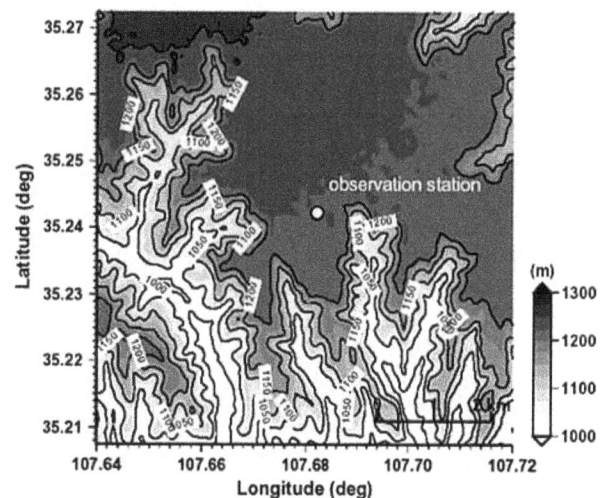

Figure 1. The topography of the experimental region. The digital elevation data were based on http://www2.jpl.nasa.gov/srtm.

(a)

(b)

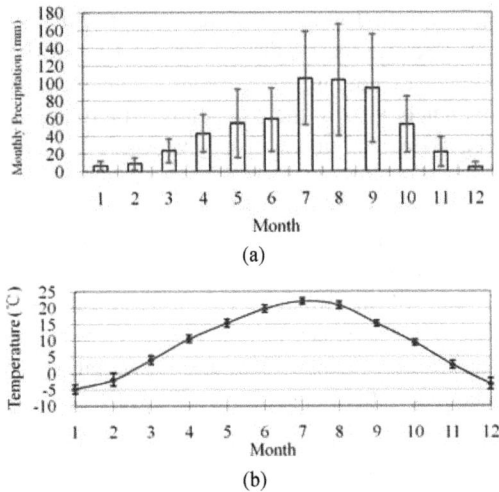

Figure 2. The monthly mean precipitation (a) and monthly mean air temperature (b) averaged during the years 1957-2001 at the experimental site (vertical bars show standard deviations).

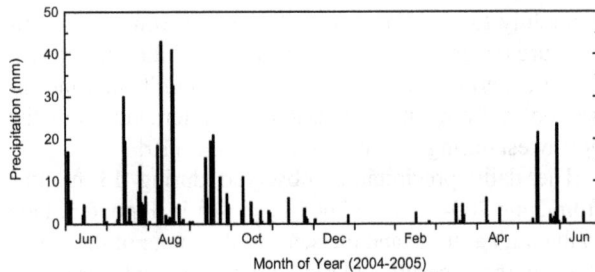

Figure 3. Daily precipitation during June 2004 to June 2005.

not much. During the remaining days of July, through August, and until the end of September, rainfall events occurred very frequently, many of which had quite prominent precipitation. The maximum precipitation occurred on August 11 and August 19, with the daily values of 42 mm and 41 mm respectively. The accumulated precipitation during the period of July 9 to September 30 was 339 mm, whereas that during June 1 to July 8 was only 36 mm. Therefore, we will use the duration of June and the first week of July to represent the dry season over the studied region, and from the second week of July until the end of September to represent the wet season. We will focus on analyzing fluxes properties during these dry and wet seasons.

3. Instrumentation

In this experiment, a flux and radiation observation system (FROS) was installed to measure turbulent fluxes and radiation components in the atmospheric surface layer. The FROS included three ultrasonic anemometer/thermometers (1210R3; GILL Instruments, Ltd., UK) mounted at 2 m, 12 m, and 32 m heights measuring turbulent quantities of wind and air temperature, as well as

three open-path infrared CO_2/H_2O gas analyzers (LI-7500; Li-Cor, Inc., USA) recording the fluctuations of CO_2 and water vapor densities at the three heights. Two shortwave radiometers (CM21; Kipp & Zonen B.V., Netherlands) and two longwave radiometers (CG4; Kipp & Zonen B.V., Netherlands) were installed at 2 m level to measure the upward/downward shortwave radiations and upward/downward longwave radiations. Soil thermometers (CS107; Campbell Scientific Inc., USA) and soil moisture sensors (TDR-615; Campbell Scientific Inc., USA) were also used to measure the soil temperature and soil moisture at the depth of 2 cm, 10 cm, 20 cm, 40 cm, and 80 cm respectively. A rain gauge (34-T; Ohtakeiki Ltd., Japan) measured the precipitation. The data logger was the type of CR5000 by Campbell Scientific, Inc., USA.

4. Methodology

4.1. Data Selection

In this study, 10 Hz raw turbulent data measured by the ultrasonic anemometer/thermometers at three heights (2 m, 12 m, and 32 m) during the observational year (June 2004 to June 2005) were used to calculate sensible and latent heat fluxes by eddy correlation method. In section 5.1, we will focus on analyzing the properties of surface heat fluxes in the wet and dry seasons of a summer during four months (June to September in 2004) within the observational year.

Because the surface parameters C_H, C_E, and β are based on the assumption of surface homogeneity, to avoid the topography complexity caused by the southeast valley, these parameters were calculated only in conditions when winds came from the northwest flat terrain. In Sections 5.2 and 5.3, we will investigate the seasonal variations of bulk transfer coefficient C_H for sensible heat and surface moisture availability β based on the data collected at 32 m height during the whole observational year (June 2004 to June 2005). In all these cases, we only selected clear days according to the precipitation data.

In previous studies, Matsushima and Kondo [14] described that C_H was little affected by the atmospheric stability within the stability range of $-0.5 < z/L < 0$. Kondo and Watanabe [7] found that β was sensitive and widely dispersed when the mean wind speed was smaller than 2.0 ms^{-1} and the solar radiation was less than 300 Wm^{-2}. Accordingly, we selected the data when 1) mean wind speed larger than 2.0 ms^{-1}; 2) solar radiation larger than 300 Wm^{-2}; 3) atmospheric stability in the range of $-0.5 < z/L < 0$.

4.2. Calculation of Surface Heat Fluxes and Surface Parameters

4.2.1. Surface Heat Fluxes

The turbulent sensible and latent heat fluxes at 2 m, 12 m,

and 32 m heights were calculated by the eddy covariance method, which can be formulated as following:

$$\begin{cases} H = \rho c_P \overline{w'\theta'} \\ LE = \rho l \overline{w'q'} \end{cases} \quad (1)$$

Here H denotes the sensible heat flux, LE the latent heat flux; c_P and ρ are the specific heat and the air density, l the latent heat for vaporization of water, w' the vertical turbulent wind speed, θ' the turbulent potential temperature, q' the turbulent specific humidity.

Firstly dynamic calibrations of the water vapor and CO_2 densities measured by the infrared CO_2/H_2O gas analyzer were conducted by a standard hygrometer and standard CO_2 gas [15]. The temperature measured by the ultrasonic anemometer/thermometer was calibrated by the water vapor density [16].

For flows over complex terrain, errors may arise when vector quantities are measured in a reference framework that is not consistent with that of the equations used to analyze them. To solve this problem, three mathematical rotations were applied here to transfer the sampled velocity data from the instrument's reference frame to the streamline reference frame according to the scheme by Kaimal and Finnigan [17]. Accounting for the effect of the variation in air density due to the transport of sensible heat flux, the WPL calibration was applied in the calculation of latent heat flux [18,19].

4.2.2. Bulk Transfer Coefficients

Sensible heat flux H and latent heat flux LE from a surface can be generally determined from the bulk transfer method [20]:

$$H = c_P \rho C_H U (T_s - T) \quad (2)$$

$$LE = l \rho C_E U (q_{sat}(T_s) - q) \quad (3)$$

Here T_s is the surface temperature, $q_{sat}(T_s)$ the saturation specific humidity at a temperature of T_s. U, T, and q represent the mean wind speed, mean air temperature, and mean specific humidity at a reference level, respectively. Here C_H and C_E are the bulk transfer coefficients for sensible heat and latent heat.

For a well-wetted surface, $C_E \approx C_H$ is expected. Since the surface moisture availability β is defined as $\beta \equiv C_E/C_H$, $\beta \approx 1$ for a well-wetted surface, and for a completely dry surface $\beta = 0$ can be expected; the values of β for land surfaces with different wetness vary between these limits.

One of the methods to determine the bulk transfer coefficients is based on the Monin-Obukhov similarity theory (MOST):

$$C_H = \frac{\kappa^2}{\left[\ln\left((z-d)/z_0\right) - \Psi_m\right]\left[\ln\left((z-d)/z_t\right) - \Psi_h\right]} \quad (4)$$

Here Ψ_m and Ψ_h are stability correction functions for momentum and heat; κ is von Kármán's constant; z_0 and z_t represent the roughness length for momentum and sensible heat, respectively; and d is the displacement height [15].

There are many formulations for the MOST stability functions in literatures [21-26]. In this study, we applied the interpolation stability functions for the ASL proposed by Brutsaert [26] as the following:

$$\phi_m(y) = \left(a + by^{m+1/3}\right)\big/\left(a + y^m\right) \quad (5)$$

$$\phi_h(y) = \left(c + dy^n\right)\big/\left(c + y^n\right) \quad (6)$$

The values assigned to the constants are $a = 0.33$, $b = 0.41$, $m = 1.0$, $c = 0.33$, $d = 0.057$, and $n = 0.78$. Equations (5) and (6) can be readily integrated into

$$\Psi(y) = \int_0^y \left[1 - \varphi(x)\right] dx/x \quad (7)$$

and then yield the stability correction functions:

$$\begin{aligned} \Psi_m(y) &= \ln(a+y) - 3by^{1/3} + \frac{ba^{1/3}}{2}\ln\left[\frac{(1+x)^2}{(1-x+x^2)}\right] \\ &\quad + 3^{1/2}ba^{1/3}\tan^{-1}\left[(2x-1)/3^{1/2}\right] + \Psi_0, y \le b^{-3} \end{aligned} \quad (8)$$

$$\Psi_m(y) = \Psi_m\left(b^{-3}\right), y > b^{-3} \quad (9)$$

$$\Psi_h(y) = \left[(1-d)/n\right]\ln\left[\left(c+y^n\right)\big/c\right] \quad (10)$$

in which $x = (y/a)^{1/3}$, $y = -(z-d)/L$, and Ψ_0 denotes a constant of integration, given by $\Psi_0 = \left(-\ln a + 3^{1/2}ba^{1/3}\pi/6\right)$ [26].

After C_H is determined from (4), as an alternative, the surface temperature T_s in Equation (2) can be replaced by the effective surface temperature for sensible heat flux T_h as:

$$T_h = \frac{H}{c_P \rho C_H U} + T \quad (11)$$

Finally from (3) the surface moisture availability β can be derived:

$$\beta = \frac{LE}{l\rho C_H U\left[q_{sat}(T_h) - q\right]} \quad (12)$$

After sensitivity tests, we found that the displacement height d had little effect on C_H and β. Thus we applied an estimated value of $d = 0.4$ m and a value of $z_0 = 0.5$ m calculated from the neutral logarithmic wind relationship based on data at 32 m height during June 2004 in determination of C_H and β. To determine z_t values, Sugita et al. [27] analyzed data over a complex surface and found that z_t varied between 10^{-14} and 10^{-2} m. This range, together with $z_0 = 0.7 - 0.8$ m ($d = 4$ m), gave a range of 4 - 32 for

$\ln(z_0/z_t)$. Because the land cover condition in the experimental area has similar heterogeneity as that in Sugita *et al.* [27], we adopted a constant value of 5 to $\ln(z_0/z_t)$ for this study and derived the corresponding value of z_t.

5. Results

5.1. Surface Heat Fluxes

Before calculating surface fluxes, the flux footprint was estimated to investigate the influence of the upwind spatial distribution of the surface emission to the vertical fluxes measured at some height. The footprint was calculated by the methods given by Horst and Weil [28] and Horst [29] which allow the estimation of the upwind distance from which most of the fluxes originate for a given measurement height and atmospheric stability. We found that for our experimental site the contributed upwind distance was approximately within 2 km for the fluxes measured at 32 m level for the stability range of $-0.5 < z/L < 0$.

5.1.1. Variations of Surface Heat Fluxes during Dry Season and Wet Season

In this section we will focus on analyzing the properties of the sensible and latent heat fluxes during the dry season (June and first week of July) and the wet season (remaining days of July, August and September in 2004) of the experimental area.

Figures 4 shows the time series of daily mean sensible and latent heat fluxes as well as net radiation from June to September in 2004. The mean sensible heat flux (denoted as H) and the mean latent heat flux (denoted as LE) showed similar fluctuation trend with the net radiation and with each other during the four months. The sensible heat flux H was close to the latent heat flux LE in June.

During July, H decreased on the whole; LE started to increase in clear days when the net radiation was also large and LE was getting larger than H gradually. The latent heat flux LE kept increasing in August and was much larger than the sensible heat flux H. The maximum of LE during August was around 200 Wm^{-2}; in August at the study site rainfalls were concentrated and the precipitation was prominent, and most of the land surface was bare soil after the wheat was harvested in June. Thus these latent heat flux values are comparable to the results of Kimura *et al.* [4]. They calculated sensible and latent heat fluxes over a bare soil surface on the Loess Plateau by a three-layer soil model, and found that the latent heat flux LE in August reached a maximum of about 200 Wm^{-2} after a rainfall. Their modeled values of sensible heat flux H had the maximum of around 200 Wm^{-2} in August over the loess soil [4], which is double of our maximum H of around 100 Wm^{-2} during August. During September LE gradually decreased to be smaller than those in August but still showed larger than the corresponding H values.

These changes of surface heat fluxes were mainly caused by the variations of the solar radiation and the precipitation during the wet season and the dry season in the region. On one side, the decreasing of the solar radiation from June to August caused the fall of the surface temperature. In consequence, less sensible heat was available to be transferred into the atmosphere by means of turbulent processes between the land surface and the atmosphere. On the other side, the predominant increasing of the precipitation during July and August (see **Figure 3**) caused the surface humidity increased largely and more latent heat was produced from the land surface; similarly, the latent heat flux decreased when the precipitation during September was relatively less compared to July and August.

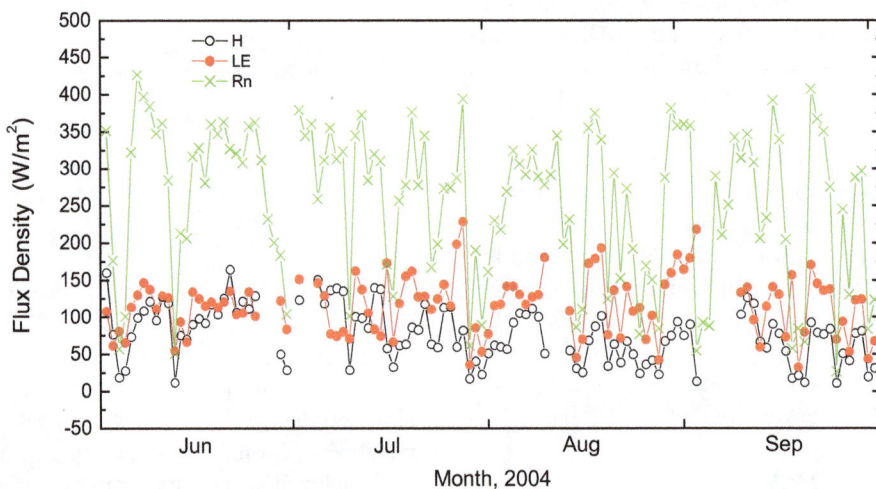

Figure 4. Seasonal variations in daily mean values of sensible (*H*) and latent (*LE*) heat fluxes during June to September 2004 (based on the sensors installed at 12 m height). Net radiation flux (Rn) is also drawn in the figure.

5.1.2. Detailed Analysis of Surface Fluxes

According to the different properties of sensible and latent heat fluxes during the dry and wet seasons, it should be meaningful to investigate the detailed characteristics of H and LE respectively in each season. For this purpose, we selected some typical days from both seasons with proper micrometeorological conditions such as no precipitation, clear sky in daytime, and mild wind speeds.

Considering the radiation components and the precipitation (see **Figure 3**), the four days from June 18 to 21 was selected to represent the dry season. Similarly, the duration between August 28 and 31, after strong rainfall events and with clear sky, was selected to represent the wet season.

We show the radiation components of each representative period in **Figure 5**. In the four representative days of dry season, the peak values of the solar radiation were around 1000 Wm^{-2}. Their maximum of 1023 Wm^{-2} occurred on June 18. However in wet season, the peak values of solar radiation were about 900 Wm^{-2} during the four representative days, with a maximum of 952 Wm^{-2} on August 29, around 70 Wm^{-2} lower than that in the dry season.

The sensible and latent heat fluxes during both representative periods of dry and wet seasons are shown in **Figure 6**. In daytime of the dry period, most of the sensible heat flux H was slightly larger than the corresponding latent heat flux LE. They were generally near zero during nighttime. However, in the wet period, LE showed largely increased in daytime, much higher than H.

After calculating ensemble averages of H and LE during each representative period, the daily variations of H and LE are shown in **Figure 7**. In dry period, the maximum of H in daytime appeared at 14:00; the value was about 225 Wm^{-2}. The peak of LE in daytime also occurred around 14:00, with a value of about 160 Wm^{-2}. During nighttime, there was small amount of downward sensible heat flux; the values were around 25 Wm^{-2}. After 10:00 in daytime, H tended to be larger than LE, and then kept larger until 15:30 in the afternoon. After that, H changed to be smaller than LE. At night LE was generally close to zero or was slightly minus.

In wet period, the maximum of H in daytime occurred at 11:30 with a value of 123 Wm^{-2}. Around 15:00, H showed a second peak of 118 Wm^{-2}. LE in daytime also showed double peak at 12:30 and 15:00, with values of 280 Wm^{-2} and 265 Wm^{-2}, respectively. In nighttime, both H and LE were near to zero or slightly minus. These four representative days were shortly after the rainy days with large precipitation. Thus the surface humidity was prominently increased. Therefore, in daytime after around 10:00, strong evaporation was produced from the land surface

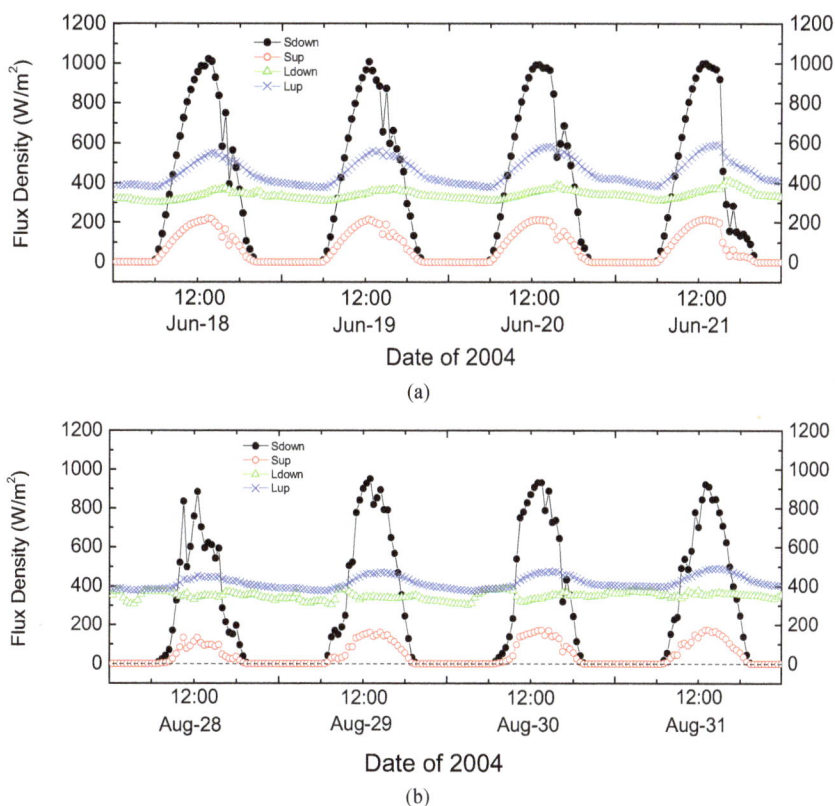

(a)

(b)

Figure 5. Time series of four radiation components during the four selected representative days in the dry period (a) and the wet period (b).

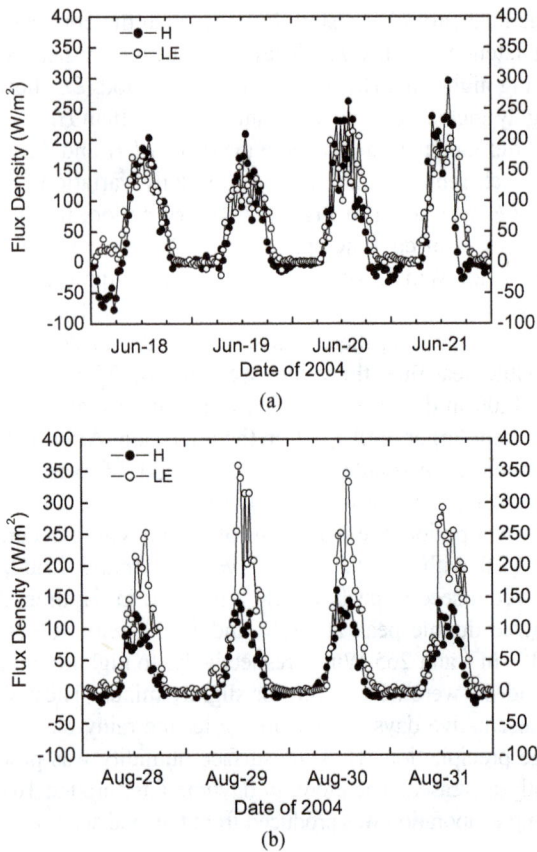

Figure 6. Time series of sensible (denoted as *H*) and latent (denoted as *LE*) heat fluxes during the four selected representative days in the dry period (a) and the wet period (b).

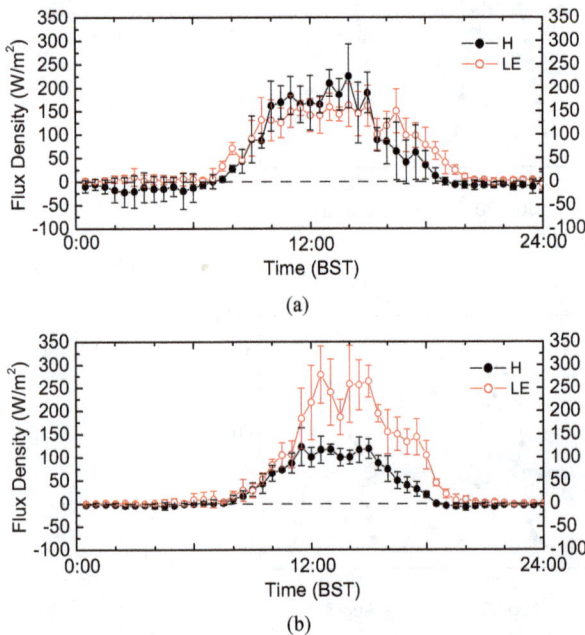

Figure 7. Daily variations of sensible (denoted as *H*) and latent (denoted as *LE*) heat fluxes averaged during the four selected representative days in the dry period (a) and the wet period (b).

due to heating effect by solar radiation. Such strong evaporation transferred large latent heat flux to the atmosphere, so that *LE* started to increase rapidly and be much higher than *H*. After sunset around 18:00, when the solar radiation tended to zero, *LE* decreased rapidly to zero.

5.2. Bulk Transfer Coefficient for Sensible Heat C_H

Figure 8(a) shows the seasonal variation of the bulk transfer coefficient for sensible heat C_H during June 2004 to June 2005. During the whole observational period, C_H did not show very obvious change with seasons, ranging from 0.004 to 0.006; these values were higher than the values ranging between 0.002 and 0.0055 derived by Kimura and Kondo [30] over paddy fields. During June to August in 2004 and April to June in 2005, C_H showed more discrete, but C_H values kept more converged during October 2004 to February 2005, although still with some scattering. This can be related to the farm season in the studied area; from spring to autumn wheat and corn were planted and harvested in turn and in winter the land surface was kept bare in the farm fields. Since the wheat was usually harvested at the end of June, from the end of June in 2004, C_H decreased sharply. Shimoyama *et al.* [31] found similar seasonal variation trends in C_H and surface roughness length from a Siberian bog, and indicated that change in C_H depended on the physical effects of surface roughness so that canopy structure had effect on seasonal variation of C_H. Kimura and Kondo [30] also observed that C_H changed with the height of canopy in paddy fields.

5.3. Surface Moisture Availability β

Figure 8(b) shows the seasonal variation of the surface moisture availability β from June 2004 to June 2005. During these 13 months, the values of β changed obviously with seasons. Higher values were typically found in June and July 2004, with large scattering; the peak value was 0.91 occurring on July 17. From October 2004, β gradually decreased, and reached the lowest value around 0.03 in February 2005. After that, β was increasing from the middle of April until the end of June in 2005. The mean value of β during the wet season of the region was 0.29; this value was much lower than the values between 0.5 - 0.8 over a paddy field derived by Kimura and Kondo [30] and higher than the values within 0.1 - 0.2 over a forest by Kondo [32]. The mean β value was 0.19 in winter; this value was comparable with the values between 0.1 - 0.2 from the forest [32] with the close roughness length of 0.3 - 1 m to that in this study. Because a higher C_H value will cause a lower β value, this forest also showed a comparable C_H value of 0.005

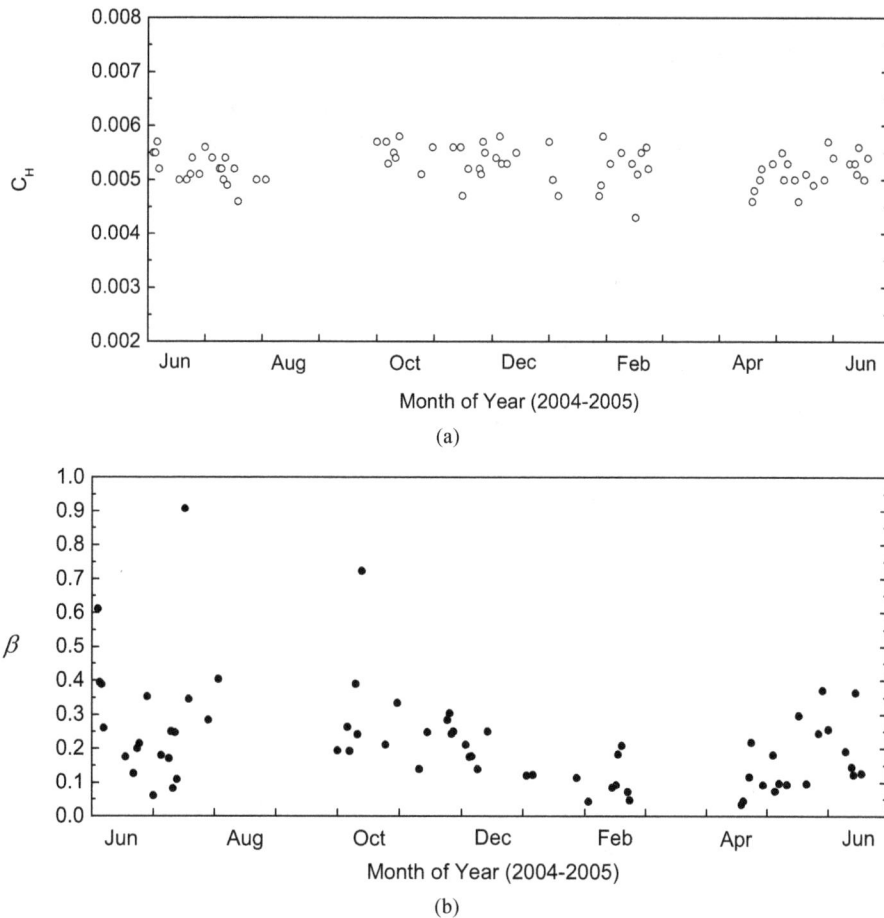

Figure 8. Seasonal variations of (a) the bulk transfer coefficient for sensible heat C_H and (b) the surface moisture availability β on the plateau during June 2004 to June 2005. Both values were evaluated only when the wind direction was from north-west (over the flat terrain).

with the C_H values in this study [32]. Kimura and Kondo [30] calculated β over a rice paddy field; and their β values decreased to around 0.7 after the irrigation periods.

Figure 9 shows the relationship between the surface moisture availability β and the soil water content measured at 10 cm depth at the experimental site. In this plot, β values show clear dependence on the surface soil water content; the three highest values of β correspond to high soil water content which were observed on the days after the raining days with large daily precipitation. These indicate the importance of soil water content in determining β; the soil water content change contributes much to the variations of surface moisture availability.

6. Conclusions and Discussions

The energy fluxes and surface parameters over a complex surface on the Loess Plateau were investigated in this study. For the flux measurement, the upwind distance of 70% flux contribution was approximately 2 km in the atmospheric stability range of neutral to weakly unstable. The dry season in the observational region occurred from

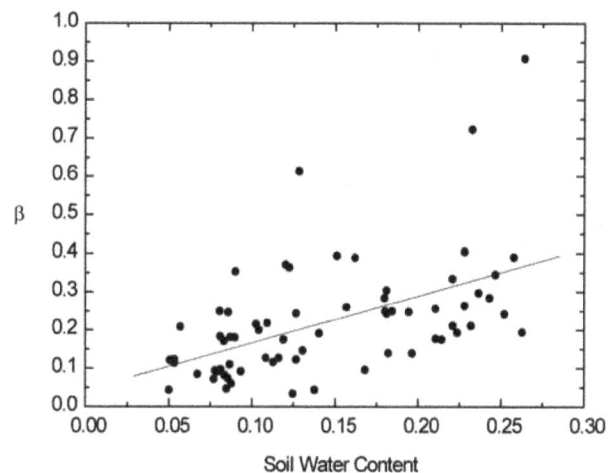

Figure 9. Relationship between the surface moisture availability and the soil water content measured at 10 cm depth (regression line: Y = 0.04344 + 1.22788X; r² = 0.27).

June to the first week of July in 2004; from the middle of July the region got into the wet season until the end of September in 2004.

In dry season the surface transferred turbulent energy to the ABL mainly by means of sensible heat flux with a maximum of around 230 Wm^{-2}. During the wet season, latent heat flux increased prominently reaching a peak around 280 Wm^{-2} and sensible heat flux decreased having a maximum half of that in dry season; the land surface mainly offered latent heat into the atmospheric surface layer.

The bulk transfer coefficient for sensible heat C_H did not change obviously with seasons and had the values between 0.004 and 0.006 during the observational year. The surface moisture availability β changed with seasons with higher values in June and July 2004 and lowest values around 0.03 in February 2005. Its peak value of 0.91 occurred on July 17 in 2004. During the wet season the mean value of β was 0.29. In the winter the mean value of β was 0.19, which was comparable with the values from a forest [32] which had a comparable roughness length as this study. The surface soil water content showed related to β; the highest values of β corresponded to the highest soil water content.

Because of the intricate topography on the Loess Plateau and the heterogeneity of the surface coverage in the experimental area, in this paper we only chose winds from the flat terrain direction to analyze the bulk heat transfer properties, more work will be meaningful to investigate the regional heat transfer and energy balance on the plateau based on synergistic approaches including remote sensing and numerical modeling.

7. Acknowledgements

This study was part of the research project of Recent Rapid Change of Water Circulation in the Yellow River and its Effects on the Environment supported by the Research Institute for Humanity and Nature, Japan. We thank all the members in our experimental team and other cooperative staffs in the Institute of Soil and Water Conservation, Chinese Academy of Sciences. We also give our special thanks to Dr. Sukanta Basu for his comments on revising this paper.

REFERENCES

[1] J. C. Wyngaard, "Scalar Fluxes in the Planetary Boundary Layer—Theory, Modeling, and Measurement," *Boundary-Layer Meteorology*, Vol. 50, No. 1-4, 1990, pp. 49-75.

[2] T. Hiyama, A. Takahashi, A. Higuchi, M. Nishikawa, W. Li, W. Liu and Y. Fukushima, "Atmospheric Boundary Layer Observations on the Changwu Agro-Ecological Experimental Station over the Loess Plateau, China," *Asia Flux Newsletter*, No. 16, 2005, pp. 5-9.

[3] R. Kimura, N. Takayama, M. Kamichika and N. Matsuoka, "Soil Water Content and Heat Balance in the Loess Plateau—Determination of Parameters in the Three-Lay-

ered Soil Model and Experimental Result of Model Calculation," *Journal of Agricultural Meteorology*, Vol. 60, No. 1, 2004, pp. 55-65.

[4] R. Kimura, M. Kamichika, N. Takayama, N. Matsuoka and X. C. Zhang, "Heat Balance and Soil Moisture in the Loess Plateau, China," *Journal of Agricultural Meteorology*, Vol. 60, No. 2, 2004, pp. 103-113.

[5] S. Liu, R. Sun, Z. Sun, X. Li and C. Liu, "Evaluation of Three Complementary Relationship Approaches for Evapotranspiration over the Yellow River Basin," *Hydrological Processes*, Vol. 20, No. 11, 2006, pp. 2347-2361.

[6] I. Tamagawa, "Turbulent Characteristics and Bulk Transfer Coefficients over the Desert in the HEIFE Area," *Boundary-Layer Meteorology*, Vol. 77, No. 1, 1996, pp. 1-20.

[7] J. Kondo and T. Watanabe, "Studies on the Bulk Transfer Coefficients over a Vegetated Surface with a Multilayer Energy Budget Model," *Journal of the Atmospheric Sciences*, Vol. 49, No. 23, 1992, pp. 2183-2199.

[8] H. K. Kafle and Y. Yamaguchi, "Effects of Topography on the Spatial Distribution of Evapotranspiration over a Complex Terrain Using Two-Source Energy Balance Model with ASTER Data," *Hydrological Processes*, Vol. 23, No. 16, 2009, pp. 2295-2306.

[9] W. P. Kustas, F. Li, T. J. Jackson, J. H. Prueger, J. I. MacPherson and M. Wolde, "Effects of Remote Sensing Pixel Resolution on Modeled Energy Flux Variability of Croplands in Iowa," *Remote Sensing of Environment*, Vol. 92, No. 4, 2004, pp. 535-547.

[10] W. Li, T. Hiyama and N. Kobayashi, "Turbulence Spectra in the Near-Neutral Surface Layer over the Loess Plateau in China," *Boundary-Layer Meteorology*, Vol. 124, No. 3, 2007, pp. 449-463.

[11] M. Nishikawa, T. Hiyama, K. Tsuboki and Y. Fukushima, "Numerical Simulations of Local Circulation and Cumulus Generation over theLoess Plateau, China," *Journal of Applied Meteorologyand Climatology*, Vol. 48, No. 4, 2009, pp. 849-862.

[12] A. Takahashi, T. Hiyama, M. Nishikawa, H. Fujinami, A. Higuchi, W. Li, W. Liu and Y. Fukushima, "Diurnal Variation of ater Vapor Mixing between the Atmospheric Boundary Layer and Free Atmosphere over Changwu, the Loess Plateau in China," *SOLA*, Vol. 4, 2008, pp. 33-36.

[13] N. Takayama, R. Kimura, M. Kamichika, N. Matsuoka and X. C. Zhang, "Climatic Features of Rainfall in the Loess Plateau in China," *Journal of Agricultural Meteorology*, Vol. 60, No. 3, 2004, pp. 173-189.

[14] D. Matsushima and J. Kondo, "A Proper Method for Estimating Sensible Heat Flux above a Horizontal-Homogeneous Vegetation Canopy Using Radiometric Surface Observations," *Journal of Applied Meteorology*, Vol. 36, No. 12, 1997, pp. 1696-1711.

[15] K. Shimoyama, T. Hiyama, Y. Fukushima and G. Inoue, "Seasonal and Inter Annual Variation in Water Vapor and Heat Fluxes in a West Siberian Continental Bog," *Journal of Geophysical Research*, Vol. 108, No. D20, 2003, p. 4648.

[16] J. C. Kaimal and J. E. Gaynor, "Another Look at Sonic Thermometry," *Boundary-Layer Meteorology*, Vol. 56, No. 4, 1991, pp. 401-410.

[17] J. C. Kaimal and J. J. Finnigan, "Atmospheric Boundary Layer Flows," Oxford University Press, New York, 1994.

[18] E. K. Webb, G. I. Pearman and R. Leuning, "Correction of Flux Measurements for Density Effects Due to Heat and Water Vapour Transfer," *Quarterly Journal of the Royal Meteorological Society*, Vol. 106, No. 447, 1980, pp. 85-100.

[19] R. Leuning and J. Moncrieff, "Eddy-Covariance CO_2 Flux Measurements using Open- and Closed-Path CO_2 Analyzers: Corrections for Analyzer Water Vapor Sensitivity and Damping of Fluctuations in Air Sampling Tubes," *Boundary-Layer Meteorology*, Vol. 53, No. 1-2, 1990, pp. 63-76.

[20] W. Brutsaert, "Evaporation into the Atmosphere: Theory, History, and Applications," Kluwer Academic Publishers, Boston, 1982.

[21] J. C. Wyngaard and O. R. Cote, "The Budgets of Turbulent Kinetic Energy and Temperature Variance in the Atmospheric Surface Layer," *Journal of Atmospheric Sciences*, Vol. 28, No. 2, 1971, pp. 190-201.

[22] J. C. Wyngaard, "On Surface-Layer Turbulence," In: D. A. Haugen, Ed., *Workshop on Micrometeorology*, American Meteorological Society, Boston, 1973, pp. 101-149.

[23] E. L. Andreas, "Two-Wavelength Method of Measuring Path-Averaged Turbulent Surface Heat Fluxes," *Journal of Atmospheric and Oceanic Technology*, Vol. 6, No. 2, 1989, pp. 280-292.

[24] R. J. Hill, "Review of Optical Scintillation Methods of Measuring the Refraction-Index Spectrum, Inner Scale and the Surface Fluxes," *Wave Random Media*, Vol. 2, No. 3, 1992, pp. 179-201.

[25] W. Brutsaert, "Stability Correction Functions for the Mean Wind Speed and Temperature in the Unstable Surface Layer," *Geophysical Research Letters*, Vol. 19, No. 5, 1992, pp. 469-472.

[26] W. Brutsaert, "Aspects of Bulk Atmospheric Boundary Layer Similarity under Free-Convective Conditions," *Reviews of Geophysics*, Vol. 37, No. 4, 1999, pp. 439-451.

[27] M. Sugita, T. Hiyama and I. Kayane, "How Regional are the Regional Fluxes Obtained from Lower Atmospheric Boundary Layer Data?" *Water Resources Research*, Vol. 33, No. 6, 1997, pp. 1437-1445.

[28] T. W. Horst and J. C. Weil, "How Far is Far Enough—The Fetch Requirements for Micrometeorological Measurement of Surface Fluxes," *Journal of Atmospheric and Oceanic Technology*, Vol. 11, No. 4, 1994, pp. 1018-1025.

[29] T. W. Horst, "The Footprint for Estimation of Atmosphere-Surface Exchange Fluxes by Profile Techniques," *Boundary-Layer Meteorology*, Vol. 90, No. 2, 1999, pp. 171-188.

[30] R. Kimura and J. Kondo, "Heat Balance Model over a Vegetated Area and Its Application to Paddy Field," *Journal of the Meteorological Society of Japan*, Vol. 76, No. 2, 1998, pp. 937-953.

[31] K. Shimoyama, T. Hiyama, Y. Fukushima and G. Inoue, "Controls on Evapotranspiration in a West Siberian Bog," *Journal of Geophysical Research*, Vol. 109, No. D8, 2004, Article ID: D08111.

[32] J. Kondo, "Meteorology of the Water Environment: Water and Heat Balance of the Earth's Surface," Asakura Shoten Press, Tokyo, 1994.

5

Aquacrop Model Calibration in Potato and Its Use to Estimate Yield Variability under Field Conditions

Antonio de la Casa[*], Gustavo Ovando, Luciano Bressanini, Jorge Martínez

Facultad de Ciencias Agropecuarias (FCA), Universidad Nacional de Córdoba (UNC), Córdoba, Argentina

ABSTRACT

AquaCrop model estimates the crop productivity decrease in response to water stress, determining the biomass (B) based on water productivity (WP) and accumulated transpiration (ΣTr); and the yield (Y) is calculated according to B and the harvest index (HI). AquaCrop was evaluated considering different WP values for 2010 late growing season to simulate crop yield of potato (*Solanum tuberosum* L.) cv. Spunta, in a commercial production field of 9 ha located in the green belt of Cordoba city (31°30'S, 64°08'W, 402 m asl), while monitoring in 2009 was used to verify the model. Canopy cover estimation by AquaCrop was adjusted using observed field data obtained from vertical digital photographs acquired at 2.5 m height. WP values of 15.8 and 31.6 (for C_3 and C_4 species, respectively) and two intermediate values 21 and 26.3 $g \cdot m^{-2}$ were considered to evaluate the model performance. While linear function between observed tuber yields and estimated by AquaCrop had always a correlation coefficient greater than 0.94 ($p < 0.001$), using WP = 26.3 and WP =31.6 $g \cdot m^{-2}$ presented overestimation, whereas with 15.8 $g \cdot m^{-2}$ had an opposite behavior, while WP = 21 $g \cdot m^{-2}$ was the value that produced the lowest estimation error. In addition, soil moisture from this estimated value of WP was highly correlated with measured water content in different areas of production field. The verification test shows that while the model slightly underestimates canopy cover, biomass was overestimated. After setting the coefficients of canopy cover development, the AquaCrop crop model estimated adequately potato yield for high production values that are less affected by lack of water, but in both years showed a tendency to overestimate the lowest yields, as was observed for other crops. Meanwhile, the dispersion between the observed and estimated yield was higher in the verification test because the sampling this year was more random.

Keywords: Potato; Canopy Cover; AquaCrop Model; Soil Water Content

1. Introduction

Cordoba Province is currently the major potato (*Solanum tuberosum* L.) producing region of Argentina, and the green belt (gb) of Cordoba city has the largest growing area nationally [1]. Spunta is the more important cultivar used, of Dutch origin and intermediate cycle, with two growing seasons in Córdoba: from August to December (semi-early) and from February to June (late) [2], with irrigated and rainfed yields ranging from 20 to 22 $Mg \cdot ha^{-1}$ and from 15 to 17 $Mg \cdot ha^{-1}$, respectively. According to estimates made in different regions of Argentina, there is a considerable difference between current and potential potato yields [3], so there are many possibilities to reduce this productive gap insofar that the limitations will be overcome [4].

The potato crop in the gb of Cordoba is produced by applying traditional agricultural practices, without considering productivity heterogeneity within each field, despite the technological advances that made possible the precision agriculture (PA). This technique consists in making agronomic practices according to the particular requirements of each sector of the field, tending to use the inputs in a rational way, as well as specific recommendations available at the site level [5]. Despite the economic importance of potato crop in the region, there is insufficient information about the PA linked to its production, even in basic issues such as yield spatial variability [6].

Integrating crop simulation models with the producer experience and PA tools, can facilitate the development of strategies for use and management practices adjusted to the condition of each sector in the field and, particularly in those sites with greater restrictions [7]. Producer skill and knowledge of spatial data contribute to identifying areas of the field, to address soil sampling. From this information, models can be applied to estimate crop yield potential and determine the particular restrictions to achieve that potential.

The PA implementation in potato crop, first requires to know the spatial variability, of both productivity as well as properties and processes responsible of yield. By making such assessment directly in a farm, it is possible not only understand broadly the productive variability, but also diagnose the current production situation and the potential deterioration of the environment faced by the productive sector of the cropped area around Cordoba city.

AquaCrop Model

AquaCrop crop model is based on FAO method that estimates crop productivity decrease in response to water stress [8]. AquaCrop has the following characteristics [9]: 1) distinguish evapotranspiration (ET) between crop transpiration (Tr) and soil evaporation (E), 2) considers a simple model of growth and senescence of canopy cover as basis for estimating Tr and to separate it from E, 3) considers that the yield obtained (Y) is a function of biomass (B) and the harvest index (HI), and 4) water stress is evaluated separately by four functions affecting growth and senescence of canopy cover, Tr and HI. By splitting the ET in E and Tr the unproductive use of E in biomass production is avoided, which is important especially during the period when the ground cover is incomplete. The AquaCrop growth engine to estimate B (biomass per unit of accumulated transpiration, $g \cdot m^{-2}$) per day is:

$$B_i = WP^* \times (Tr_i / ETo_i) \qquad (1)$$

where WP^* is the water productivity normalized by the evaporative demand of atmosphere, which is defined by the reference evapotranspiration (ETo), and the CO_2 atmospheric concentration. This normalization produces water productivity coefficients which tend to be relatively constant under particular climatic conditions [10,11].

As in other crop models, AquaCrop structures the soil-plant-atmosphere system by including 1) the soil, by incorporating water and nutrients budgets, 2) the plant, through their processes of growth, development and yield, and 3) the atmosphere, with its thermal regime, rainfall, evaporative demand and carbon dioxide concentration. While it is a generic model, presents specific parameters for different crops, and some of them have a conservative character [9].

Crop models need to be analyzed in its predictive behavior, and adjusted or calibrated prior to more extensive use, particularly when considering conditions of genotype, environment and specific management. In this sense, AquaCrop was been configured and tested in corn [12], and also subject to validation under irrigated and water deficit conditions [13]; parameterized and tested in irrigated and rainfed cotton [14]; compared with other crop models to estimate sunflower growth under different water regimes [15]; and to evaluate the quinoa (*Chenopodium quinoa* Willd.) yield response to water availability [16].

Unlike other crop models that use the leaf area index as biophysics variable through which the crop interacts with the atmosphere, AquaCrop employs canopy cover and, based thereon, structures the dynamic of growth, water consumption and crop productivity [17]. The model estimates the canopy cover development from an integrated set of three exponential functions throughout the crop cycle, first the coverage value is calculated in the absence of restrictions, which then is adjusted according to the water stress or actual soil nutritional conditions. In most studies of AquaCrop calibration and verification is to be noted that are lacking foliage coverage measurements and, for this reason, assessment of model behavior is implemented with estimated coverage data.

Because the recent appearance of AquaCrop, information about the model performance is relatively scarce, and it is not known if the model has been put operative for potato yet. Although the model includes a set of coefficients to represent crop bioclimatic requirements and tolerances, the global geographic expansion of potato causes a wide range of behaviors and productive responses [18]. Only in Argentina, for all production regions [4] established the possibility of four growing seasons, each one characterized according to their different climatic conditions, soil type, cultural practices and yield level. Thus, the availability of a potato model adapted to local and regional conditions should have a strong impact to project the crop expansion into new regions or analyze any restrictions that should be overcome in the currently production areas. Moreover, it is interesting to evaluate the model performance in the context of productive variability which may present at field conditions [6], where differences of climatic, soil and technological environments are less pronounced and demand more to the predictive ability of the simulation tool.

Within these general guidelines, this work considered two objectives: 1) calibrate the AquaCrop model using data of canopy cover obtained through digital photographs to represent the potato behavior under the environmental conditions of Córdoba green belt, Argentina, 2) analyze the model performance to evaluate production variability in a commercial field in two years under contrasting weather conditions.

2. Material and Methods

2.1. Description of the Study Area

The work was performed in a 9 ha potato (*Solanum tu-*

berosum L., cv. Spunta) field of commercial production, during late autumn cycle in 2009 and 2010. The property is located in the green belt of Cordoba city, Argentina (31°30'S, 64°08'W, 402 m asl). The soil is classified as Haplustol entic series Manfredi, fine silty, mixed, thermic [19], without limitations for irrigated agriculture.

2.2. Crop Management

According to regular crop practices, the planting was carried out with a density of 6 pl·m^{-2}. The 2009 growing season of potato was carry on from February 9 to May 29. The crop was fertilized with 260 kg·ha^{-1} NPK (20-14-3) at planting with the supplement of Mg and S and, in addition, was watered by furrows on 2 moments: at 58 (07/04/2009) and 71 (20/04/2009) days after planting, applying 25 mm at each date. The experiment consisted of a rectangular grid of 5 × 5 sampling sites spaced 47 m in the N-S direction and 44 m following the irrigation furrows (E-W).

The late cycle of 2010 was extended from 16 February to 29 May. The crop was fertilized at planting with 250 kg·ha^{-1} NPK (20-17-3) and the addition of Mg and S, and later with 22-0-0 and the supplement of S, Ca and Mg. Under these conditions of high fertilization level, productive losses due to nutrient deficiency were not considered in the simulation. Meanwhile, during the crop cycle received a surface irrigation of 20 mm at 43 days after planting (31/03/2010). The sampling grid in this year was square, with 3 rows × 3 columns spaced every 70 m.

2.3. Field Measurements

2.3.1. Canopy Cover

In 2009 and with an average frequency of 12 days, a vertical photography (at 1.2 m height) was taking to calculate the canopy cover (CC) near (to the west) of each grid node. In 2010, at each node of the grid an approximately 4 m^2 sampling was performed acquiring nine vertical photographs (taken at 2.5 m above the crop) in different positions equidistant to obtain the mean value of CC by

date with an average frequency of 12 days. In both years, the crop cover was obtained according to [20] and [21], from a digital procedure which determines the presence of vegetation above ground by performing the colorimetric decomposition of the image in the visible range [22]. For the purposes of classification, in each pixel of the image is performed the green (**g**) and red (**r**) bands ratio, and vegetation was considered when **g/r** was greater than 1.05 and soil in the contrary case. [23] determined that the ground cover calculated in this way do not differ from those obtained by the maximum likelihood method and is easier to implement.

2.3.2. Total Biomass

In 2009, two sectors, one between nodes 12 and 13 and the other between the nodes 21 and 22, with the same frequency than for crop cover, 2 samples of 1 m^2 of total biomass were taken to obtain total dry matter.

2.3.3. Crop Yield

In 2009 all tubers were harvested in 1 m^2 of each node. The yield of 2010 was obtained collecting all tubers in 1 m^2 of 6 places equidistant from each node. To express crop yield as dry matter, a ratio dry/fresh weight of 0.2 was used [4].

2.4. Information Used by AquaCrop

The description of the data is performed according to file structure required to run AquaCrop model.

2.4.1. Soil Information

According to the profile description for the soils of the region [24], the file was integrated with three soil horizons whose limits of availability water are shown in **Table 1**. Using Soil Water characteristics program [25], and considering the low organic matter due continuous tillage, the field capacity value (FC) was reduced in A horizon, while the abundant fertilization justifies wilting point values (WP) slightly higher compared to those presented by [24].

Table 1. Soil properties used to represent the soil modal green belt (gb) of Cordoba City. These data constitute the file soil used by AquaCrop model for late potato season in 2009 and 2010.

Horizon	Texture	Depth	Saturation	FC	WP	TAW	Ksat	tau
		m	m^3·m^{-3}			mm·m^{-1}	mm·d^{-1}	
A	Silty Loam	0 - 0.23	0.46	0.30	0.14	160	150	0.50
AC	Silty Loam	0.24 - 0.46	0.46	0.30	0.14	160	150	0.50
C	Loam	>0.47	0.46	0.27	0.12	150	250	0.60

Reference: FC: Field capacity; WP: Wilting point; TAW: Total available water; Ksat: Saturation coefficient; tau: Drainage coefficient.

2.4.2. Climate

Meteorological data were obtained from an automatic monitoring station located 9 km from the site. The station provides hourly records of solar radiation, precipitation, temperature and relative humidity. From hourly records of temperature and humidity, daily maximum and minimum values were obtained of both elements. Similarly, daily radiation and precipitation data were accumulated. With the daily values of solar radiation, maximum and minimum temperature and maximum and minimum relative humidity the reference evapotranspiration (ETo) was calculated by the Penman-Monteith method, using ETo-Calc v3.1 software [26].

Weather files were created for each year from daily values of rain, reference evapotranspiration, maximum and minimum temperature and atmospheric CO_2 concentration, according to the program default file (AquaCrop considers 369.47 ppmv as reference level, which is the average of the atmospheric concentration of CO_2 in 2000 at Mauna Loa observatory, Hawaii).

2.4.3. Crop Management

According to potato production in the gb of Cordoba city, the model considered in both years a proportion of ground covered by stubble of only 4%, which practically does not limit the evaporation rate from the soil. Furthermore, the ground was systemized for furrow irrigation so that restricted flow is not considered, nor are water embankments. In order to estimate the surface runoff, a curve number 65 was used [27]. While not considered limitations on fertility, irrigation water supply was simulated incorporating water amounts listed above for each of the dates.

2.4.4. Model Parameters

Coefficients and conservative parameters (generics) that AquaCrop uses to represent crop performance against water stress conditions were modified slightly so that the crop was considered moderately tolerant to water stress for canopy expansion and stomatal closure, keeping unchanged the remaining (moderately tolerant to early canopy senescence and stress by aeration).

2.5. Calibration and Validation

The AquaCrop calibration strategy to reproduce the particular potato production conditions of the gb of Cordoba was implemented considering observed ground cover data throughout the cycle against those estimates by the model. Model calibration for late season potato crop was made from spatially distributed data observed in 2010, due this year the sampling was more intensive, while the 2009 data obtained under less intensive sampling conditions, were used for validation. Furthermore, while in 2010 the potato crop had more favorable conditions for growth, so that canopy cover was relatively uniform and reached maximum values above 90%, in 2009 the "top necrosis" disease (caused by the Tomato Spotted Wilt Virus—TSWV) restricted ground cover in different sectors across the field and thus increased the spatial yield variability.

The canopy cover (CC) is a relevant biophysical parameter that represents the ability of a crop for intercepting solar radiation, as well as to discriminate ETo between crop transpired water (Tr) from soil evaporated water. To obtain the CC that expresses a potential growth value, we proceeded according to [9], who represents the development of the cover by coupling three exponential equations:

$$CC = CC_0 e^{CGC \times t} \qquad (2)$$

where: CC is the canopy cover at time t, expressed as a fraction ground cover; CC_0 is the initial size of the canopy (at t = 0) in fraction, and CGC is the growth rate of the canopy in fraction per day, which is a constant value for a crop under optimal growth conditions and is modulated by water stress effect.

This function represents the exponential growth of the crop during the first instance after emergence, when growth is proportional to existing size of CC. At the moment that the plants begin to shade each other, development progresses according to second stage, where CC follows an exponential decay rate, according to:

$$CC = CC_x - (CC_x - CC_0) \times e^{-CGC \times t} \qquad (3)$$

where: CC_x is the maximum coverage for optimal growing conditions.

As the crop approaches maturity, CC begins to show a decline phase due to general leaf senescence. The model describes this third stage according to following expression:

$$CC = CC_x \left[1 - 0.05 \left(e^{\frac{CDC \times t}{CC_x}} - 1 \right) \right] \qquad (4)$$

where: CDC is the canopy decline coefficient (in fraction per day reduction), and t is the time from the onset of senescence.

Due to the absence of local or regional determinations of water productivity (WP) in potato, the calibration consisted in obtain its value using AquaCrop basic information, considering firstly the range of typical model values for C_3 and C_4 crops metabolism.

3. Results and Discussion

3.1. Year 2010

Figure 1(a) shows canopy cover (CC) variation during

the late potato growing season of 2010 in one of the 9 sectors of the commercial field. While dots show observations of CC determined from digital photographs at different times of the crop cycle, the continuous line corresponds to the values estimated by AquaCrop for weather, soil and crop management conditions of late potato growing season in the Cordoba gb. The development coefficients of CC that were used, obtained by trial and error, are presented in **Table 2**.

The simulated function describes the canopy cover observed very closely throughout the crop cycle, so the linear fit between the two variables, as shown in **Figure 1(b)**, presents a slope very close to one, while the value of the intercept approaches zero. This analysis was extended to the rest of field and the results obtained are presented in **Table 2**. As in the Sector 1.1, the remaining nodes analyzed have linear functions that not differ statistically from the identity function.

In order to obtain a characteristic value of water productivity (WP) for environmental and technological conditions of Córdoba green belt, the final yield of potato crop cultivar Spunta was estimated considering different values. Then a relationship was established between the

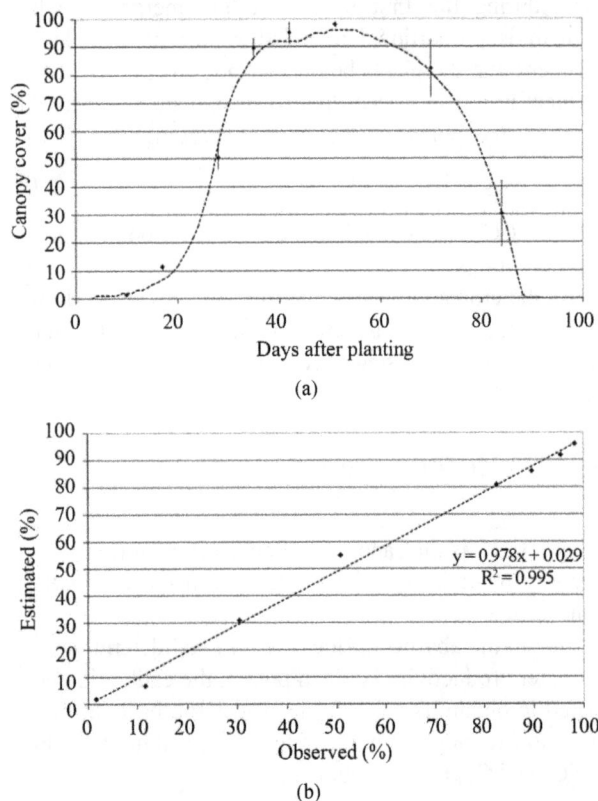

Table 2. Development coefficients of canopy cover model for potato crop used by AquaCrop, and regression (a: intercept, b: slope) and determination (R^2) coefficients for the relationship between observed and estimated canopy cover for each node in 2010.

Node	CC_0	CC_x	CGC	CDC	a	b	R^2
1.1.	0.4	98	0.19	9.0	0.029	0.978	0.995
1.2.	0.4	99	0.20	8.0	−0.608	1.007	0.989
1.3.	0.4	99	0.20	8.0	−1.778	1.019	0.990
2.1.	0.3	99	0.18	8.0	−2.614	0.985	0.991
2.2.	0.4	98	0.19	8.4	−0.912	0.987	0.993
2.3.	0.4	98	0.19	8.3	−0.687	0.983	0.995
3.1.	0.3	97	0.19	7.8	−1.540	1.006	0.996
3.2.	0.3	96	0.20	8.0	2.292	0.980	0.991
3.3.	0.3	97	0.19	7.7	0.395	0.980	0.992

References: Node: row.column. CC_0: initial size of canopy (in t = 0); CC_x: maximum canopy cover for optimal growing conditions; CGC: growth rate of the canopy in fraction per day; CDC: canopy decline coefficient in reduced fraction per day.

(a)

$$y = 0.978x + 0.029$$
$$R^2 = 0.995$$

(b)

Figure 1. (a) Canopy cover variation observed (dots) and estimated by AquaCrop (dash line) with respect to days after planting in node 1.1 in 2010. Vertical bars indicate +/− one standard deviation; (b) Relationship between observed and estimated canopy cover.

final tuber yields observed and estimated by AquaCrop for different levels of WP, as shown in **Figure 2**.

The adjustment function between the observed and calculated values in all cases presents a determination coefficient equal to 0.9 (p < 0.001), but with WP = 26.3 $g \cdot m^{-2}$ and WP = 31, 6 $g \cdot m^{-2}$ occurs a marked overestimation, and when employing 15.8 $g \cdot m^{-2}$ the behavior is opposite, meanwhile using WP = 21 $g \cdot m^{-2}$ produces the lowest estimation error. As is the case with radiation use efficiency (RUE) values [28], the water productivity in potato crop also presents intermediate values between those characteristic for C_3 and C_4 species.

Evaluations about radiation use efficiency in potato assigned this crop the highest value within the C_3, even similar to C_4 species like corn, levels that have also been reported for cv Spunta under environmental conditions Córdoba green belt [29] [30]. Based on this argument, and according to the results obtained here, it is considered appropriate to use a value of water productivity of 21 $g \cdot m^{-2}$, intermediate between those proposes by the model for C_3 and C_4 species. Anyway, overestimation of yields increases when observed productivity are lower, suggesting, like the first evaluations in other crops [13], that AquaCrop behavior is more suitable under greater water supply conditions, but tends to estimate incorrectly the biomass or yield under water stress conditions [31].

Having established estimated conditions of CC similar to the measured, the difference in coverage that occurs in different sectors of the commercial field should explain, according to the logic of the model, the productivity levels achieved at each site particularly. The different sizes of crop transpiration surface also should be reflected in a particular soil moisture condition during the growing season, specially the water consumption (evaporation and transpiration), because the replenishment (rain and irrigation) is assumed uniform in the field. Considering soil moisture as a control variable, and analyzing the degree of correspondence between observed and estimated by AquaCrop, **Figure 3** shows the variation experienced by the soil water content along the growing season in 1.1 node.

The water content variation estimated by AquaCrop for the layer between 0.15 and 0.35 m, represents this dynamic in an appropriate way for the node 1.1, while the last measured value has a more significant underestimation. The analysis of the relationship between observed and estimated values was extended to the other nodes in

the field, obtaining the linear regression and correlation coefficients presented in **Table 3**.

The estimated and observed values of soil moisture in each sector present a significant linear relationship (p < 0.01), with the intercept being different from zero and the slope greater than 1 which, except for node 3.3, shows some inconsistency. The spatial distribution of errors between the mean tuber yield observed and estimates by AquaCrop for the nodes of the sampling grid are shown in **Figure 4**. It shows that there is a trend of the error to decrease in the field, being consistently higher the error obtained on the nodes located on row 3 (north of the field).

Similarly, as shown in **Table 4**, the relationship between potato yields and observed moisture condition in each sector presents a correlation coefficient always positive, that has statistical significance (p < 0.05) in 2 of 6 sampling dates.

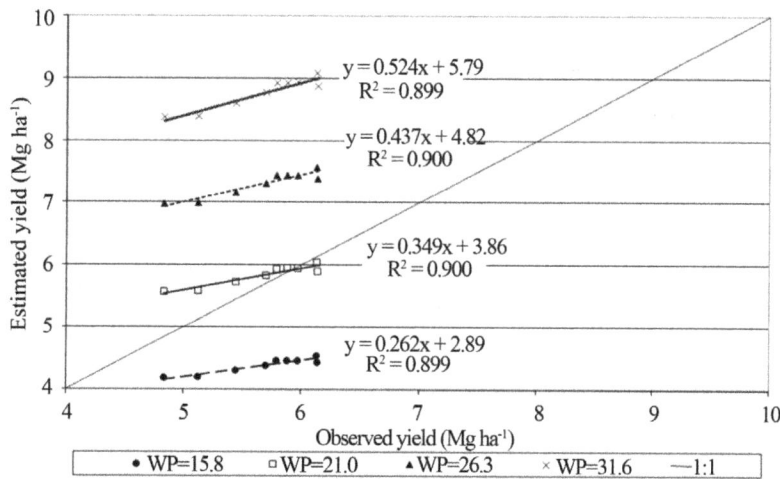

Figure 2. Relationship between observed tuber yield (dry matter) in different nodes of a production field of Córdoba green belt monitored in 2010, and those estimated by AquaCrop considering different water productivity values (WP, $g \cdot m^{-2}$).

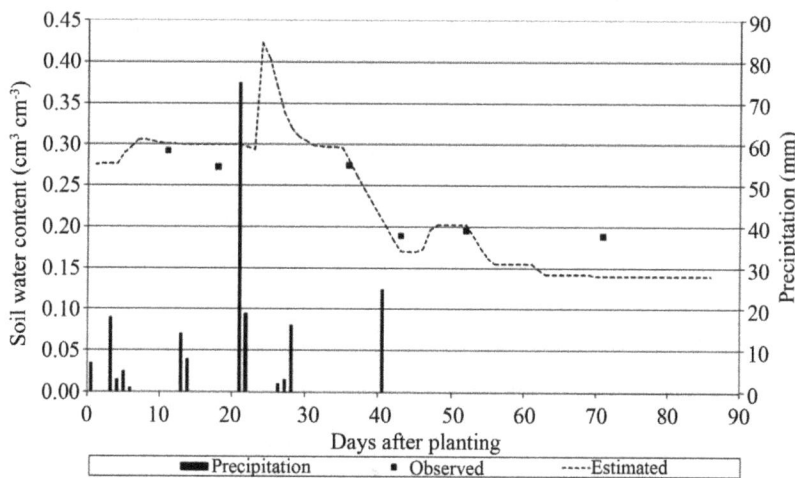

Figure 3. Soil water content observed (0.20 - 0.30 m) and estimated (0.15 - 0.35 m) by AquaCrop for the node 1.1. in 2010 in relation to days after planting. Also presented daily precipitation records in bar graph.

Table 3. Coefficients of regression (a: intercept; b: slope) and determination (R^2) for the relationship between soil water content observed in 0.20 - 0.30 m layer and estimated by AquaCrop in 0.15 - 0.35 m layer, for each of 9 nodes in 2010.

Node	n	a	b	R^2
1.1.	6	−9.198	1.381	0.925
1.2.	6	−13.160	1.522	0.886
1.3.	6	−15.051	1.655	0.924
2.1.	6	−13.751	1.567	0.923
2.2.	6	−9.134	1.363	0.930
2.3.	6	−8.074	1.406	0.943
3.1.	6	−5.395	1.262	0.886
3.2.	6	−5.622	1.313	0.896
3.3.	6	−0.083	1.100	0.928
All	54	−6.468	1.298	0.864

Table 4. Coefficients of correlation (r), t value and its probability (p) for the linear relationship between the soil moisture content and tuber yield in different dates of growing season.

Date	r	t	df	p
Feb-26	0.5318	1.6614	7	0.1406
Mar-05	0.6911	2.5296	7	0.0393
Mar-23	0.2896	0.8005	7	0.4497
Mar-30	0.5519	1.7512	7	0.1234
Apr-08	0.5638	1.806	7	0.1139
Apr-27	0.7058	2.636	7	0.0336

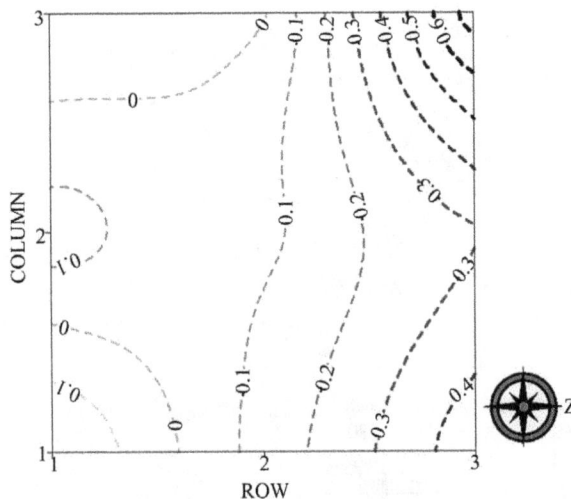

Figure 4. Spatial variation of the error (Mg·ha⁻¹) between observed and estimated yields with WP = 21 g·m⁻² in 2010.

3.2. Year 2009

The potato crop was monitored in the same field of commercial production of 2010, and was used as a control test to verify the predictive performance of AquaCrop. Sampling was less intensive as canopy cover was determined from a single photograph obtained only on the nodes of grid sampling. Furthermore, due to the impact of the Black Death, a disease that was so variable in the field, the canopy cover presented a higher variability compared to 2010. The variables used to evaluate Aqua-Crop, additionally to CC, were, first, the biomass produced from measurements taken during the growth season into two sectors of the potato field (between nodes 1.2. and 1.3. and between nodes 2.1. and 2.2.) and, in second term, the yield of tubers that were harvested on each of the 25 nodes.

With respect to dry matter production, **Figure 5** shows the variation of canopy cover and the biomass obtained along crop season in two sectors of the field.

It shows a strong contrast between the two sectors monitored, with maximum coverage values below 40% in the 1.2. - 1.3. sector and that exceeds 80% in 2.1. - 2.2., what characterizes the greater heterogeneity of crop production in 2009. Consistent with the difference between canopy coverage, total crop biomass also shows a significant difference between the two sites.

Figure 6 shows the correlation and regression analysis between CC and biomass values observed and estimated by AquaCrop for both sectors simultaneously (with initial conditions of soil moisture at field capacity). The results are contradictory since, while the regression line of canopy cover approximates the identity function, Although the dispersion is somewhat high ($R^2 = 0.81$) and a trend toward underestimation for higher values, the figure for dry matter has a slope of 1.4, with a somewhat lower level of dispersion, which means that the model systematically overestimates measurements. As seen in the canopy cover values of **Figure 5**, the incidence of "top necrosis" disease produced a marked retardation of cover growth between 30 and 40 days after planting, which explains the increase of dispersion between the observed canopy cover and estimated by model.

The aptitude of AquaCrop to estimate potato yield was verified in the first instance assimilating canopy cover estimated by the model in the absence of growth restrictions and the observations obtained in the sampling grid. For this purpose, the development coefficients of canopy cover were determined using the model, solving its value by trial and error based on a linear fit function between observed and estimated, with the results shown in **Table 5**. Although the R^2 values obtained are high and exceeds 0.9 in 20 of 25 cases, and the slopes do not differ significantly than 1, assuring the similarity between calculated and observed, in other cases the observed curves of canopy cover presented significant alteration making difficult to achieve greater approximation.

Observed and estimated canopy cover variation throughout the growing season for nodes 2.4. and 4.4. of 2009 is shown in **Figure 7**.

The relationship between observed tuber yield in each 1 m^2 plots in the nodes of the sampling grid and those calculated with AquaCrop is shown in **Figure 8**. The estimates were made considering initial moisture content at field capacity (FC), to avoid the influence of this factor so that the difference in yield in the field is determined only by changes in coverage.

Both for 2009 and 2010 late potato growing season, the results confirm that the duration of canopy cover (coverage integration along the cycle) is a good indicator of the productive capacity of the potato crop [30,32], so that its monitoring during the crop cycle can generate useful information to plan or to establish eventually precision agriculture, or to establish areas of located management [33]. Because of the less intensive sampling and probably as a consequence of the "top necrosis" disease, in 2009 there was a significant variability in the commercial field, which is only partially explained by the model. However, it is considered that this result can be more linked to spatial variability analysis of the production field because, unlike 2010, is performed directly at level of unit area. Similarly, while it is considered necessary produce further evidence to confirm the water use efficiency or water productivity by the potato cv Spunta, estimates produced by AquaCrop seem to be consistent to evaluate the spatial variability of the productivity under cultivation regime of Córdoba green belt.

4. Conclusions

Once the coefficients of canopy cover development are adjusted, AquaCrop estimated adequately crop yield of

Table 5. Coefficients of canopy cover development used by AquaCrop model to estimate potato coverage and coefficients of regression (a: intercept, b: slope) and determination (R^2) between observed and estimated values at the nodes of the sampling grid in 2009.

Node	CC$_0$	CC$_x$	CGC	CDC	a	b	R^2
1.1.	0.02	55	0.16	6.0	1.597	1.066	0.958
1.2.	0.02	25	0.15	3.5	−0.766	0.999	0.606
1.3.	0.02	54	0.16	6.0	−0.810	1.107	0.982
1.4.	0.02	18	0.13	2.0	0.682	1.039	0.874
1.5.	0.02	30	0.15	3.0	−2.564	1.104	0.816
2.1.	0.02	88	0.16	10.0	1.380	1.058	0.958
2.2.	0.02	88	0.18	8.0	1.766	1.025	0.909
2.3.	0.02	90	0.17	11.0	3.009	1.061	0.897
2.4.	0.02	83	0.18	8.0	1.010	1.120	0.917
2.5.	0.02	86	0.18	8.0	−0.333	1.031	0.965
3.1.	0.02	66	0.15	8.0	1.252	0.838	0.886
3.2.	0.02	26	0.14	2.5	0.613	1.067	0.922
3.3.	0.02	72	0.16	8.0	−2.049	1.004	0.916
3.4.	0.02	60	0.15	6.5	2.166	1.051	0.918
3.5.	0.02	52	0.16	4.5	−0.278	0.993	0.953
4.1.	0.02	30	0.15	4.0	−0.322	1.056	0.964
4.2.	0.02	60	0.16	7.5	0.817	1.104	0.955
4.3.	0.02	66	0.16	8.0	1.752	1.025	0.974
4.4.	0.02	45	0.16	4.5	1.356	1.029	0.945
4.5.	0.02	60	0.16	7.0	1.328	0.992	0.933
5.1.	0.02	96	0.19	9.0	0.209	1.129	0.914
5.2.	0.02	87	0.17	8.0	2.251	1.010	0.974
5.3.	0.02	80	0.18	8.0	−1.882	1.143	0.951
5.4.	0.02	84	0.17	8.0	1.165	0.999	0.944
5.5.	0.02	72	0.17	7.0	2.071	1.050	0.899

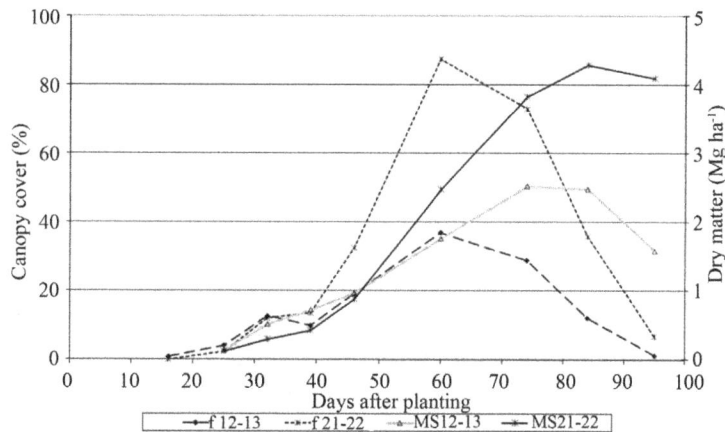

Figure 5. Total dry matter (DM) and canopy cover variation according to the days after planting in two sectors of field production in 2009.

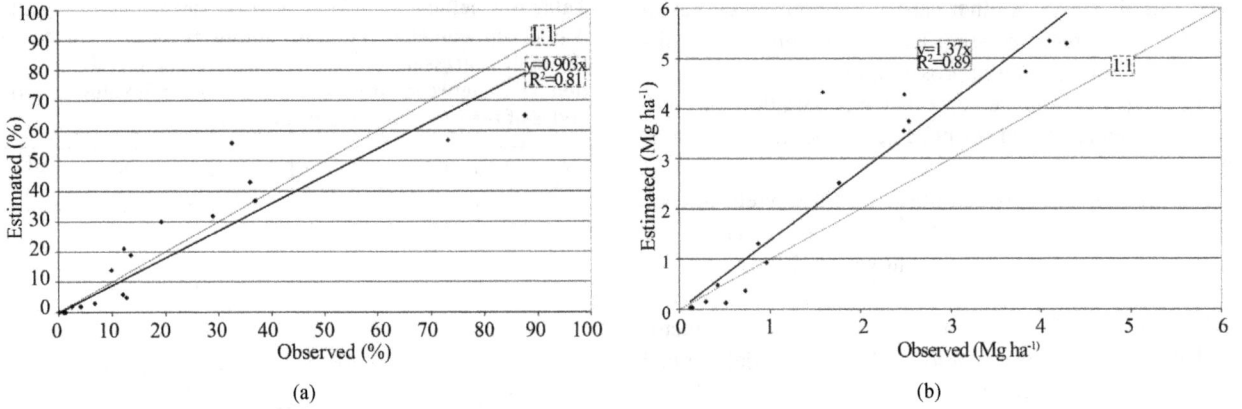

(a)   (b)

Figure 6. Relationship between observed canopy cover (a) and biomass (b) and those estimated by AquaCrop in sectors 12 - 13 and 21 - 22 of the production field in 2009. The dashed line is the 1:1 function.

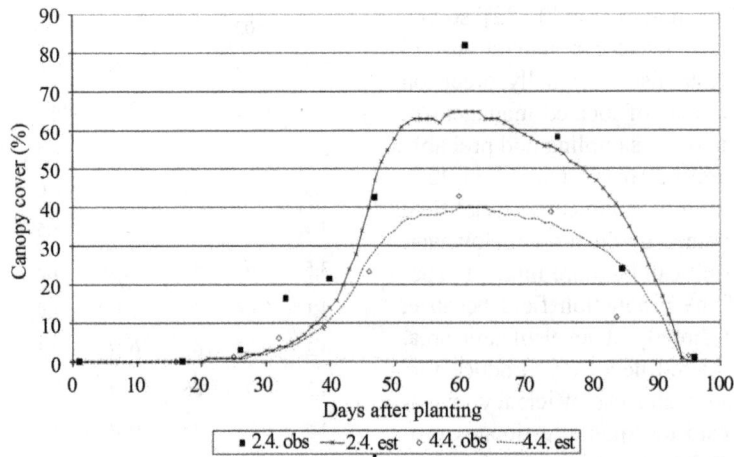

Figure 7. Canopy cover observed and estimated by AquaCrop in relation to days after planting in 2 nodes (2.4 y 4.4) in 2009.

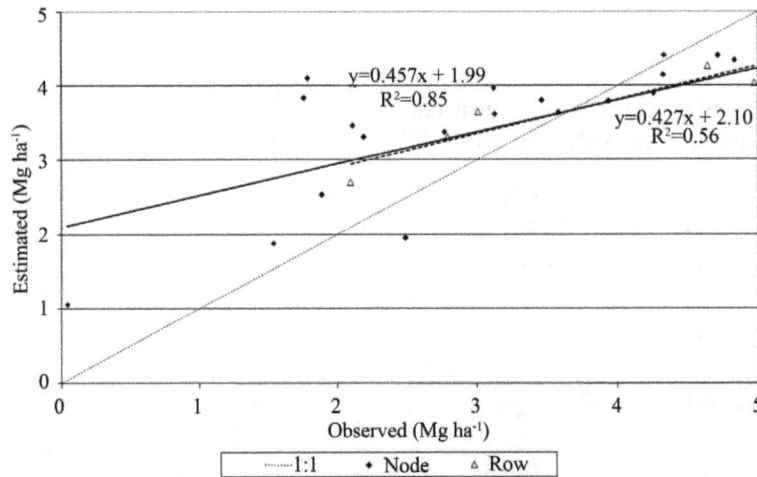

Figure 8. Tuber yields (dry weight) observed and estimated with AquaCrop at 25 nodes and the mean value per row in 2009.

potato, particularly for high production levels. For the environmental, cultural and technological conditions of Córdoba greenbelt, water productivity in potato was 21 g·m^{-2}, intermediate value for species of C_3 and C_4 me-tabolism. This water productivity level proposed to represent potato crop cv Spunta under autumn season conditions of Córdoba greenbelt, can be considered acceptable as a first approximation insofar the productivity levels calculated are similar to those observed.

AquaCrop estimated adequately the evolution of soil water content of the potato crop during the growing season, as well as the spatial distribution of productivity was simulated consistent with the observed values. The model performance can be improved by adjusting the differences of the input data (irrigation, drainage, bulk density) in different sectors of the commercial field.

REFERENCES

[1] M. Mosciaro, "Caracterización de la Producción y Comercialización de Papa en Argentina," Área de Economía y Sociología Rural, EEA-INTA, Balcarce, 2004.

[2] D. O. Caldiz and P. C. Struik, "Survey of Potato Production and Possible Yield Constraints in Argentina," *Potato Research*, Vol. 42, No. 1, 1999, pp. 51-71.

[3] J. Kadaja and H. Tooming, "Potato Production Model Based on Principle of Maximum Plant Productivity," *Agricultural and Forest Meteorology*, Vol. 127, No. 1-2, 2004, pp. 17-33.

[4] D. O. Caldiz, F. J. Gaspari, A. J. Haverkort and P. C. Struik, "Agro-Ecological Zoning and Potential Yield of Single or Double Cropping of Potato in Argentina," *Agricultural and Forest Meteorology*, Vol. 109, No. 4, 2001, pp. 311-320.

[5] F. J. Pierce and P. Novak, "Aspects of Precision Agriculture," *Advances in Agronomy*, Vol. 67, 1999, pp. 1-85.

[6] E. A. Po, S. S. Snapp and A. Kravchenko, "Potato Yield Variability across the Landscape," *Agronomy Journal*, Vol. 102, No. 3, 2010, pp. 885-894.

[7] Y. M. Oliver, M. J. Robertson and M. T. F. Wongbet, "Integrating Farmer Knowledge, Precision Agriculture Tools, and Crop Simulation Modelling to Evaluate Management Options for Poor-Performing Patches in Cropping Fields," *European Journal of Agronomy*, Vol. 32, 2010, pp. 40-50.

[8] J. Doorenbos and A. H. Kassam, "Yield Response to Water, FAO Irrigation and Drainage Paper No. 33," Food and Agriculture Organization of United Nations, Rome, 1979.

[9] P. Steduto, T. C. Hsiao, D. Raes and E. Fereres, "Aqua-Crop—The FAO Crop Model for Predicting Yield Response to Water: I. Concepts and Underlying Principles," *Agronomy Journal*, Vol. 101, No. 3, 2009, pp. 426-437.

[10] C. B. Tanner and T. R. Sinclair, "Efficient Water Use in Crop Production," In: H. M. Taylor, *et al.*, Ed., *Limitations to Water Use in Crop Production*, ASA, CSSA, and SSSA, Madison, 1983.

[11] P. Steduto, T. C. Hsiao and E. Fereres, "On the Conservative Behavior of Biomass Water Productivity," *Irrigation Science*, Vol. 25, No. 3, 2007, pp. 189-207.

[12] T. C. Hsiao, L. K. Heng, P. Steduto, B. Rojas-Lara, D. Raes and E. Fereres, "AquaCrop—The FAO Crop Model for Predicting Yield Response to Water: III. Model Parameterization and Testing for Maize," *Agronomy Journal*, Vol. 101, No. 3, 2009, pp. 448-459.

[13] L. K. Heng, S. R. Evett, T. A. Howell and T. C. Hsiao, "Calibration and Testing of FAO AquaCrop Model for Rainfed and Irrigated Maize," *Agronomy Journal*, Vol. 101, No. 3, 2009, pp. 488-498.

[14] H. J. Farahani, G. Izzi, P. Steduto and T. Y. Oweis, "Parameterization and Evaluation of AquaCrop for Full and Deficit Irrigated Cotton," *Agronomy Journal*, Vol. 101, 2009, pp. 469-476.

[15] M. R. Todorovic, R. Albrizio, L. Zivotic, M. T. Abi Saab, C. Stöckle and P. Steduto, "Assessment of AquaCrop, CropSyst, and WOFOST Models in the Simulation of Sunflower Growth under Different Water Regimes," *Agronomy Journal*, Vol. 101, No. 3, 2009, pp. 509-521.

[16] S. Geerts, D. Raes, M. Garcia, R. Miranda, J. A. Cusicanqui, C. Taboada, J. Mendoza, R. Huanca, A. Mamani, O. Condori, J. Mamani, B. Morales, V. Osco and P. Steduto, "Simulating Yield Response to Water of Quinoa (*Chenopodium quinoa* Willd.) with FAO-AquaCrop," *Agronomy Journal*, Vol. 101, No. 3, 2009, pp. 499-508.

[17] D. Raes, P. Steduto, T. C. Hsiao and E. Fereres, "Aqua-Crop—The FAO Crop Model to Predict Yield Response to Water: II Main Algorithms and Software Description," *Agronomy Journal*, Vol. 101, No. 3, 2009, pp. 438-447.

[18] P. L. Kooman, M. Fahem, P. Tegera and A. J. Haverkort, "Effects of Climate on Different Potato Genotypes. 1. Radiation Interception, Total and Tuber Dry Matter Production," *European Journal of Agronomy*, Vol. 5, 1996, pp. 193-205.

[19] B. Jarsún, J. Gorgas, E. Zamora, H. Bosnero, E. Lovera, A. Ravelo and J. Tassile, "Los Suelos de Córdoba," Agencia Córdoba Ambiente e Instituto Nacional de Tecnología Agropecuaria, EEA Manfredi, Córdoba, 2006.

[20] F. J. Adamsen, P. J. Pinter Jr., E. M. Barnes, R. L. La Morte, G. W. Wall, S. W. Leavitt and B. A. Kimball, "Measuring Wheat Senescence with a Digital Camera," *Crop Science*, Vol. 39, No. 3, 1999, pp. 719-724.

[21] A. A. Gitelson, Y. J. Kaufman, R. Stark and D. Rundquist, "Novel Algorithms for Remote Estimation of Vegetation Fraction," *Remote Sensing of Environment*, Vol. 80, No. 1, 2002, pp. 76-87.

[22] Y. Li, D. Chen, C. N. Walker and J. F. Angus, "Estimating the Nitrogen Status of Crops Using a Digital Camera," *Field Crops Research*, Vol. 118, No. 2, 2010, pp. 221-227.

[23] A. de la Casa, G. Ovando, L. Bressanini, Á. Rodríguez and J. Martínez, "Determinación de la Fracción de Suelo Cubierta con el Follaje de Papa a Partir del Cociente Entre Bandas de Fotografías Digitales," Actas de la XIII

Reunión Argentina y VI Latinoamericana de Agrometeorología, Bahía Blanca, Buenos Aires, 2010.

[24] J. D. Dardanelli, O. A. Bachmeier, R. Sereno and R. Gil, "Rooting Depth and Soil Water Extraction Patterns of Different Crops in a Silty Loam Haplustoll," *Field Crop Research*, Vol. 54, 1997, pp. 29-38.

[25] K. E. Saxton and W. J. Rawls, "Soil Water Characteristics by Texture and Organic Matter for Hydrologic Solutions," *Soil Science Society of America Proceedings of Annual Conference*, Seattle, 2004.

[26] D. Raes, "ETo Calculator v3.1," Land and Water Digital Media Series No. 36, Food and Agriculture Organization of United Nations, Rome, 2009.

[27] Soil Conservation Service, "Estimation of Direct Runoff from Storm Rainfall," In: *National Engineering Handbook*, Soil Conservation Service, USDA, Washington DC, 1964.

[28] T. R. Sinclair and R. C. Muchow, "Radiation Use Efficiency," *Advances in Agronomy*, Vol. 65, 1999, pp. 215-265.

[29] A. de la Casa, G. Ovando, L. Bressanini, Á. Rodríguez and J. Martínez, "Uso del Índice de Área Foliar y del Porcentaje de Cobertura del Suelo Para Estimar la Radiación Interceptada en Papa," *Agricultura Técnica (Chile)*, Vol. 67, 2007, pp. 78-85.

[30] A. de la Casa, G. Ovando, L. Bressanini, J. Martínez and Á. Rodríguez, "Eficiencia en el Uso de la Radiación en Papa Estimada a Partir de la Cobertura del Follaje," *Agriscientia*, Vol. 26, 2011, pp. 21-30.

[31] S. R. Evett and J. A. Tolk, "Introduction: Can Water Use Efficiency Be Modeled Well Enough to Impact Crop Management?" *Agronomy Journal*, Vol. 101, No. 3, 2009, pp. 423-425.

[32] N. S. Boyd, R. Gordon and R. C. Martin, "Relationship between Leaf Area Index and Ground Cover in Potato under Different Management Conditions," *Potato Research*, Vol. 45, No. 2, 2002, pp. 117-129.

[33] A. Hornung, R. Khosla, R. Reich, D. Inman and D. G. Westfall, "Comparison of Site-Specific Management Zones: Soil-Color-Based and Yield-Based," *Agronomy Journal*, Vol. 98, No. 2, 2006, pp. 407-415.

6

Spatial and Temporal Variation of Normalized Difference Vegetation Index (NDVI) and Rainfall in the North East Arid Zone of Nigeria

Christiana F. Olusegun[1]*, Zachariah D. Adeyewa[2]
[1]Department of Physical and Chemical Sciences, Elizade University, Ilara-Mokin, Nigeria
[2]Department of Meteorology, Federal University of Technology, Akure, Nigeria

ABSTRACT

This study examines the spatial and temporal variation of onset and cessation of rainfall and greenness in the North East Arid Zone of Nigeria. Onset and cessation of greenness dates were determined from mean monthly time series of Normalized Difference Vegetation Index (NDVI) using Advance Very High Resolution Radiometer (AVHRR) data for five meteorological stations in the zone for a period of nineteen years (1981-1999). Lowest growing days of six weeks were observed in Nguru (12.53 N, 10.28 E, alt. 343 m), Potiskum (11.42 N, 11.02 E, alt. 415 m) and Maiduguri (11.51 N, 13.05 E, alt. 354 m), while Yola (12.28 N, 9.14 E, alt. 174 m) and Bauchi (10.17 N, 9.49 E, alt. 609 m) have growing days of 15 and 16 weeks respectively. Highest rate of greenness of 0.18/month was observed in Maiduguri while the lowest rate of green-up of 0.07/month was observed in Bauchi. Similarly, highest rate of senescence (0.08/month) was observed in Bauchi while lowest rate of senescence (0.04/month) was observed in Nguru.

Keywords: NDVI; Onset of Greenness; Cessation of Greenness; Rainfall

1. Introduction

The close relationship between rainfall and the growth of vegetation has made it possible to use Normalized Difference Vegetation Index (NDVI) data as a proxy for precipitation variation over land. The Normalized Difference Vegetation Index (NDVI) can be calculated from satellite imagery and is generally recognised as a reliable index of ground vegetation cover [1]. Satellite data processed into Normalized Difference Vegetation Index (NDVI) can be used to indicate onset and cessation of greenness, rate of green-up and senescence, growing days and growing season. Most previous regional-scale studies were based on time series of remotely sensed indicators of vegetation greenness, mainly the Normalized Difference Vegetation Index (NDVI), and rainfall measurements from ground stations. NDVI has been extensively and qualitatively used to infer changes in vegetation response to rainfall in seasonally arid regions [2]. Le Houérou [3] concluded that the monthly NDVI could be best explained by a linear correlation with rainfall. The correlation was achieved when

the rainfall of the preceding two months was also included. Quantitative variations of satellite derived vegetation index over Africa were also provided by Adeyewa [4]. His analysis of anomalies in the NDVI showed that significant reductions in the vegetative activities are in the western, eastern and southern part of Africa, while the western and central part witnessed increased activities for different time period between the years 1981-1999. NDVI is a good indicator of the ability for vegetation to absorb photosynthetically active radiation and has been widely used by researchers to estimate green biomass [5], leaf area index [6] and patterns of productivity [7] because the internal mesophyll structure of healthy green leaves strongly reflects NIR radiation, and leaf chlorophyll and other pigments absorb a large proportion of the red VIS radiation [8,9].

This study is focused on the North East Arid Zone of Nigeria which lies between 12 N and the Niger border and 10 E meridian and the Cameroon border. The climate of the North East Arid Zone of Nigeria is semi-arid in nature because the ratio of average rainfall to average potential evapotranspiration is between 0.20 and 0.45

---

*Corresponding author.

[10]. The vegetation in North East Arid zone of Nigeria is mainly Sudan or Sahel Savannah and it has also been confirmed that crop yields in this zone is dependent on rainfall amounts and distribution [11-13]. The annual rainfall in semi-arid regions has been observed to be declining for more than five decades now, resulting in significant decrease in the length of the growing season in this region. Inconsistent rainfall amount and distribution always resulted in very low crop yield on an annual basis [14]. However, accurate determination of onset of greenness is important for high crop yield. It is therefore imperative for us to determine the time lag between onset of greenness and onset of rainfall. Another peculiar feature of this zone is the short rainy season between June/July and September/October immediately followed by a long dry season. Therefore, this research work is aimed at determining onset and cessation of rainfall and greenness, time lag between onset and cessation of rainfall and greenness, growing days and growing seasons besides rate of green-up and rate of senescence in the North East Zone of Nigeria.

2. Materials and Methods

The Normalized difference Vegetation Index (NDVI) is calculated as the difference between near infrared (NIR) and visible reflectance value normalized over the sum of both [15].

NDVI   is therefore estimated as: $\dfrac{NIR - VIS}{NIR + VIS}$

where VIS is the land surface reflectance in the visible wavelength (0.58 - 0.68 micrometer) and NIR is the land surface reflectance in the near infrared wavelength (0.725 - 1.1 micrometer).

Hulme [16], classified the numerous approaches towards the determination of growing season determination into two categories. The first category is the determination of growing season using standardized meteorological variables such as the magnitude of annual rainfall [16-19] and evapotranspiration is one of the categories. The second category is when the growing season is determined from absolute daily or monthly total rainfall [20-26]. Growing seasons in the North East Arid zone of Nigeria were determined in this research using NDVI.

Monthly rainfall data archived at The Nigerian Meteorological Agency, Oshodi, Lagos, which covers the time period of January 1981 through to December 1999, was used for analysis in this research. The NDVI data were extracted from the Pathfinder AVHRR Land (PAL) Global 8 km, 10-day composite, Normalized Difference Vegetation Index (NDVI) product archived at the Goddard Earth Sciences, Distributed Active Centre (GES-DAAC). The data covers the time period from January 1981 through December 1999. The PAL 10-day data were originally created from the Pal 8-km daily product

using a temporal re-sampling method based on maximum NDVI values. The Pal 8 km daily data were spatially re-sampled based on maximum NDVI values, from the Advanced Very High Resolution Radiometer (AVHRR), Global Area Coverage (GAC) data, which has a nominal resolution of 4km. The data were analyzed for five stations The rainfall synoptic stations used to represent the North East Arid Zone are Nguru (12.53 N, 10.28 E, alt. 343 m), Potiskum (11.42 N, 11.02 E, alt. 415 m) and Maiduguri (11.51 N, 13.05 E, alt. 354 m), Yola (12.28 N, 9.14) and Bauchi (10.17 N, 9.49 E) for the years 1981-1999.

Data Analysis

Onset of greenness was empirically determined using Equation (1)

$$\begin{aligned} &\text{Onset of greenness} \\ &= 1/4\left(\text{NDVI max} - \text{NDVI min}\right) + \text{NDVI max} \end{aligned} \quad (1)$$

Cessation of greenness was also empirically obtained by determining the time the maximum NDVI had decreased by 30%.

In determining onset and cessation of rainfall Ilesanmi [17], method was adopted. Onset occurs when cumulative rainfall reaches 8% of the total rainfall while cessation occurs when cumulative rainfall reaches 90% of the total rainfall. Growing season was determined by considering the length of time between onset date of greenness and cessation date of greenness. The rate of green-up and rate of senescence was determined using Equations (2) and (3)

$$\text{Rate of greenup} = \frac{\text{NDVI}_{max} - \text{Onsetofgreenness}}{\text{Growingdays}} \quad (2)$$

$$\text{Rate of senescence} = \frac{\text{NDVI}_{max} - \text{Cessationofgreenness}}{\text{Brown days}}$$

$$(3)$$

The NDVI data used in this work were made ready by further analyzing the Pathfinder AVHRR Land (PAL) data through a previous study byAdeyewa [4]. Program codes were written for all the thirty stations with the center latitude and longitude, start latitude, stop latitude, start longitude and stop longitude for each station. The steps of making the data ready have been described by Adeyewa [4]. These are some excerpts of the analytical steps taken:

Read data in binary form (unzipped data) from the compressed data.

Convert raw data to scaled format in order to obtain the NDVI data.

Combine the three 10-day composites available or each month into a singular monthly dataset using the maximum observed NDVI for each pixel in the particular station.

Compute annual mean of the NDVI from the monthly

Spatial and Temporal Variation of Normalized Difference Vegetation Index (NDVI) and Rainfall in the North East Arid Zone of Nigeria

51

records.

Estimate monthly and annual averages of NDVI for five stations in Nigeria.

3. Results and Discussion

Figure 1(a) shows the rainfall and NDVI variation at Potiskum. Onset of rainfall occurs in the fourth week of the month of May while onset of greenness was observed in the fourth week of June. Onset of greenness occurs in June as a result of the northward movement of the Inter tropical Convergence Zone (ITCZ) which then brings about the availability of sufficient rainfall necessary for

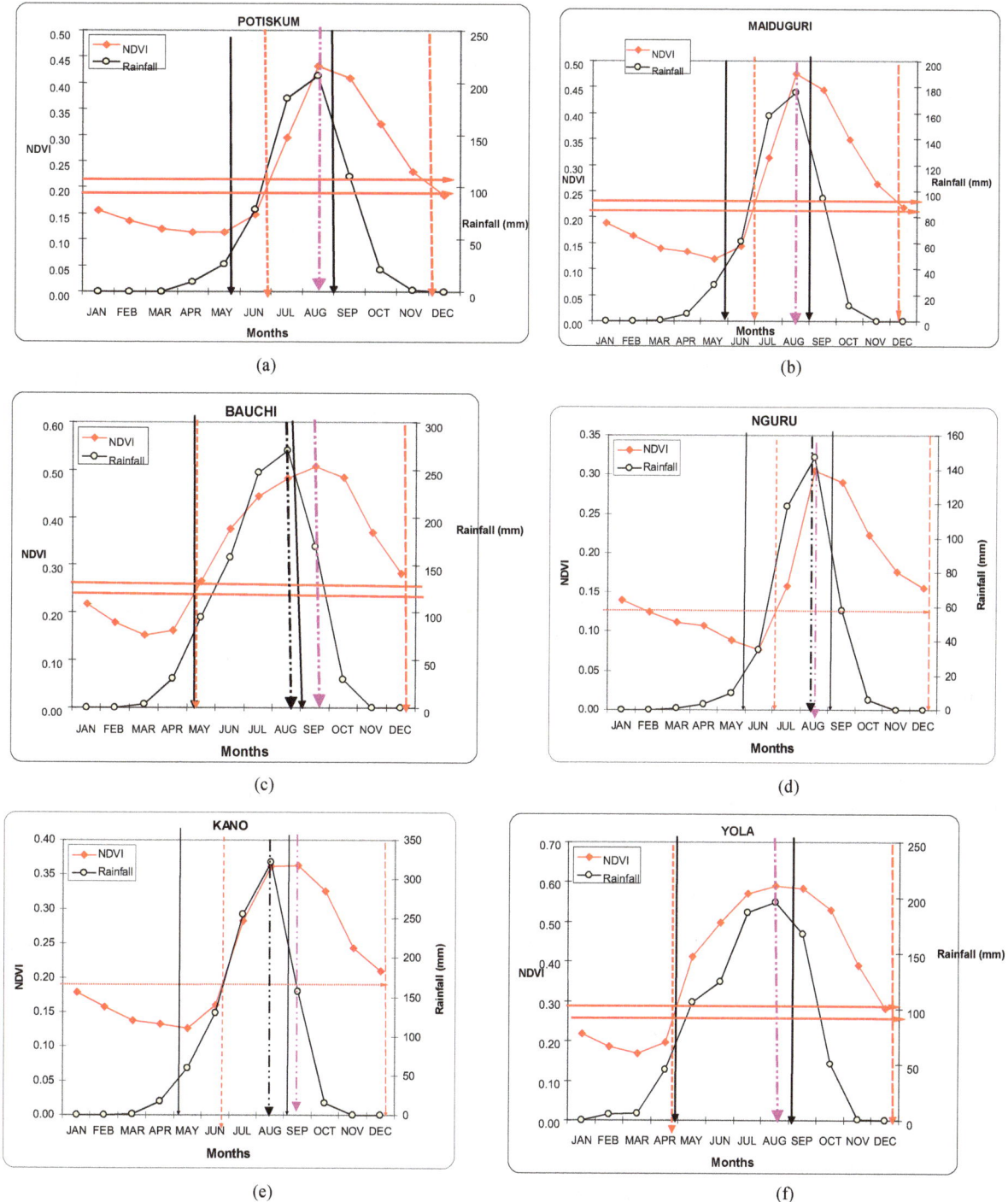

Figure 1. (a)-(f) Graphical Determination of onset and cessation of rainfall and green-up, maximum NDVI and maximum rain. Onset and cessation of rainfall (continuous solid line), Onset and cessation of greenness (short dashed line), Maximum NDVI and Maximum Rainfall (long dash dot).

the emergence of greenness. However, the time lag between onset of rainfall and greenness was observed to be
four weeks. Cessation of rainfall occurs in the first week
of September while cessation of greenness was observed
in the first week of December. The time lag between
cessation of rain and cessation of greenness is twelve
weeks. Maximum rainfall and NDVI occur simultaneously in August. Growing days is one month two weeks.
Growing season spans from June to December. The rate
of green-up was calculated to be 0.16 while the rate of
senescence was 0.07.

Determined onset of rainfall and onset of greenness
occur in the fourth week of May and first week of July
respectively as shown in **Figure 1(b)** for Maiduguri station, the time lag between onset of rainfall and onset of
greenness is five weeks. Cessation of rainfall and greenness occurs in the first week of September and December
respectively. Time lag between cessation of rainfall and
greenness is twelve weeks. The peak of rainfall and
NDVI was observed in August. Growing days is one
month two weeks while growing season spans from June
to December. The rate of green-up is 0.18 while rate of
senescence is 0.07.

Also onset of rainfall and greenness for Bauchi was
obtained from the variation pattern of rainfall and NDVI
shown in **Figure 1(c)**. Onset of rainfall and greenness
occur in the first week of May. No time lag was observed
between onset of rainfall and greenness in this station.
Cessation of rainfall occurs in the first week of September while cessation of greenness was observed in the
fourth week of December. Time lag between cessation of
rainfall and cessation of greenness is fifteen (15) weeks.
Maximum rainfall and NDVI occur in August and September respectively and the time lag being one month.
Growing days is four months. The growing season spans
from May to December and the rate of green-up is 0.07
while the rate of senescence is 0.08.

At Nguru, onsets of rainfall occur in the fourth week
of May while onset of greenness was observed in the first
week of July as shown in **Figure 1(d)**. The time lag between onset of rainfall and onset of greenness is one
month one week. Cessation of rainfall occurs in the first
week of September while cessation of greenness was
observed in the fourth week of December and the lag is
fifteen (15) weeks. NDVI and rainfall peaks in August.
Growing days is one and a half month. Growing season
spans from July to December. The rate of green-up is
0.11 while the rate of senescence is 0.04.

At Kano Station, onset of rainfall and greenness was
observed in the first week of May and fourth week of
June respectively for Kano station as depicted in **Figure
1(e)**. The time lag between onset of rainfall and onset of
greenness is seven weeks. Cessation of greenness was
observed in the fourth week of December wile cessation

rainfall was observed first week of September. Time lag
between cessation of rainfall and cessation of greenness is
fifteen (15) weeks. Inter-Annual variation of rainfall and
NDVI attains its maximum level in August and September respectively and the lag is one month. Growing days
is two months three weeks. Growing season spans from
June to December. The rate of green-up and senescence is
0.06.

Determined onset of rainfall and greenness in Yola
was observed in the fourth week of April while cessation
of rainfall and greenness occur in the first week of September and fourth week of December respectively as
shown **Figure 1(f)**. Time lag between cessation of rainfall and cessation of greenness is fifteen (15) weeks.
Rainfall and NDVI peaks in August. Growing days is
three months three weeks. Growing season spans from
April to December. Rate of green-up is 0.08 while the
rate of senescence is 0.07.

Accurate timing of onset and cessation of rainfall can
be determined using onset and cessation of greenness
analysis obtained from analysis of NDVI data. In most
parts of North East Arid Zone of Nigeria, false onset of
rain and early cessation of rain is prevalent as a result of
the haphazard pattern of rainfall distribution [27-33].

Comparison between rate of green-up and senescence
is shown in **Figure 2**. It was observed that Maiduguri has
the highest rate of green-up. This implies that acceleration of photosynthetic activity is high in this area when
compared to other areas in this zone. This may be due to
the presence of high temperature regimes mostly observed in this area at most times throughout the year. This
supports the fact that the sun is a prominent factor in
photosynthesis. Bauchi has the highest rate of senescence.
This implies that rate of decreasing photosynthetic activity is highest in this zone. Hence, plants are forced to
stop growing in this region at a shorter time interval
more than other region in the North East arid Zone.

4. Conclusion

This research work has been able to reveal the ability of
remotely sensed satellite derived NDVI in the monitoring
of greenness of vegetation in North East Arid zone of

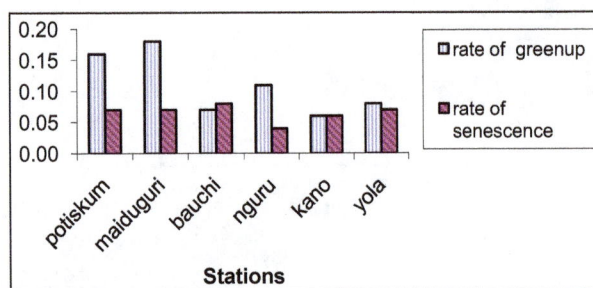

**Figure 2. Comparison of rate of green-up and senescence
the North East arid zone.**

Nigeria. It has also revealed to us that the long-term yearly pattern of NDVI is dependent on the climatic characteristic of each station in the North East Arid zone of Nigeria. The time lag existing between onset of greenness and onset of rainfall if disseminated to the farmers in this region will help enhance crop yield and food security. In all the stations considered in this research, NDVI peaks in August/September thus confirming the unimodal distribution of rainfall in the North East Arid Zone of Nigeria. The study also established that the growing season in this region though short spanning between May/June to September/October can be maximally utilised if predetermined. Deforestation must be discouraged in this zone and afforestation project should be carried out. Finally, if the accurate timing of growing season is disseminated to the farmers promptly then the use of irrigation practice will be adopted when necessary and genetically improved drought resistant crop can be planted in the North East arid zone of Nigeria.

REFERENCES

[1] T. Hess, W. Stephens and G. Thomas, "Modelling NDVI from Decadal Rainfall Data in the North East Arid Zone of Nigeria," *Journal of Environmental Management*, Vol. 48, 1996, pp. 249-261.

[2] E. F. Lambin, P. Cashman, A. Moody, B. H. Parkhurst, M. H. Pax and C. B. Schaaf, "Agricultural Production Monitoring in the Sahel Using Remote Sensing: Present Possibilities and Research Needs," *Journal of Environmental Management*, Vol. 38, No. 4, 1993, pp. 301-322.

[3] A. D. Malo and S. E. Nicholson, "A Study of Rainfall and Vegetation Dynamics in the African Sahel Using Normalized Difference Vegetation Index," *Journal of Arid Environments*, Vol. 19, No. 1, 1990, pp. 1-24.

[4] Z. D. Adeyewa, "Satellite Derived Vegetation Index over Africa," PGD Dissertation, African Regional Centre for Space Science and Technology Education, Obafemi Awolowo University, 2003.

[5] S. D. Prince, C. J. Tucker and F. Rosental, "Satellite Remote Sensing of Primarchy Production: Comparison of Results for Sahelian Grassland 1981-1988," *International Journal of Remote Sensing*, Vol. 12, No. 6, 1991, pp. 1301-1312.

[6] G. Asar, E. T. Kanemasu, G. P. Miller and R. L. Weiser, "Light Interception and Leaf Area Estimates from Measurements of Grass Canopy Reflectance," *IEEE Transactions on Geosciences and Remote Sensing*, Vol. 24, 1986, pp. 76-82.

[7] S. N. Goward, D. G. Dye, S. Turner and S. Liang, "The University of Maryland Improved Global Vegetation Index," *International Journal of Remote Sensing*, Vol. 15, No. 17, 1994, pp. 3365-3395.

[8] J. A. Gausman, C. J. Tucker and P. J. Sellers, "Global Monthly AVHRR Climatology," Boulder NOAA NESDIS,

National Geophysical Data Centre (CD Rom), 1995.

[9] P. J. Sellers and D. S. Schimel, "Remote Sensing of the Land Biosphere and Biogeochemistry in the EOS Era: Science Priorities, Methods and Implementation—EOS Land Biosphere and Biogeochemical Panels," *Global Planet Change*, Vol. 7, No. 4, 1993, pp. 279-297.

[10] H. N. Le Houérou, "Climate Change, Drought and Desertification," *Journal of Arid Environments*, Vol. 34, No. 2, 1996, pp. 133-185.

[11] J. M. Kowal and D. T. Knabe, "An Agro Climatological Atlas of the Northern States of Nigeria," Ahmadu Bello University Press, Zaria, 1972.

[12] N. Peacock and J. Hemrich, "Light and Temperature Responses in Sorghum," Agrometeorology of Sorghum and Millet in the Semi-Arid Tropics, 42, ICRISAT Centre, 1984.

[13] I. J. Ekpoh, "Estimating the Sensitivity of Crop Yields to Potential Climate Change in North-Western Nigeria," *Global Journal of Pure and Applied Sciences*, Vol. 5, No. 3, 1999, pp. 303-308.

[14] M. J. Mortimore, "Adopting to Drought: Farmers, Famines and Desertification in West Africa," Cambridge University Press, Cambridge, 1989.

[15] J. C. Eidenshink, "The 1990 Conterminous US AVHRR Data Set," *Photogrammetric Engineering and Remote Sensing*, Vol. 58, No. 6, 1992, pp. 809-813.

[16] M. Hulme, "Secular Changes in Wet Season Structure in Central Sudan," *Journal of Arid Environment*, Vol. 13, No. 1, 1987, pp. 31-46.

[17] O. O. Ilesanmi, "An Empirical Formulation of the Onset, Advance and Retreat of Rainfall in Nigeria," *Tropical Geography*, Vol. 34, No. 1, 1972, pp. 17-34.

[18] P. Benoit, "The Start of the Growing Season in Northern Nigeria," *Agricultural Meteorology*, Vol. 18, No. 2, 1977, pp. 91-99.

[19] A. B. Shaw, "An Analysis of Rainfall Regimes on the Coastal Region of Guyana," *Journal of Climatology*, Vol. 7, No. 3, 1986, pp. 291-302.

[20] M. W. Walter, "Length of the Rainy Season in Nigeria," *Nigerian Geographical Journal*, Vol. 10, 1967, pp. 123-128.

[21] E. G. Davey, F. Mattie and S. I. Solomon, "An Evaluation of Climate and Water Resources for the Development of Agriculture in the Sudan Savannah Zone of West Africa," World Meteorological Organization (WMO) Special Environment Report No. 9, Geneva, 1976.

[22] R. D. Stern, M. D. Dennett and D. J. Garbutt, "The Start of the Rains in West Africa," *Journal of Climatology*, Vol. 1, No. 1, 1981, pp. 59-68.

[23] R. D. Stern, M. D. Dennett and I. C. Dale, "Analysis of Daily Rainfall Measurements to Give Agronomically Useful Results. 1. Direct Methods," *Experimental Agri-*

culture, Vol. 18, No. 3, 1982, pp. 223-236.

[24] O. J. Olaniran and G. N. Summer, "A Study of Climatic Variability in Nigeria Based on the Onset, Retreat and Length of the Rainy Season," *International Journal of Climatology*, Vol. 9, No. 3, 1989, pp. 253-269.

[25] N. Singh, "The Duration of the Rainy Season over Different Parts of India," *Theoretical and Applied Climatology*, Vol. 37, No. 1-2, 1986, pp. 51-62.

[26] R. Ananthakrishnan and M. R. Soman, "Onset of the South West Monsoon over Kerala: 1901-1980," *Journal of Climatology*, Vol. 8, No. 3, 1998, pp. 283-296.

[27] M. V. L. Sivakumar, "Predicting Rainy Season Potential from the Onset of Rains in Southern Sahelian and Sudanian Climatic Zones of West Africa," *Agricultural and Forest Meteorology*, Vol. 42, No. 4, 1988, pp. 295-305.

[28] D. O. Adefolalu, "Rainfall Trends in Nigeria," *Journal of Theoretical and Applied Climatology*, Vol. 37, No. 4, 1986, pp. 205-219.

[29] R. C. N. Anyadike, "Seasonal and Annual Rainfall Variations over Nigeria," *International Journal of Climatology*, Vol. 13, No. 5, 1993, pp. 567-580.

[30] T. Aondover and W. Ming-Ko, "Changes in Rainfall Characteristics of Northern Nigeria," *International Journal of Climatology*, Vol. 18, No. 11, 1998, pp. 1261-1271.

[31] I. J. Ekpoh, "Rainfall and Peasant Agriculture in Northern Nigeria," *Global Journal of Pure and Applied Sciences*, Vol. 5, No. 1, 1999, pp. 123-128.

[32] A. Dai, P. Lamb, K. E. Trenberth, M. Hulme, P. D. Jones and P. Xie, "The Recent Sahel Drought Is Real," *International Journal of Climatology*, Vol. 24, No. 11, 2004, pp. 1323-1331.

[33] K. C. Anyanwale, "Climate Dynamics of the Tropics," KAP, Dordrecht, 2007, p. 488.

7

Stable Boundary Layer Height Parameterization: Learning from Artificial Neural Networks

Wei Li

Department of Marine, Earth, and Atmospheric Sciences, North Carolina State University, Raleigh, USA

ABSTRACT

Artificial neural networks (ANN) are employed using different combinations among the surface friction velocity u_*, surface buoyancy flux B_s, free-flow stability N, Coriolis parameter f, and surface roughness length z_0 from large-eddy simulation data as inputs to investigate which variables are essential in determining the stable boundary layer (SBL) height h. In addition, the performances of several conventional linear SBL height parameterizations are evaluated. ANN results indicate that the surface friction velocity u_* is the most predominant variable in the estimation of SBL height h. When u_* is absent, the secondly important variable is the surface buoyancy flux B_s. The relevance of N, f, and z_0 to h is also discussed; f affects more than N does, and z_0 shows to be the most insensitive variable to h.

Keywords: Artificial Neural Network; Large-Eddy Simulation; Stable Boundary Layer Height

1. Introduction

Parameterizations of the stable boundary layer (SBL) height are often critical in many practical problems, such as air pollutant dispersion modeling [1-3], weather modeling, and climate modeling. Due to the complex relationship between the mean profiles and the turbulence in the stable boundary layer [4,5], it is less straightforward to determine the SBL height from wind speed and temperature profiles compared to observing the convective boundary layer height from mean profiles. During recent several decades, a number of model formulations of SBL depth have been developed based on various datasets [6-11]; these formulations are mainly in forms of linear equations or complex non-linear multi-limit equations. Most of these schemes are surface flux-dominated models; they are summarized by Zilitinkevich et al. [11].

Various linear relationships [6-8] were formulated by considering the SBL height relying on one or two variables among the earth's rotation (f), surface friction velocity (u_*), surface buoyancy flux (B_s), and free-flow stability (N) which have been identified as the key physical processes that govern the SBL height. However, no consensus was achieved from these linear models.

By employing the turbulent kinetic energy (TKE) budget equation, Zilitinkevich and Mironov [10] derived two diagnostic multi-limit equations for the equilibrium depth of SBL that contains all the four aforementioned

variables and several unknown coefficients, and that can hold in both the general case and the limiting cases. Furthermore, Zilitinkevich et al. [11] proposed two refined Ekman-layer height equations from the momentum balance equations and validated their model against observations.

The above parameterizations of SBL height have been evaluated by many researchers based on various observational datasets [11-14] and controlled numerical simulations [15-17]. Using datasets over grassland, cooler ocean surface and snow cover, Vickers and Mahrt [13] evaluated a variety of surface flux-based SBL height formulations; they summarized that the existing formulations generally perform poorly and often overestimate the depth of SBL.

The aforementioned multi-limit equations were further discussed and evaluated by Zilitinkevich and Baklanov [12]; they found that the Ekman-layer height equations by Zilitinkevich et al. [11] produced more accurate estimation of SBL height than that by the equations in Zilitinkevich and Mironov [10]. Kosovic and Lundquist [17] compared several parameterizations to calculate SBL height using large-eddy simulations of moderately stable boundary layers and they demonstrated that the gravity waves in the free atmosphere do affect the height of SBL.

Steeneveld et al. [14] evaluated the performance of the Zilitinkevich and Mironov [10] multi-limit equations against four observational datasets over different terrains; they found that the multi-limit equations underestimate the SBL height, especially for shallow boundary layers, and no unique parameter that sets for these equations can be determined. Alternatively, Steeneveld et al. [14] developed a formulation based on formal dimensional analysis using the same quantities as in the multi-limit equations; this new formulation showed to be more robust and it significantly reduced the model bias for shallow boundary layer heights compared with the multi-limit equations, and a unique parameter set was found for this new equation. Furthermore, Steeneveld et al. [18] proposed another alternative equation by an inverse interpolation of the eddy diffusivities for each boundary layer prototype [16] instead of interpolating the height scales for each prototype in the equations by Zilitinkevich and coworkers when applying the definition of Ekman layer depth. This equation reduced the bias of the predicted SBL height compared to their equations. However, the formulation derived from formal dimensional analysis [14] still shows better performance than this equation [18].

Therefore, the above studies have enlightened us that some formulations which involve less variables can perform better than the complex multi-limit formulations for the SBL height estimation. In this paper, we will take an unconventional approach to figure out which variables are indeed essential for an optimum representation of the SBL height. Artificial neural networks will be employed to model this non-linear phenomenon based on a large-eddy simulation output dataset.

2. Data Description

The data used in this study are the output from 68 runs of large eddy simulation (LES). They are idealized simulations, by setting different values for the parameters of geostrophic wind G, cooling rate C of air temperature, surface roughness length z_0, initial boundary layer height H, Brunt-Vaisala frequency N in the free atmosphere above the SBL, and Coriolis parameter f, to represent typical low-level jet scenarios in the stable boundary layer.

The LES computational domain employed for this study extends 800 meters in the lateral direction and 795 meters in the vertical direction. The grid size in both y and z directions is 10 meters; at each grid point, the three dimensional wind velocity and air temperature are generated as a time series; the time step for the LES flow fields is 0.1 sec [19]. The 5-min mean values of the simulated data are output for the boundary layer height studies. From each LES run, a total of 144 time steps are obtained representing 12 hours of simulation between 1800 to 0600 LST.

Each simulation starts from a neutral profile. As shown by the wind profiles, the damping layer of wind speed starts from 550-m level around, we only use the

data from the surface to 550-m height to analyze the stable boundary height.

There have been many definitions of SBL height with different emphasis on properties of turbulence [20], heat flux [21], mean wind speed [22], mean temperature [23], and other effects [16,24,25]. In this study, we will employ the definition of the height of the lowest maximum of the wind speed [22,26], often referred to as the low-level jet height, as the targeted SBL height h in the parameterization. The LES wind profiles show that low-level jets usually start developing after 0000 LST. Therefore, only the data between 0000 - 0600 LST in each simulation are selected to represent typical low-level jets.

3. Resampling Strategy

In this study, at each time step in each LES run, from the wind speed profile, a SBL height h can be estimated as the height of the maximum wind speed, i.e., the low-level jet height. Thus from each simulation there are totally 72 output values of 6 hours for surface friction velocity u_*, surface buoyancy flux B_s, and boundary layer height h. The geostrophic wind G, the Coriolis parameter f, the Brunt-Vaisala frequency N in the free atmosphere, and the surface roughness length z_0 are set as various values for each LES run.

We will firstly combine the data from all the 68 LES sruns together to form a total dataset, and then apply repeated random sub-sampling cross-validation algorithm, to evaluate five conventional linear formulations of stable boundary layer height. This method randomly partitions the dataset into training and validation data by a certain percentage (60% training and 40% validation in this study). For each such split, the model is fit to the training data, and assesses the predictive accuracy using the validation data. This procedure is repeated for 100 times in this study.

The variables from the training data are used in the conventional linear formulations to calculate their included constants first. Then substituting the variables in the validation data, together with the derived constants into the linear equations, the SBL height h can be predicted. Consequently based on the difference between the predicted and the targeted values of h, the statistical quantities such as root mean squared error (RMSE), median absolute error (MAE), median bias error (MBE), and index of agreement (IOA) are calculated. Therefore the performance of these linear formulations can be evaluated from the distribution of the statistical errors.

4. Performance of Conventional SBL Height Parameterizations Based on LES Data

Most conventional formulations for the equilibrium

depth of the stably stratified boundary layer are surface-flux-dominated formulations. In these parameterizations, the turbulent heat flux and momentum flux at the surface are required to be estimated. Rossby and Montgomery [27] proposed that the surface friction velocity u_* and Coriolis parameter f mainly affect the stable boundary layer depth:

$$h = C_n \frac{u_*}{f} \qquad (1)$$

Then Zilitinkevich [7] considered the surface heat flux together with the influences of surface friction and the earth's rotation such that

$$h = C_{sr} \frac{u_*^2}{\left(-fB_s\right)^{1/2}} \qquad (2)$$

where B_s is the surface buoyancy flux. Pollard $et\ al.$ [8] were the first to suggest that N, the free-flow stratification above the stable boundary layer, would affect the boundary layer height:

$$h = C_{ir} \frac{u_*}{\left(fN\right)^{1/2}} \qquad (3)$$

Kitaigorodskii [6] proposed that only the surface heat flux and momentum flux are dominated on the depth of the boundary layer without the effect of earth rotation

$$h = -C_s \frac{u_*^3}{B_s} \qquad (4)$$

When only considering surface momentum flux and the overlying free-flow stratification, Kitaigorodskii and Joffre [9] obtained

$$h = C_i \frac{u_*}{N} \qquad (5)$$

In this study, we evaluate the performance of the above five parameterizations against controlled large eddy simulation output. **Figure 1** shows the statistical errors of RMSE, MAE, MBE, and IOA between the predicted and the target values of stable boundary layer

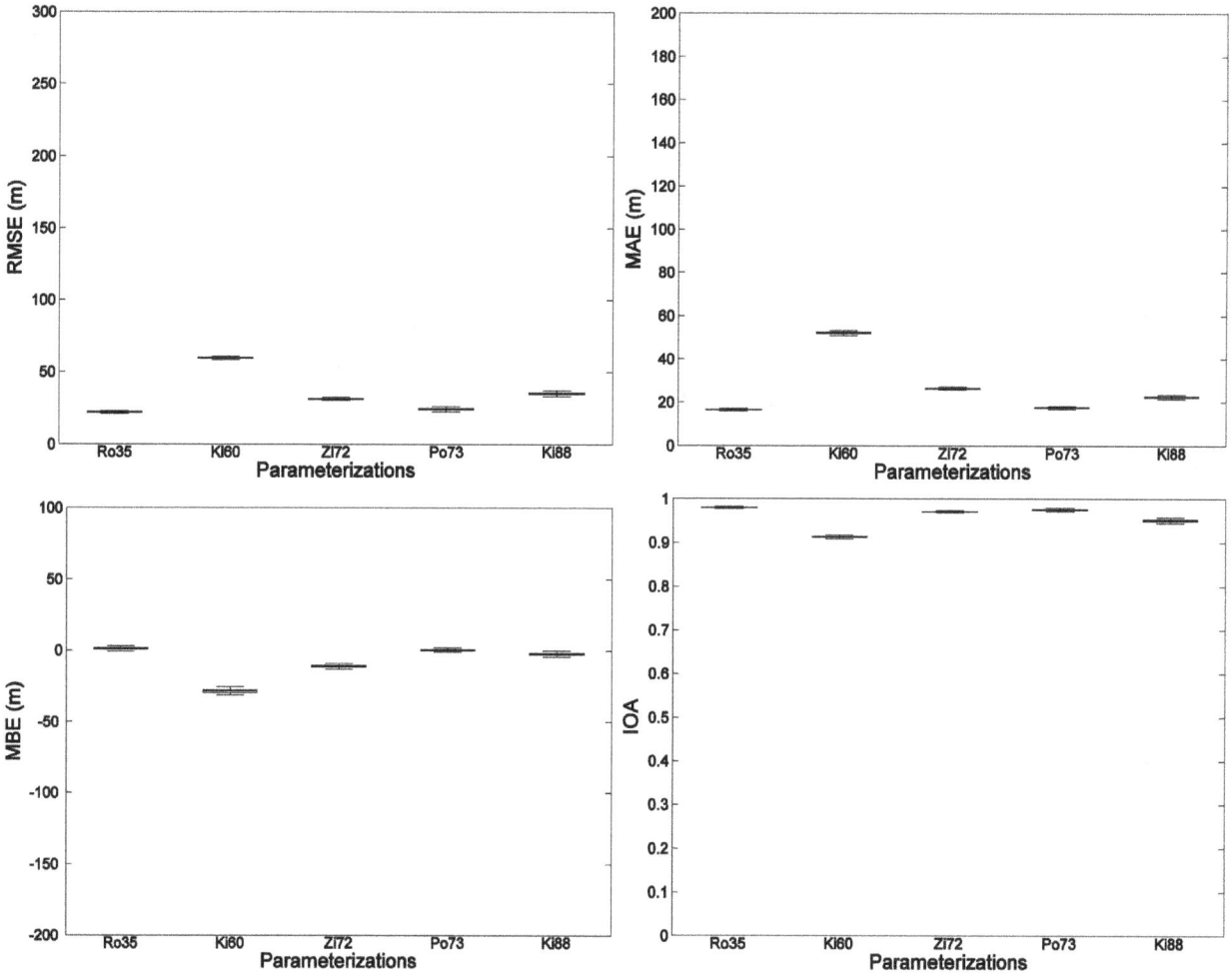

Figure 1. Performance of conventional linear parameterizations.

height h based on these five formulations.

The formulation by Rossby and Montgomery [27] when only considering influences of u_* and f shows the best performance with minimum values of RMSE (25 m) and MAE (18 m) among others. The Pollard *et al.* [8] scheme performs secondly best with a RMSE of 30 m and a MAE of 20 m when N is taken into account. Then Zilitinkevich [7] formulation has slightly larger RMSE than that by Pollard *et al.* [8] when N is replaced by B_s. When f is not involved in the formulations [Equations (4) and (5)], the models biases are largely increased with Equation (4) being the worst case.

Therefore, the performance of these conventional linear parameterizations of SBL height against LES data indicates that the surface friction velocity u_* and Coriolis parameter f are proved to be the two most predominant variables to determine h and the free-flow stability N shows more important than the surface buoyancy flux B_s.

5. Artificial Neural Networks

Artificial neural networks (ANN) are composed of simple interconnected neurons. Unlike other statistical techniques, ANN usually makes no prior assumptions regarding the data distribution, and can model extremely non-linear relationships and be trained to be accurately representative for new and unseen data [28]. Typically, a neural network can be trained based on a comparison between the output and the target until the network output matches the target, so that a particular input leads to a specific target output.

Artificial neural networks have been applied to perform complex functions in various fields. In recent years, a significant number of ANN applications have been developed in different fields of geosciences such as satellite remote sensing, meteorology, oceanography, numerical weather prediction, and climate studies [29].

The work flow for the general neural network design process has the following three primary steps:

5.1. Collect and Prepare the Data

After the sample data have been collected, they need to be preprocessed and to be divided into three subsets before they are used to train the network. The first subset is the training set, which is used for computing the gradient and updating the network weights and biases. The second subset is the validation data which will determine the stopping criteria. The error on the validation set is monitored during the training process. The validation error normally decreases during the initial phase of training, as does the training set error. However, when the network begins to over fit the data, the validation error typically begins to rise after some iterations. The network weights and biases are saved when the validation error reaches

the minimum at the certain iteration, which gives the optimum iteration number for the network training. The third subset in the data is the testing data, the error on which is not used during the training process, but used to compare different models. Typically, a poor division of the dataset happens when the testing error reaches a minimum at a significantly different iteration number from that at which the validation error reaches its minimum.

5.2. Create, Configure and Initialize the Network

After the data have been prepared, the next step is to select the network architecture and create the network. In this paper, the most predominant network architecture of multilayer perceptron is utilized to map the relationships between the stable boundary layer height h and the associated variables u_*, B_s, N, f, and z_0. We create a two-layer feedforward network (see **Figure 2**). The sum of the weighted inputs and the bias forms the input to the transfer function in the network; then the transfer function generates the output vector [28,30]. Multilayer networks often use the log-sigmoid transfer function.

After configuring the network object and also initializing the weights and biases of the network, the network is ready for training.

5.3. Train and Validate the Network

The process of training a neural network involves adjusting the values of the weights and biases of the network to optimize the network performance, that is, to find the combination of weights which can result in the smallest error [28]. The default performance function for feedforward networks is the mean square error (MSE) between the network outputs and the targets.

Gradient descent is the simplest optimization algorithm; the gradient is calculated using the technique of

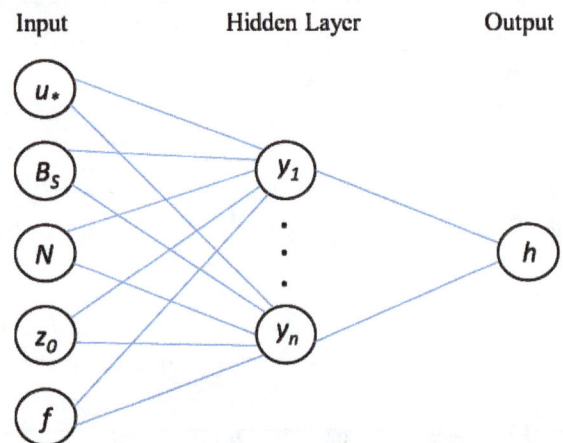

Figure 2. Illustration of the ANN architecture for modeling stable boundary layer height.

back propagation algorithm, which involves performing computations backward through the network and is the most computationally straightforward algorithm for training the multilayer perceptron [28,31]. The gradient descent algorithm updates the network weights and biases in the direction in which the performance function decreases most rapidly.

Once the training procedure is complete, the network validation can be done by checking the training record and creating a regression plot between the outputs of the network and the targets.

6. ANN Modeling Strategy of Stable Boundary Layer Height

In this study, we will apply ANN to predict the stable boundary layer height. ANN development for simulating the SBL height requires identification of the input and output variables. From the knowledge of conventional parameterizations, the SBL height h can be expressed as a function of surface friction velocity u_*, surface buoyancy flux B_s, Brunt-Vaisala frequency N in the free atmosphere above the SBL, Coriolis parameter f, and surface roughness length z_0.

Figure 2 shows the illustration of the two-layer perceptron ANN architecture in modeling the SBL height h. The architecture of the network depends on the number of input and output variables. For the present problem, the number of input nodes in the two-layer perceptron network varies from the minimum of 1 to the maximum of 5; the input variables are randomly combined among the five associated variables u_*, B_s, N, f, and z_0. **Table 1** shows the 31 variable combinations of inputs used in the network. Only one output node exists in this problem.

However, the number of nodes in the hidden layer is not fixed. For each combination of input variables, we test the network by changing the neuron number in the hidden layer from 1 to 20 in turn to find a best number of hidden nodes. At each specific hidden nodes number, the network is repeated to run 100 times to calculate a statistical median value of the errors. For each run, the dataset are randomly partitioned into three subsets of training, validation, and testing data with the proportion of 30%, 30%, and 40%, respectively.

ANNs are developed using MATLAB Neural Network Toolbox (version 2010). The back propagation algorithm is applied for training. The training is performed using the default values for the parameters of the network.

7. Performance of ANN-Based Parameterizations Based on LES Data

From previous studies, the basic variables that govern the SBL height h are the surface friction velocity u_*, the surface buoyancy flux B_s, the free-flow stability N, the

Table 1. Input variable combinations in ANN modeling.

Combination #	u_*	B_s	N	f	z_0
31	√	√	√	√	√
30	√	√	√	√	
29	√	√	√		√
28	√	√	√		
27	√	√		√	√
26	√	√		√	
25	√	√			√
24	√	√			
23	√		√	√	√
22	√		√	√	
21	√		√		√
20	√		√		
19	√			√	√
18	√			√	
17	√				√
16	√				
15		√	√	√	√
14		√	√	√	
13		√	√		√
12		√	√		
11		√		√	√
10		√		√	
9		√			√
8		√			
7			√	√	√
6			√	√	
5			√		√
4			√		
3				√	√
2				√	
1					√

Coriolis parameter f, and the surface roughness length z_0. In this study, these five variables u_*, B_s, N, f, and z_0 were randomly combined into 31 different combinations of inputs for the neural network in the order of binary numbers (see **Table 1**) to investigate which combination of variables is the optimum input in determining the stable boundary layer height h. **Figure 3** shows the statistical errors of RMSE, MAE, MBE, and IOA calculated between the output value and the target value of SBL height h in the network based on the 31 input combinations utilizing the LES output data.

We divided the 31 combinations into 4 groups. From the first group of combinations (combination number 1 to 7 in **Table 1**) where the variables u_* and B_s are not involved in the input, the model performance is very poor. The values of RMSE are as high as 75 m; all the IOA values are lower than 0.3.

In the second group of combinations (combination number 8 to 15 in Table 1), when the surface buoyancy flux B_s are included in the model input, the RMSE decrease to be around 50 m, the IOA are largely increased up to 0.9. This indicates that buoyancy flux is a very important variable in predicting h when friction velocity is absent in the input.

When the third (combination number 16 to 23 in **Table 1**) and the fourth (number 24 to 31) groups of combinations are input into the network, where the surface friction velocity u_* is involved in each input, the model performance is largely improved again to be the best among all the variable combinations of input. The RMSE are very close to each other based on these 16 set of inputs, with a mean value of 20 m, and their IOA reach to be as high as 0.98. This strongly proves that the surface friction velocity u_* is the most dominant variable in determining the SBL height.

From the second group to the fourth group of variable combinations, keeping the way of combining other three variables (N, f, z_0) as that in the first group, when adding u_* with B_s in the input variables, the model performance is largely improved. From the second group to the third group, when B_s is replaced by u_* in the input, the model performance is largely improved similarly. This indicates that the variable B_s has less relevance compared to u_* in predicting h, even if B_s has been proved to be an important one. This can also be verified from that when changing the input combinations from the fourth group to the third group, where B_s is removed from the variable combinations, the model performance is not changed obviously.

The median RMSE based on the variable combination of (u_*, B_s, N, f, z_0) is the minimum of the RMSE values based on all combinations. This means that when all the

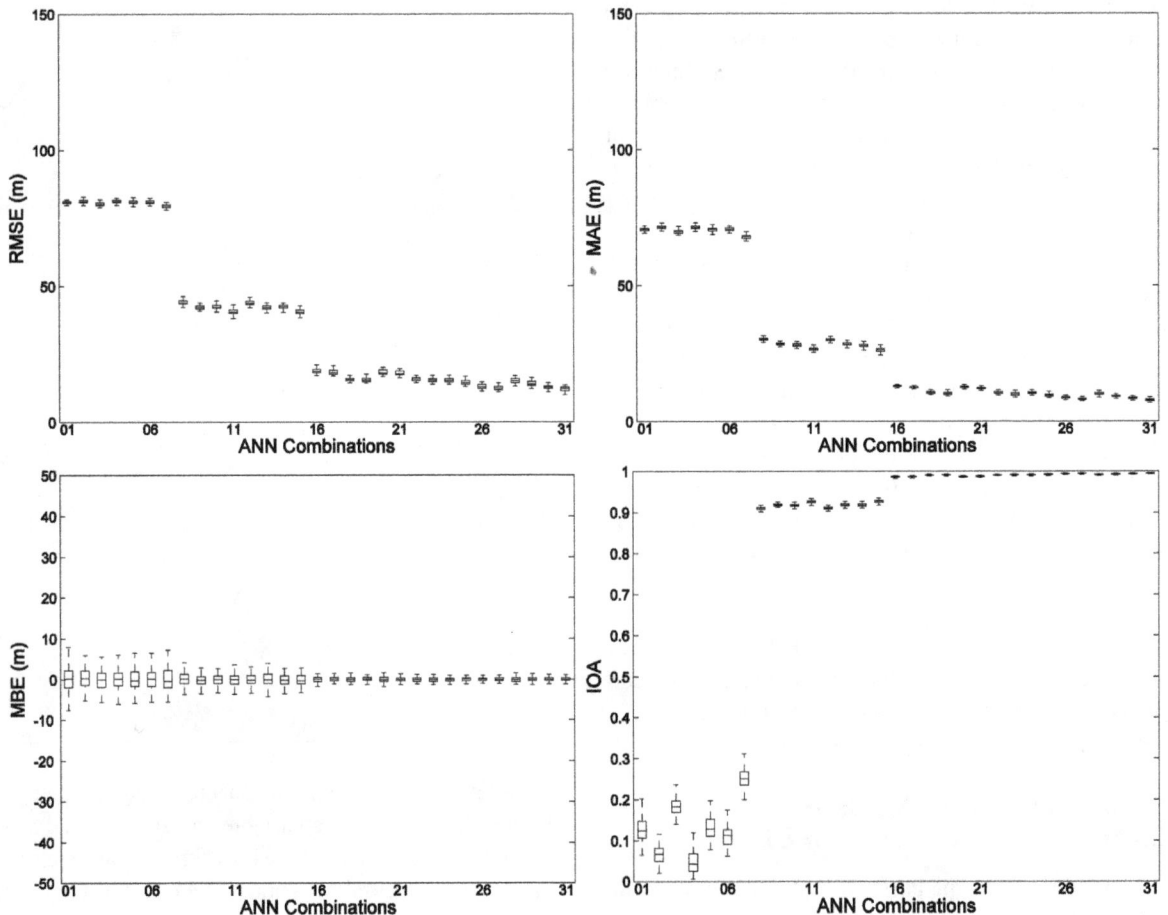

Figure 3. Performance of ANN-based parameterizations using 31 different variable combinations as inputs.

five relevant variables u_*, B_s, N, f, and z_0 are taken into account in the model input, the model performs best.

In addition, within the third and fourth groups, when the coriolis parameter f is added into the combinations, the RMSE values are decreased; when replacing f by N in the combinations, RMSE are increased; when f and N are both involved, the RMSE values are decreased again. Therefore, the coriolis parameter f shows to be more important for h prediction compared to the inversion strength N.

In further details, within both the third group and the fourth group, the values of RMSE are changing in pairs. The two median RMSE values of each pair are almost equal; the difference between the two input variable combinations within each pair is that the surface roughness length z_0 is included or not. This demonstrates that the surface roughness length z_0 is not obviously affective to h determination.

Overall, our results based on ANN shows that the strongest relation of h exists with u_*, and then with B_s secondly. Koracin and Berkowicz [32] ever proposed a simple empirical estimate of $h = 700u_*$. This formula was evaluated by Steeneveld et al. [14] by observational data and it performs well over their datasets. In the alternative formulation using dimensional analysis by Steeneveld et al. [14], they also concluded that h relies on u_* most strongly, and then on B_s in the very stable limit, then on N while f and z_0 are less relevant for their data; when the Coriolis parameter f is omitted, the SBL height h for moderately stable conditions is proportional to u_*/N, and for high stability conditions h is proportional to the length scale $\sqrt{|B_s|/N^3}$.

Kosovic and Lundquist [17] also used LES study to explore various parameterizations of SBL height. They found that from a small-domain simulation, the SBL height formulation proposed by Kosovic and Curry [15] in which the strength of the inversion N was not involved (except for u_*, B_s and f) appeared to match the SBL height defined by turbulent shear stress better than the scheme by Zilitinkevich et al. [11] did in which N was involved, due to the simulation domain was too small to include the effects of gravity waves that developed above the SBL. They also conducted medium-resolution LES with the domain size sufficiently large to resolve the overlying gravity waves; the results showed that the formulation [11] with N involved consequently gave a better prediction of SBL height.

In the conventional formulations, the friction velocity u_* and Coriolis parameter f are proved to be the two most predominant variables in determining h, whereas the surface buoyancy flux B_s is not an important factor and even has less influence than the free-flow stability N. **Table 2** compares the model performance given by the statistical errors of RMSE, MBE, MAE and IOA based

Table 2. Performance of two linear parameterizations and two ANN-based parameterizations using corresponding variable combinations as inputs.

Parameterization	RMSE	MBE	MAE	IOA
(u_*, f)	15.59	−0.07	10.54	0.991
(u_*, B_s)	15.26	−0.07	10.48	0.991
Ro35	22.15	0.95	16.51	0.980
Ki60	59.53	−28.56	52.16	0.913

on the conventional formula of $h = C_n u_*/f$ by Rossby and Montgomery [27] (Ro35 in **Figure 1**) and based on the input variable combination of (u_*, f) in ANN. Their performance is quite good and they have close RMSE values of 22 m and 16 m; both have very high IOA values nearing to 1. The cases for the conventional formula $h = -C_s u_*^3/B_s$ by Kitaigorodskii [6] (Ki60 in **Figure 1**) and the variable combination of (u_*, B_s) in ANN are also listed in **Table 2**. Although Ki60 results in a very high IOA value of 0.91, its RMSE is as large as 60 m, which is much larger than the RMSE of 15 m for the combination (u_*, B_s) in ANN. As discussed in section 4, this Ki60 formulation performs worst among the five conventional formulations (**Figure 1**); this might be due to the cube of u_* in the formulation and needs to be further explored.

8. Conclusions

In this paper, we tried to find out which variables are essential and can achieve the best performance in determining the stable boundary layer height h based on large-eddy simulation data. Artificial neural networks were applied to investigate which combination of variables is the optimum input. We also evaluated the performance of conventional linear formulations for the SBL height prediction, in which the surface friction velocity u_* and the Coriolis parameter f were proved to be the two most predominant variables to determine h and the free-flow stability N showed to be more important than the surface buoyancy flux B_s for h prediction.

The results based on ANN showed that the strongest relation of h exists with the surface friction velocity u_*. The secondly dominant variable relevant to h is the surface buoyancy flux B_s. The Coriolis parameter f shows to be more important compared to the inversion strength N. The surface roughness length z_0 is not an affective variable to h.

9. Acknowledgements

We thank Dr. Sukanta Basu in the Department of Marine, Earth, and Atmospheric Sciences at North Carolina State University for providing the LES dataset and codes in this study.

REFERENCES

[1]   P. Seibert, F. Beyrich, S. E. Gryning, S. Joffre, A. Rasmussen and P. H. Tercier, "Review and Intercomparison of Operational Methods for the Determination of the Mixing Height," *Atmospheric Environment*, Vol. 34, No. 7, 2000, pp. 1001-1027.

[2]   J. A. Salmond and I. G.McKendry, "A Review of Turbulence in the Very Stable Boundary Layer and Its Implications for Air Quality," *Progress in Physical Geography*, Vol. 29, No. 2, 2005, pp. 171-188.

[3]   G. J. Steeneveld and A. A. M. Holtslag, "Meteorological Aspects of Air Quality," *Air Quality in the 21st Century*, Nova Science Publishers, New York, 2009, pp. 67-114.

[4]   W. Brutsaert, "Radiation, Evaporation and the Maintenance of the Turbulence under Stable Conditions in the Lower Atmosphere," *Boundary-Layer Meteorology*, Vol. 2, No. 3, 1972, pp. 309-325.

[5]   L. Mahrt, "Modeling the Height of the Stable Boundary Layer," *Boundary-Layer Meteorology*, Vol. 21, No. 1, 1981, pp. 3-19.

[6]   S. A. Kitaigorodskii, "On the Computation of the Thickness of the Wind-Mixing Layer in the Ocean," *Izvestiya AN SSSR, Geophysical Series*, Vol. 3, 1960, pp. 425-431.

[7]   S. S. Zilitinkevich, "On the Determination of the Height of the Ekman Boundary Layer," *Boundary-Layer Meteorology*, Vol. 3, No. 2, 1972, pp. 141-145.

[8]   R. T. Pollard, P. B. Rhines and R. Thompson, "The Deepening of the Wind-Mixed Layer," *Geophysical Fluid Dynamics*, Vol. 3, 1973, pp. 381-404.

[9]   S. A. Kitaigorodskii and S. M. Joffre, "In Search of Simple Scaling for the Heights of the Stratified Atmospheric Boundary Layer," *Tellus*, Vol. 40A, No. 5, 1988, pp. 419-443.

[10] S. S. Zilitinkevich and D. V. Mironov, "A Multi-Limit Formulation for the Equilibrium Depth of a Stably Stratified Boundary Layer," *Boundary-Layer Meteorology*, Vol. 81, No. 3-4, 1996, pp. 325-351.

[11] S. S. Zilitinkevich, A. Baklanov, J. Rost, A. S. Smedman, V. Lykosov and P. Calanca, "Diagnostic and Prognostic Equations for the Depth of the Stably Stratified Ekman Boundary Layer," *Quarterly Journal of the Royal Meteorological Society*, Vol. 128, No. 1, 2002, pp. 25-46.

[12] S. S. Zilitinkevich and A. Baklanov, "Calculation of the Height of Stable Boundary Layers in Practical Applications," *Boundary-Layer Meteorology*, Vol. 105, No. 3, 2002, pp. 389-409.

[13] D. Vickers and L. Mahrt, "Evaluating Formulations of the Stable Boundary Layer Height," *Journal of Applied Meteorology*, Vol. 43, No. 11, 2004, pp. 1736-1749.

[14] G. J. Steeneveld, B. J. H. Van de Wiel and A. A. M. Holtslag, "Diagnostic Equations for the Stable Boundary Layer Height: Evaluation and Dimensional Analysis," *Journal of Applied Meteorology and Climatology*, Vol. 46, No. 2, 2007, pp. 212-225.

[15] B. Kosovic and J. A. Curry, "A Large Eddy Simulation Study of a Quasi-Steady, Stably Stratified Atmospheric Boundary Layer," *Journal of the Atmospheric Sciences*, Vol. 57, No. 8, 2000, pp. 1052-1068.

[16] S. S. Zilitinkevich and I. Esau, "The Effect of Baroclinicity on the Equilibrium Depth of Neutral and Stable Planetary Boundary Layers," *Quarterly Journal of the Royal Meteorological Society*, Vol. 129, No. 595, 2003, pp. 3339-3356.

[17] B. Kosovic and J. K. Lundquist, "Influences on the Height of the Stable Boundary Layer," *16th Symposium on Boundary Layers and Turbulence, American Meteor Society*, Portland, 23 June 2004, Preprints.

[18] G. J. Steeneveld, B. J. H. Van de Wiel and A. A. M. Holtslag, "Notes and Correspondence: Comments on Deriving the Equilibrium Height of the Stable Boundary Layer," *Quarterly Journal of the Royal Meteorological Society*, Vol. 133, No. 622, 2007, pp. 261-264.

[19] C. Sim, S. Basu and L. Manuel, "On Space-Time Resolution of Inflow Representations for Wind Turbine Loads Analysis," *Energies*, Vol. 5, No. 7, 2012, pp. 2071-2092.

[20] D. H. Lenschow, X. S. Li, C. J. Zhu and B. B. Stankov, "The Stably Stratified Boundary Layer over the Great Plains. I. Mean and Turbulent Structure," *Boundary-Layer Meteorology*, Vol. 42, No. 1-2, 1988, pp. 95-121.

[21] S. J. Caughey, J. C. Wyngaard and J. C. Kaimal, "Turbulence in the Evolving Stable Layer," *Journal of the Atmospheric Sciences*, Vol. 36, No. 6, 1979, pp. 1041-1052.

[22] J. W. Melgarejo and J. W. Deardorff, "Stability Functions for the Boundary Layer Resistance Laws Based upon Observed Boundary Layer Heights," *Journal of the Atmospheric Sciences*, Vol. 31, No. 5, 1974, pp. 1324-1333.

[23] T. Yamada, "Prediction of the Nocturnal Surface Inversion Height," *Journal of Applied Meteorology*, Vol. 18, No. 4, 1979, pp. 526-531.

[24] J. R. Garratt and R. A. Brost, "Radiative Cooling Effects within and above the Nocturnal Boundary Layer," *Journal of the Atmospheric Sciences*, Vol. 38, No. 12, 1981, pp. 2730-2746.

[25] R. K. Newsom and R. M. Banta, "Shear-Flow Instability in the Stable Nocturnal Boundary Layer as Observed by Doppler Lidar during CASES-99," *Journal of the Atmos-*

pheric Sciences, Vol. 60, No. 1, 2003, pp. 16-33.

[26] F. Beyrich, "Sodar Observations of the Stable Boundary-Layer Height in Relation to the Nocturnal Low-Level Jet," *Meteorologische Zeitschrift*, Vol. 3, No. 1, 1994, pp. 29-34.

[27] C. G. Rossby and R. B. Montgomery, "The Layer of Frictional Influence in Wind and Ocean Currents," *Papers in Physical Oceanography and Meteorology*, Vol. 3, No. 3, 1935, pp. 1-101.

[28] M. W. Gardner and S. R. Dorling, "Artificial Neural Networks (the Multilayer Perceptron): A Review of Applications in the Atmospheric Sciences," *Atmospheric Environment*, Vol. 32, No. 14-15, 1998, pp. 2627-2636.

[29] V. M. Krasnopolsky, "Neural Network Emulations for Complex Multidimensional Geophysical Mappings: Applications of Neural Network Techniques to Atmospheric and Oceanic Satellite Retrievals and Numerical Modeling," *Reviews of Geophysics*, Vol. 45, No. 3, 2007, Article ID: RG3009.

[30] M. Gevrey, I. Dimopoulos and S. Lek, "Review and Comparison of Methods to Study the Contribution of Variables in Artificial Neural Network Models," *Ecological Modelling*, Vol. 160, No. 3, 2003, pp. 249-264.

[31] D. E. Rumelhart, G. E. Hinton and R. J. Williams, "Parallel Distributed Processing: Explorations in the Microstructure of Cognition, Vol. 1," MIT Press, Cambridge, 1986.

[32] D. Koracin and R. Berkowicz, "Nocturnal Boundary-Layer Height: Observations by Acoustic Sounders and Predictions in Terms of Surface-Layer Parameters," *Boundary-Layer Meteorology*, Vol. 43, No. 1-2, 1988, pp. 65-83.

8

Solar Activity, Solar Wind and Geomagnetic Signatures

Jean-Louis Zerbo[1,2,3*], Frédéric Ouattara[4], Christine Amory Mazaudier[2],
Jean-Pierre Legrand[5], John D. Richardson[6]

[1]Université Polytechnique de Bobo-Dioulasso, Bobo-Dioulasso, Burkina Faso
[2]LPP-Laboratoire de Physique des Plasmas/UPMC/Polytechnique/CNRS,
Saint-Maur-des-Fossés, France
[3]Laboratoire d'Energies Thermiques Renouvelables (L.E.T.RE), Université de Ouagadougou,
Ouagadougou, Burkina Faso
[4]Ecole Normale Supérieure de l'Université de Koudougou, Koudougou, Burkina Faso
[5]Institut National des Sciences de l'Univers (INSU), Paris, France
[6]Kavli Institute for Astrophysics and Space Research, Massachusetts Institute of
Technology, Cambridge, USA

ABSTRACT

The present study investigates solar events through geomagnetic activity and physical processes on the Sun: 1) Quiet activity (QA) related to the slow solar wind, 2) Recurrent activity (RA) related to high and moderate speed solar wind streams from coronal holes, 3) Shock activity (SA) identified by observations of SSCs and 4) Unclear activity (UA) which contains all activity not covered by the first three cases. For recent cycles, we analyze and emphasize some important results: Quiet activity is predominant for cycle 23, comprising 40% of the total activity and over 80% of the activity near solar minimum. Shock and recurrent activity contributions to total geomagnetic activity are largest in cycle 20. The most fluctuating events are observed during cycles 21 and 22. Throughout solar cycle 23, the contribution, from each type of activity, differs from recent solar cycles, with larger percentages of quiet and recurrent activity and less unclear activity. These percentages are similar to those in solar cycles observed in the late 1800s. Since 1963, solar wind data are available. We analyze the distribution of the solar wind velocity for each geomagnetic class of activity and find that: 1) Within each activity type aa does not depend on V, 2) Approximately 80% of the solar wind has V < 450 km/s for QA and 80% of the solar wind has V > 450 km/s for RA, 3) SA and UA both have 60% of the solar wind V > 450 km/s. We found the following conditions for all four solar cycles: 1) For QA 95% of solar wind speeds are in the range 399 ± 69 km/s, 2) For RA 95% of the solar wind speeds range from 582 ± 110 km/s, 3) For SA 95% of the solar wind velocities are order of 482 ± 101.4 km/s, and 4) For UA 95% of solar wind speeds are 480 ± 85.82 km/s. These results confirm the classification scheme that QA reflects slow wind effects, RA effects high wind stream and UA answers to the fluctuations between high wind stream (~60%) and slow wind (~40%). The study shows that high wind stream (~60%) and slow wind (~40%) are both registered for SA.

Keywords: Solar Wind; Geomagnetic Activity; Solar Activity; Geomagnetic Index aa

1. Introduction

Geomagnetic activity may be divided into four classes: quiet, recurrent, shock, and fluctuating/unclear [1-7]. Legrand and Simon [1] and Simon and Legrand [2] proposed the first classification of geomagnetic activity for the period 1868-1977 using pixel diagram. Pixel diagrams represent the geomagnetic data as a function of solar activity for each solar rotation (27 days) and give an overview of the geoeffectiveness of solar events. Pre-

vious pixel diagrams were built using the the geomagnetic aa index defined and explained by Mayaud [8-10], the times of SSCs (Sudden Storm Commencements) and the strong correlation between the aa index and solar wind established by Svalgaard [11]. Important works [12, 13] show that geomagnetic activity has two components. The first component varies in phase with the sunspot cycle and the second begins immediately after the maximum of the sunspot cycle. Following Legrand and Simon [1]'s works, Ouattara and Amory-Mazaudier [6] investigate the time variation of solar activity from 1968 to

---

*Corresponding author.

1996. Very recent study [7] extends and improves the classification of geomagnetic activity, and we use it to investigate the time profiles of geomagnetic activity during solar cycles 20, 21, 22, and 23. We also study the statistical distribution of solar wind speeds in each class of geomagnetic activity for cycles 20 to 23.

2. Data and Methodology

2.1. Data

The times of sudden substorm commencements (SSC), which are rapid increases in the magnetic field observed at the ground per day are taken from http://isgi.latmos.ipsl.fr/. The solar wind speed and the international sunspot number (SSN) are obtained from http://omniweb.gsfc.nasa.gov/form/dx1.html.

2.2. Geomagnetic Classification Methodology

Legrand and Simon [1] justified their use of the aa index to classify geomagnetic activity on two features: 1) The contribution of solar wind shocks to geomagnetic activity can be determined from ground-based observations of SSCs; time periods with SSCs are defined to have shock activity (SA). 2) The strong correlation between the aa index [8-10] and the solar wind speed [11]. This correlation allows the data to be divided into three classes of geomagnetic activity: 1) low-speed solar wind, 2) variable-speed solar wind and 3) high speed solar wind. Zerbo J.-L. et al. [7] consider these criteria, the contributions of magnetic clouds, and the contribution of moderate speed wind streams to improve the classifications [1]. We divide the solar wind into classes using the aa index and then use the results to study correlations with solar wind parameters. **Figure 1** illustrates the pixel diagrams of the aa indices. The top panel illustrates the year 1974,

a very active year [7], and the bottom panel the year

2009, a very geomagnetic quiet year at the minimum of the sunspot cycle. The pixel diagrams display the daily averages of aa as a table. Each horizontal line contains 27 days corresponding to a 27-day Bartels solar rotation with two days of overlap on each end. The number in each square is the mean daily value of the aa index and the squares are color coded based on these values using the color bar shown in **Figure 1**. Circles show the days when SSCs were observed. This diagram, similar to a Bartels diagram, simplifies the identification of geomagnetic phenomena. From the pixel diagrams, the class of activity, QA, SA, RA or UA, for each day is determined. The 1974 pixel diagram illustrates the different classes of activity. The use of these diagrams is best explained by Zerbo J.-L. et al. [7]. 1) Quiet activity days are defined as days when aa < 20 nT (colors white and blue on **Figure 1**). The three other classes (shock activity, recurrent activity and fluctuating activity) constitute the disturbed geomagnetic activity classes which occur on days when aa ≥ 20 nT. These classes are distinguished as follows:

2) **Recurrent (stream)** activity corresponds in the pixel diagrams to days where aa ≥ 20 which repeat at the same solar longitude for at least two consecutive solar rotations. This class is driven by fast solar wind from coronal holes which persist for more than one solar rotation. SSCs are not observed during the main phases of storms driven by recurrent streams. An example of RA is shown in **Figure 1** (top panel).

3) **Shock activity** is defined to occur on days when SSCs are observed and up to 3 days after the shock passage if aa remains > 20 nT. SA is driven by CMEs on the Sun which often produces high solar wind speeds. An example of shock activity which persists for three days is shown in **Figure 1** (top panel).

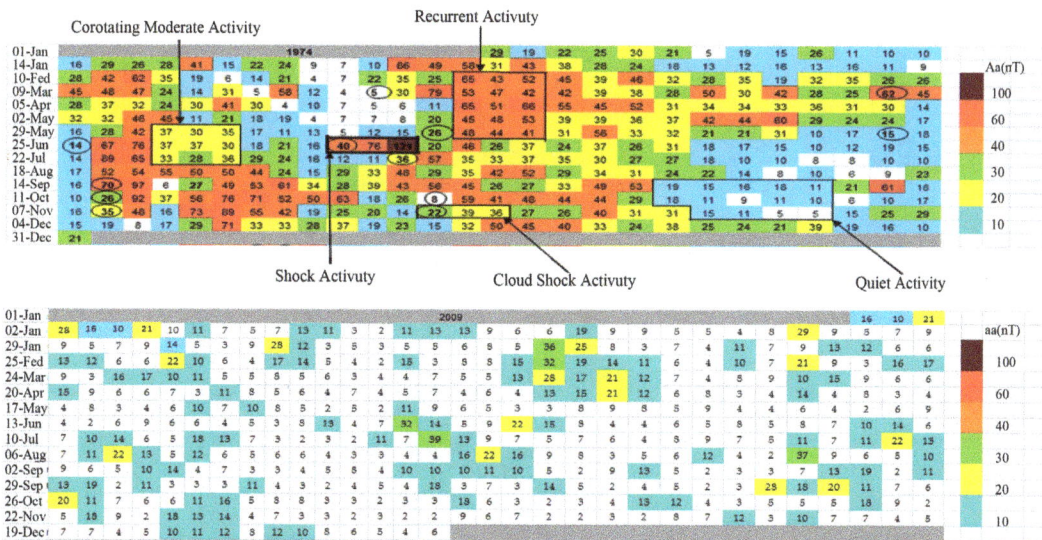

Figure 1. Pixel diagrams for the years 1974 (panel a) and 2009 (panel b) constructed with aa indices.

4) *Unclear activity* includes all times which do not fit the criteria for the other three classes and is thus a mixed class of disturbances. This activity results from variable moderate and high-speed solar wind and may be related to the fluctuations of the heliospheric neutral sheets.

3. Results

3.1. Geomagnetic Classes, Their Activity Variation during the Solar Cycles 20, 21, 22, and 23

In this section, we study the four latest solar cycles using aa indices and solar wind data. **Figure 1** shows the large difference in solar activity between the beginning of the declining phase in 1974 and very quiet recent solar minimum in 2009.

Figure 2 extends the Legrand and Simon [1]'s work to include 32 more years of data. It shows the yearly percentage of occurrence of each class of geomagnetic activity from 1868 through 2009. **Figures 2(a)-(e)** show long time variations of QA, RA, SA, FA, and SSN respectively (classification of Legrand and Simon, 1989). QA decreases with time from 1868 until solar cycle 22 and then strongly increases at the end of solar cycle 23.

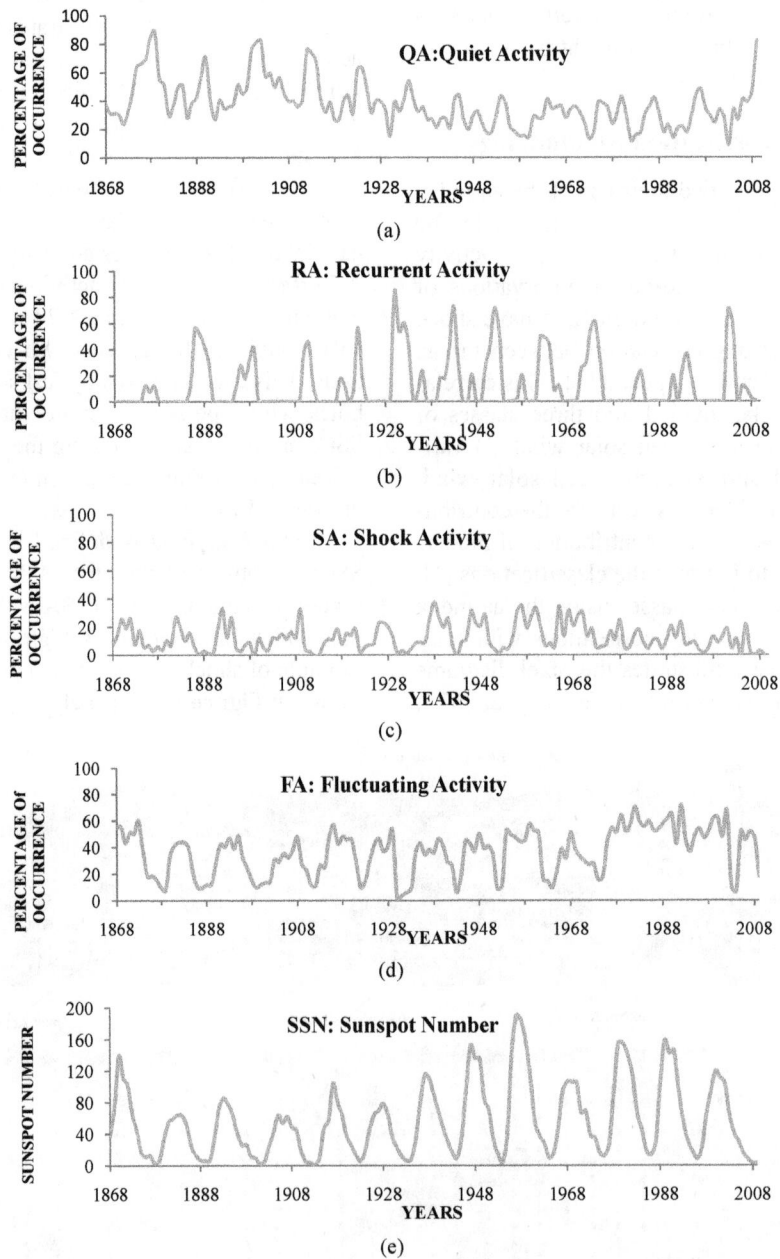

Figure 2. Long-term trends of solar activity classified using the aa index from 1868 to 2009. Quiet activity (a) Recurrent activity (b) Shock activity (c) and Fluctuating activity (d).

The recurrent activity RA in solar cycle 23 is one of the four highest since 1868. The shock activity SA is low compared to other recent cycles.

Cycle 23 is different from recent solar cycles in that RA and QA are more frequent and SA is less frequent. This cycle is more similar to the cycles observed in the late 1800s. Likewise, cycle 22 seems to be similar to the cycles observed in the beginning of the 1900s. These results are perhaps indicative of longer-scale solar changes.

Figure 3 shows the percentages of each class of geomagnetic activity for each solar cycle year per year using the most recent geomagnetic activity [7]. From this figure, it is easy to remark that all four types of geomagnetic activity coexist most of the time over these four solar cycles.

Figure 3(a) shows the profiles of the four classes during the solar cycle 20. Quiet activity (QA) makes the highest contribution in 1975 and 1976 (39% and 38%) and the lowest in year 1973 (13%). Recurrent activity (RA) exhibits the most important contribution to geomagnetic activity in 1973 and 1974 (61% and 74%). The shock activity (SA) has the highest percentages in 1965, 1969 and 1970 (51%, 39%, and 38%). Unclear activity (UA) has the lowest contributions in 1964, 1965, and 1974 (0%).

The time profiles of the different geomagnetic classes are plotted in **Figure 3(b)** for solar cycle 21. QA is most important in 1976, 1981 and 1986 (~37%) and makes its lowest contribution from 1982 to 1984 (16%). SA shows a large occurrence frequency in 1978 and 1982 (~30%). As to UA, it is important most of the time with more occurrences in 1977, 1978 and 1985(~40%).

Figure 3(c) plots the four classes during solar cycle 22. QA is most important in 1987 and 1996 (42% and 47%). RA presents the highest contributions from 1993 to 1994 (30% - 44%). SA is important in 1989 and 1991 (~30%). For UA, the occurrence is ≥27% over the entire solar cycle 22. In **Figure 3(d)**, the time profiles of each class are plotted for solar cycle 23. In 1996, QA was observed 45% of the time. The most SA in cycle 23 occurs during the maximum and declining phases with very little near solar minimum. QA and UA are observed in all solar cycle phases. Almost all RA is observed in the declining phase. This figure shows that UA has a fairly constant time profile with an average occurrence rate of about (40%), with lower values in the declining phase in 2003 (9%), 2004 (4%), and 2005 (0%). The largest UA value is 57% and occurs near the maximum phase in 1999. RA is the dominant activity in 2003 (~70%) and is still important in 2004 (~63%). QA is most frequent in 1997 and 2009 when it comprises 48% and 91% of the activity, respectively. The QA frequency minimum of 9% occurred in 2003 near solar maximum.

These results show that the most UA occurs during the ascending phase, SA occurs mainly near solar maximum and most RA is observed during the declining phase, consistent with the results of previous works [1,6,7].

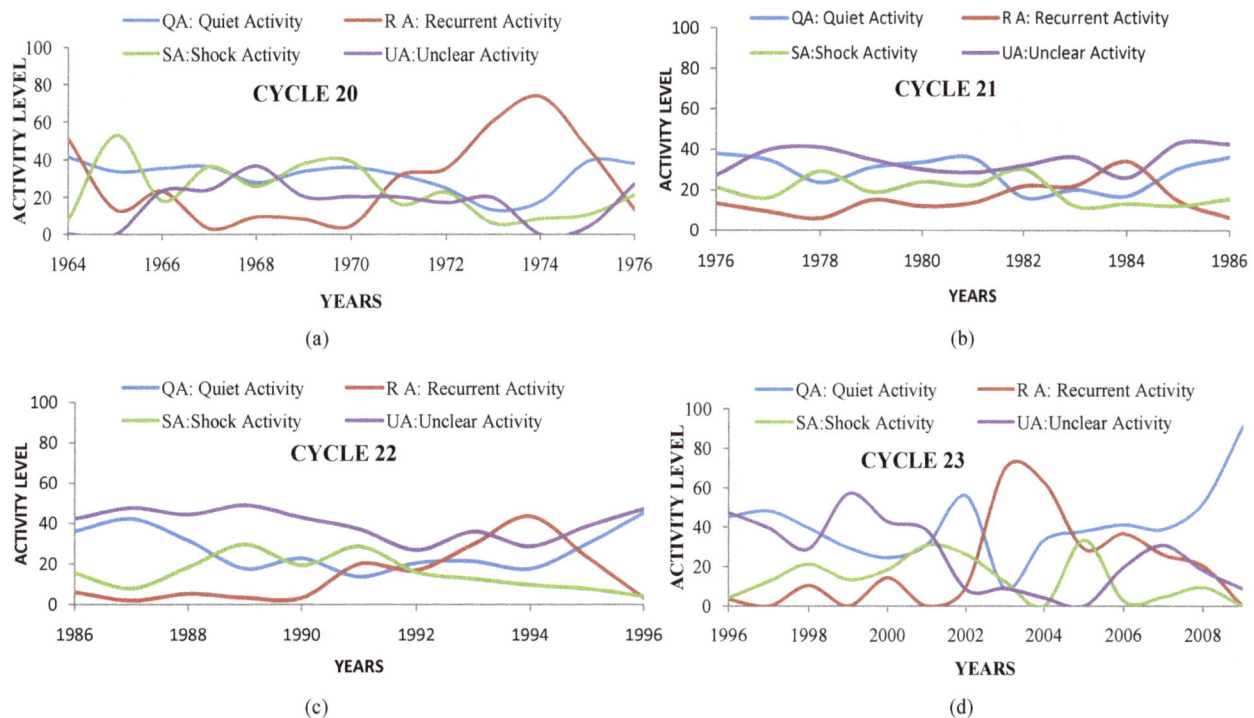

Figure 3. Percentage of each class of activity for: (a) Cycle 20; (b) Cycle 21; (c) Cycle 22; (d) Cycle 23.

3.2. Solar Wind Characteristic and Geomagnetic Activity during the Four Latest Solar Cycles 20, 21, 22, and 23

Time scales for solar variation range from solar flare timescales of several seconds, to the solar rotation period of 27 days for corotating streams, to 11 years (sunspot cycle), to 22 years (Hale cycle). Hourly solar wind data and three-hourly geomagnetic data are available for only a few sunspot cycles, so we can only analyze time scales from days to years. **Figures 4-7** show the correlation between the daily-averaged aa index and solar wind speed for the four classes of geomagnetic activity during the cycles 20, 21, 22, and 23.

Figures 4 shows the daily aa index as a function of solar wind speed and the range of solar wind speeds observed during periods with quiet geomagnetic activity, respectively for the solar cycle 20 (panel a), solar cycle 21 (panel b), solar cycle 22 (panel c), and the solar cycle 23 (panel d). **Figure 5**, devoted to recurrent geomagnetic activity, present the same morphology as **Figure 4**. Similarly, **Figure 6** shows the link between geomagnetic activity and solar wind during periods with shock activity. **Figure 7** shows the statistical distribution of the solar wind for unclear activity. It emerges from all the plots (**Figure 4**) that there is no linear correlation between the aa index and solar wind speed for the combined data. It means that within each activity type aa does not depend only on V. Nevertheless, all the panels from the **Figure 4-7** show interesting statistics:

1) **Figure 4** shows that more than 77% of the solar wind speeds for QA are less than 450 km/s for all four solar cycles. This result agrees with [1,7] where the quiet magnetic activity is fixed; 2) For RA, **Figure 5** shows that 79% - 98% of solar wind speeds are more than 450 km/s; 3) In **Figure 6**, 59% - 61% of the wind speeds are over 450 km/s for SA over all four solar cycles; 4) 59% - 70% of the wind speeds are more than 450 km/s for UA, **Figure 7**. The last three classes form the disturbed magnetic activity groups [1,7]. These results show the plausible role plays by the interplanetary magnetic field, in particular the southward component, within the groups of solar activity. That is in phase with the work of Feynman and Crooker [14] who conclude that aa~BV**2.

This study is the first to discuss of the solar wind distribution for each geomagnetic class. To classify the geomagnetic activity, Legrand and Simon [1] and Zerbo J.-L. *et al.* [7] based their solar wind condition on the correlation made by Svalgaard [11] to estimate and fix the value of solar wind: V < 450 km/s for the quiet activity and V ≥ 450 km/s for the three other classes. Our study will help to understand the links between solar and geomagnetic events.

4. Discussion and Conclusion

The solar cycle 23 is one of the longest cycles since 1868 (13 years) and the geomagnetic activity in this cycle presents some interesting features:

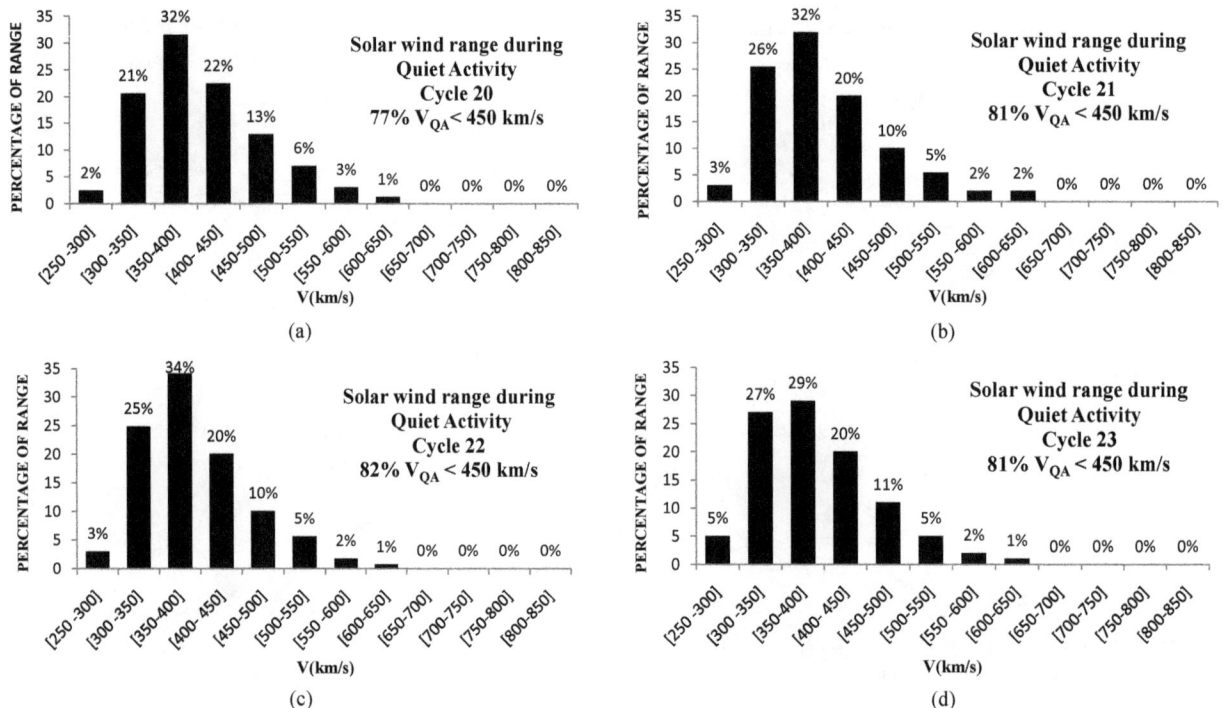

(a)

(b)

(c)

(d)

Figure 4. The range of solar wind speeds for quiet activity: (a) Solar cycle 20; (b) Solar cycle 21; (c) Solar cycle 22 and (d) Solar cycle 23.

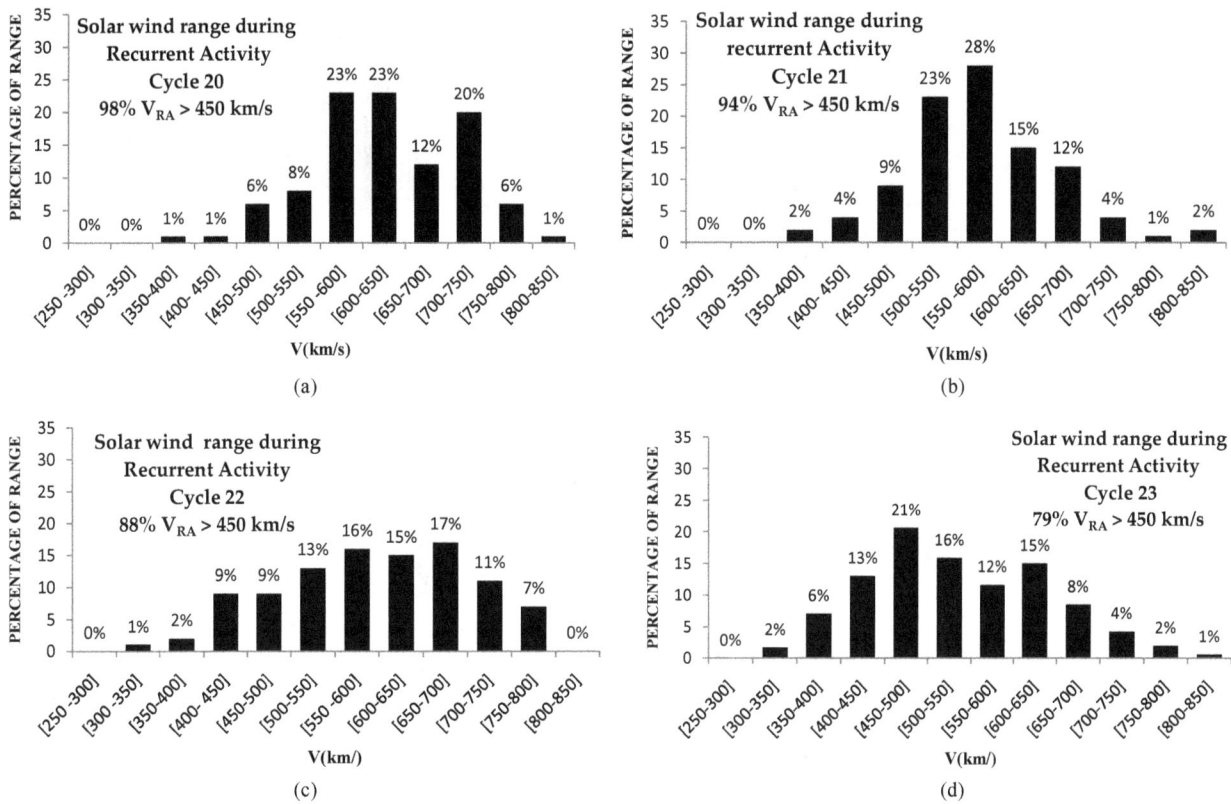

Figure 5. The range of solar wind speeds for recurrent activity: (a) Solar cycle 20; (b) Solar cycle 21; (c) Solar cycle 22 and (d) Solar cycle 23.

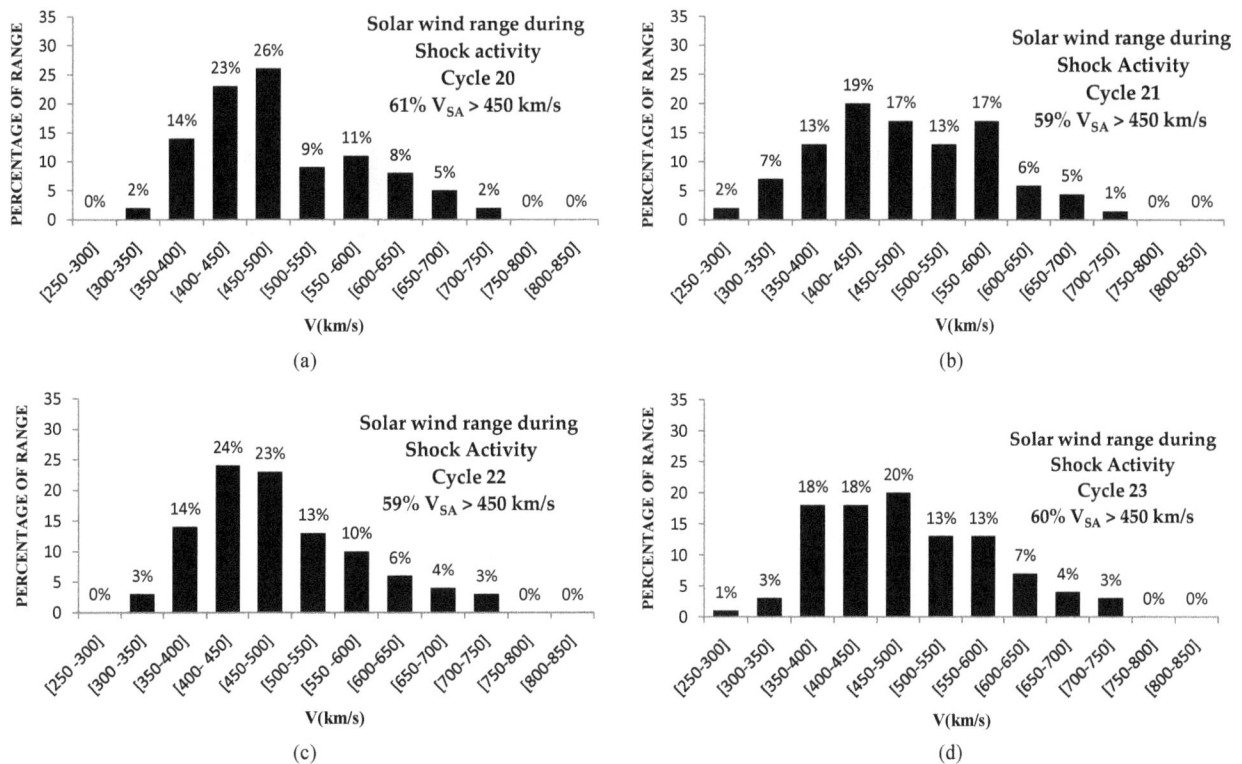

Figure 6. The range of solar wind speeds for shock activity: (a) Solar cycle 20; (b) Solar cycle 21; (c) Solar cycle 22 and (d) Solar cycle 23.

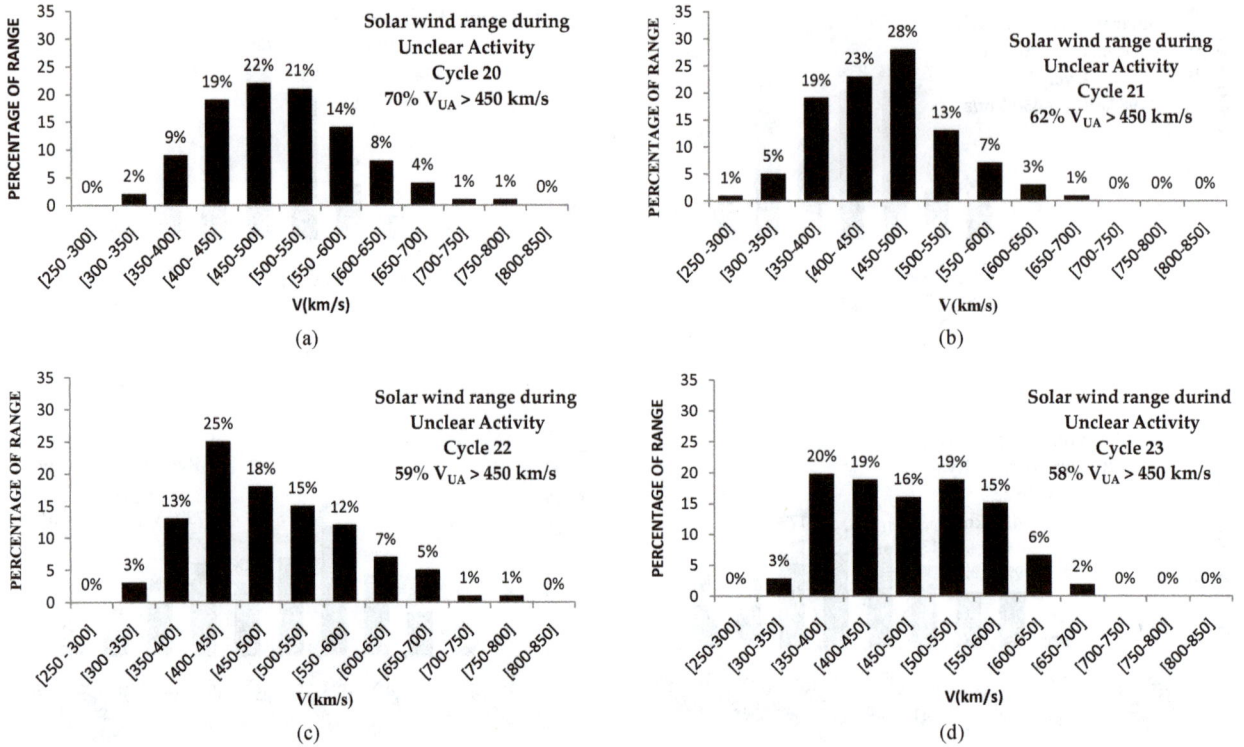

Figure 7. The range of solar wind speeds for unclear activity: (a) Solar cycle 20; (b) Solar cycle 21; (c) Solar cycle 22 and (d) Solar cycle 23.

QA is the most frequent geomagnetic class in solar cycle 23 with the highest percentages (91%) near solar minimum. UA is the second most frequent with the highest occurrence rate (57%) at the ascending phase of the solar cycle. SA is the least frequent class of activity and occurs mainly at the maximum and in the declining phase.

Figure 8 summarizes the contribution of each type of geomagnetic activity to the global activity during cycles 20, 21, 22, and 23. The high frequency of QA during cycle 23 in comparison with the other cycles may be due to the sudden increase of quiet magnetic days at the end of this cycle as shown in **Figure 2**. Cycles 20 and 23 have the highest frequency of recurrent events (RA). The highest level of shock activity (SA) is observed during cycle 21. Cycle 23 is largely under the influence of unclear events (UA).

We note that RA is caused by high wind speed streams ejected by coronal holes and RA is observed mainly in the declining phase of the sunspot solar cycle when the poloidal solar magnetic field component is maximal. The statistical study points out very interesting wind conditions for each class of geomagnetic activity: 1) For QA 95% of wind speed is about 399 ± 69 km/s, 2) For RA 95% of the wind speed is about 582 ± 110 km/s, 3) for SA 95% of the solar wind velocity is order of 482 ± 101.4 km/s, and 4) for FA 95% of wind speed is about 480 ± 85.82 km/s. These results which give in average the order of magnitude of wind speed into each geomagnetic class

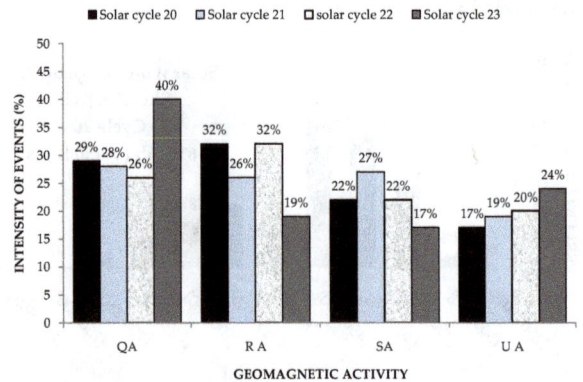

Figure 8. Global contribution of the four classes of geomagnetic activity during solar cycles 20, 21, 22, and 23.

are useful and prove the fact that RA is the effect of high solar wind stream; the QA is the effect of slow solar wind, and moderate and variable solar wind speeds are recorded during shock events. It means that SA doesn't reflect the solar winds' effects but the contribution of shock wave to geomagnetic activity. In addition, except for UA, the solar wind speed distributions are remarkably similar for the four solar cycles in each class of geomagnetic activity. Thus, the character of QA, RA, and SA does not change significantly from cycle to cycle.

Our study shows how similar slow solar wind distributions are for the four cycles. This result points out clearly that the process accelerating the slow wind does not change even though the solar wind flux is decreasing. Another interesting result is the drop in solar wind speed during RA in cycle 23 compare to the other cycles. The decrease in the RA speeds, with the most probably speeds down to 450 - 500 km/s from 550 - 600 km/s, shows the particularity of cycle 23.

Other major conclusions are that approximately 80% of solar wind V < 450 km/s for QA and about 80% of solar wind V > 450 km/s for RA, SA and UA present similar proportion with 60% of the solar wind V > 450 km/s.

These new results open the way to a more physical interpretation of the Sun-Earth relationship needed to understand geomagnetic activity and space dependences on solar activity. The study of the statistics of each geomagnetic class shows the need to refine the geomagnetic classification scheme to account for solar wind conditions.

This work shows that the magnetic signatures given by the aa index correspond to large scale physical process in the sun earth system: high speed solar wind related to coronal hole, CME, slow solar wind. With the use of this aa magnetic index, built since 1868, we can see the strong dependence on solar wind parameters and the effects of toroidal field in interplanetary medium [5,15-17]. Our present study could contribute to reconstructing the main features of solar events until 1868.

5. Acknowledgements

The authors thank all the members of LPP/CNRS/UPMC for their welcome. They thank Paul and Gérard VILA for their advice and collaboration.

The authors thank the NGDC data centre for providing the aa indices and the ACE data center for providing the solar wind velocity and IMF components. We express many thanks to Coopération Française and Burkina Faso for their financial help. JDR was supported by NASA's Wind mission. We are obliged to the anonymous referees with help us to improve science. Special thanks to ACS editors.

REFERENCES

[1] J. P. Legrand and P. A. Simon, "Solar Cycle and Geomagnetic Activity: A Review for Geophysicists. Part I. The Contributions to Geomagnetic Activity of Shock Waves and of the Solar Wind," *Annales Geophysicae*, Vol. 7, No. 6, 1989, pp. 565-578.

[2] P. A. Simon and J. P. Legrand, "Solar Cycle and Geomagnetic Activity: A Review for Geophysicists Part II. The Solar Sources of Geomagnetic Activity and Their Links with Sunspot Cycle Activity," *Annales Geophysicae*, Vol. 7, No. 6, 1989, pp. 579-594.

[3] I. G. Richardson, H. V. Cane and E. W. Cliver, "Sources of Geomagnetic Activity during Nearly Three Solar Cycles (1972-2000)," *Journal of Geophysical Research*, Vol. 107, No. A8, 2002, pp. 107-118.

[4] I. G. Richardson, E. W. Cliver and H. V. Cane, "Sources of Geomagnetic Activity over the Solar Cycle: Relative Importance of Coronal Mass Ejections, High-Speed Streams, and Slow Solar Wind," *Journal of Geophysical Research*, Vol. 105, No. A8, 2000, pp. 200-213.

[5] F. Ouattara, C. Amory-Mazaudier, M. Menvielle, P. Simon and J.-P. Legrand, "On the Long Term Change in the Geomagnetic Activity during the XXth Century," *Annales Geophysicae*, Vol. 27, No. 5, 2009, pp. 2045-2051.

[6] F. Ouattara and C. Amory-Mazaudier, "Solar-Geomagnetic Activity and Aa Indices toward a Standard," *Journal of Atmospheric and Solar-Terrestrial Physics*, Vol. 71, No. 17-18, 2009, pp. 1736-1748.

[7] J. L. Zerbo, C. Amory-Mazaudier, F. Ouattara and J. Richardson, "Solar Wind and Geomagnetism, toward a Standard Classification 1868-2009," *Annales Geophysicae*, Vol. 30, 2012, pp. 421-426.

[8] P. N. Mayaud, "Une Mesure Planétaire d'Activité Magnétique Basée sur deux Observatoires Antipodaux," *Annales Geophysicae*, Vol. 27, 1971, p. 71.

[9] P. N. Mayaud, "A Hundred Series of Geomagnetic Data, 1868-1967," *IAGA Bulletin*, Vol. 33, Zurich, 1973, p. 251.

[10] P. N. Mayaud, "Deviation, Meaning, and Use of Geomagnetic Indices," *Geophysical Monograph Series*, Vol. 22, Washington DC, 1980, p. 154.

[11] L. Svalgaard, "Geomagnetic Activity: Dependence on Solar Wind Parameters," In: J. B. Zirker, Ed., *Coronal Holes and High Speed Wind Streams*, Colorado Association University Press, Boulder, 1977, pp. 371-432.

[12] P. A. Simon and J. P. Legrand, "A Two Component Solar Cycle," *Solar Physics*, Vol. 131, 1992, pp. 187-209.

[13] J. P. Legrand and P. A. Simon, "Toward a Model of Two-Component Solar Cycle," *Solar Physics*, Vol. 141, 1991, pp. 391-410.

[14] J. Feynman and N. U. Crooker, "The Solar Wind at the Turn of the Century," *Nature*, Vol. 275, No. 5681, 1978, pp. 626-627.

[15] L. Svalgaard and E. W. Cliver, "Interhourly Variability Index of Geomagnetic Activity and Its Use in Deriving Long-Term Variation of Solar Wind Speed," *Journal of Geophysical Research*, Vol. 112, No. A10, 2007, Article ID: A10111.

[16] J. P. Legrand and P. A. Simon, "Some Solar Cycle Phenomena Related to the Geomagnetic Activity from 1868 to 1980, I. The Shock Events, or the Interplanetary Expansion of the Toroidal Field," *Astronomy and Astrophysics*, Vol. 152, No. 2, 1985, pp. 199-204.

[17] P. N. Mayaud, "The aa Indices: A 100-Year Series Characterizing the Magnetic Activity," *Journal of Geophysical Research*, Vol. 77, No. 34, 1972, pp. 6870-6874.

9

Effects of Sky Conditions Measured by the Clearness Index on the Estimation of Solar Radiation Using a Digital Elevation Model

Marcelo de Carvalho Alves[1], Luciana Sanches[2], José de Souza Nogueira[3], Vanessa Augusto Mattos Silva[3]

[1]Soil and Rural Engineering Department, Federal University of Mato Grosso, Mato Grosso, Brazil
[2]Department of Sanitary and Environmental Engineering, Federal University of Mato Grosso, Cuiabá, Brazil
[3]Department of Physics, Federal University of Mato Grosso, Mato Grosso, Brazil

ABSTRACT

This study evaluated the effects of sky conditions (measured by the clearness index, K_T) on the estimation of solar radiation and its components. Solar radiation was calculated by a digital elevation model derived from the Shuttle Radar Topography Mission (SRTM). The calculated radiation was parameterized and validated with measured solar radiation from two stations inside the urban perimeter of the city of Cuiabá, Brazil, during 2006 to 2008. The measured solar radiation varied seasonally, with the highest values in December-March and the lowest in June-September. Comparisons between calculated and measured values for two sites in Cuiabá demonstrate that the model is accurate for daily Rg estimates under clear sky conditions based on Root Mean Square Error, Mean Bias Error and Willmott's index. However, under partially cloudy and cloudy sky conditions the model was not able to provide robust estimates. Spatially, the highest values of incident Rg occurred on strands with North, Northeast and Northwest orientations and were lowest on those oriented to the South, Southeast and Southwest.

Keywords: Sky Cover; Spatial; Radiation; Mapping Solar; Geographic Information Systems

1. Introduction

Air quality in the urban areas is assessed through a combination of various meteorological factors [1], including solar energy, the main driver of life on Earth. Various authors [2-4] have noted that solar energy influences all the physical, chemical and biological processes operating in terrestrial ecosystem due to its role in energy and water vapor balance. For example, solar energy plays a controlling role in plant growth, soil heating and evapotranspiration. Moreover, knowledge of the spatial and temporal dynamics of solar radiation reaching the Earth's surface is also very important for estimating current and future basic energy sources [5,6].

There are three groups of factors that determine the interaction of solar radiation with Earth's atmosphere and surface: a) the geometry of the Earth's surface, revolution and rotation(declination, latitude, solar hour angle), terrain (elevation, albedo, slope and surface orientation, shadows); b) attenuation of the atmosphere(absorption,

scattering) by gases(air molecules, ozone, CO_2 and O_2) and solids, and; c) attenuation by liquid particles (aerosols, including non-condensed water) and clouds (condensed water) [7].

Knowledge of local variability in solar radiation is fundamental to studies about climate and has applications from eteorology, industry, agriculture, architecture, engineering and studies of water resources. This is because solar radiation affects air temperature, air movement and water availability for plants [8].

Variation in solar radiation as a function of the climate can be related to sky conditions (e.g. cloudiness). In general, sky conditions are difficult to predict, but can be categorized in terms of the weather parameters prevailing under clear sky, partially cloudy and overcast conditions. These descriptive terms can be quantitatively defined in terms of climatic variables (e.g. cloud cover, solar radiation) and are useful for energy-efficient building projects [9]. In Cuiabá, the sky cover conditions are influenced by

Effects of Sky Conditions Measured by the Clearness Index on the Estimation of Solar Radiation Using a Digital Elevation Model

73

several factors, including cloudiness and atmospheric combustion products and others. For example, [10] observed that sky cover in Cuiaba, Brazil, was influenced by a range of factors, including burning, mainly in the beginning of the dry season, with maximum intensity in September.

Continuous data on solar radiation at a given location are often difficult to obtain due to practical constraints. Thus, several models for solar radiation estimation based on meteorological variables (precipitation, temperature and sunshine) measured at conventional stations have recently been developed [11,12].

Mathematical models for radiation balance estimation should consider the influences of sky cover in areas with biomass burning activities dependent on the time of the year [10]. The concentration of aerosols will depend on the proximity of outbreaks of fire and weather conditions that influence the transport and dispersion of particulate matter in the atmosphere. The major absorbents, water vapour and aerosols, are particularly variable over the time, and when combined with changes in elevation they may significantly influence the spatial pattern of attenuation [13].

Burning events in central-western Brazil release a high concentration of aerosols into the atmosphere causing a significant attenuation of solar radiation not currently parameterized by models [14-16].

In this context, this study aims to evaluate the influence of sky coverage on global solar radiation estimation through a SRTM digital elevation model for the urban region of Cuiabá in central Brazil.

2. Material and Methods

2.1. Location and Description of the Studied Area

The study area, known locally as *Baixada Cuiabana*, is located in the urban region of Cuiaba city, Mato Grosso state, between latitude 15°10' - 15°50'S and longitude 54°50'W - 58°10'W. It has an average elevation of 165 m above sea level, ranging from 146 to 250 m, in a depression, surrounded on the north and east by the Chapada dos Guimarães city. Two meteorological stations were installed in this area at points characterized by different constructive features.

The first point was named Center station and was located in the center of Cuiabá city, (15°36'1"S, 56°5'29"W, 187 m altitude). This area is completely urbanized with sparse individual vegetation. The second point, named CPA (*Centro Político Administrativo*) station, was located at the perimeter of city in the Policy Centre Administrative of Cuiabá (15°33'59"S, 56°4'30"W, 239 m altitude). The station was surrounded by a few buildings and was close to a small lake. The straight line distance

between the stations was 4 km (**Figure 1**).

The climate is classified as Aw (tropical dry and wet), according to Köppen's classification, with two well defined seasons: a dry season (autumn-winter) and a wet season (spring-summer).The accumulated precipitation in the region is 1500 mm·year^{-1} [17].

2.2. Experimental Data for Validation

Data acquired at the two stations were used for validation purpose. The solar radiation (Rg) was measured using a pyranometer installed in both stations (WM 918, Davis Instruments, Hayward, California, USA). The Centerstation pyranometer was in stalled 4.20 m height, and CPA station pyranometer was installed at10.50mheight, for safety reasons. The acquired data were stored every 30 minutes by a data acquisition and data storage system (Data logger and Vantage Pro 2, Davis Instruments, USA), from August 2006 to June 2008.

Upper and lower limits of 0 and 1300 Wm^{-2}, respectively, were adopted in order to avoid compromising the quality of the estimates.

2.3. Estimation of the Clearness Index and Sky Cover

We used the clearness index (K_T) to describe cloud cover that is characterized according to attenuation levels. The K_T is an indicator of the relative clearness of the atmosphere [18] and is defined as the ratio of direct incident solar radiation (Rg) to the extraterrestrial solar atmosphere (Ro) (MJ m^{-2}·day^{-1}).

The classification of the sky cover was based on [19]. A cloudy sky was defined in the range $0 < K_T < 0.3$, a partially cloudy sky between $0.3 \leq 0.65 \leq K_T$ and a clear sky between $0.65 < K_T < 1.0$. Irradiation in the upper atmosphere (Ro) (MJ·m^{-2}·day^{-1}) was calculated by Equation (1).

$$Ro = 1367E_o\left(\frac{\pi}{180}W_s \sin\varphi \cdot \sin\delta + \cos\varphi \cdot \cos\delta \cdot \sin W_s\right)$$

(1)

where, E_o is the correction factor of eccentricity of the orbit (Equation (2)), W_s is the solar angle (degrees) (Equation (4)), f is the local latitude (degrees) and d is the solar declination (degrees) (Equation (5)).

$$E_o = 1.000110 + 0.034221\cos\Phi + 0.00128\sin\Phi + 0.000719\cos 2\Phi$$

(2),

where, Φ is defined by Equation (3) according to the Julian day (dJ)

$$\Phi = \frac{2\pi(dJ - 1)}{365.242}$$

(3)

$$W_s = \arccos(tg\varphi \cdot tg\delta)$$

(4)

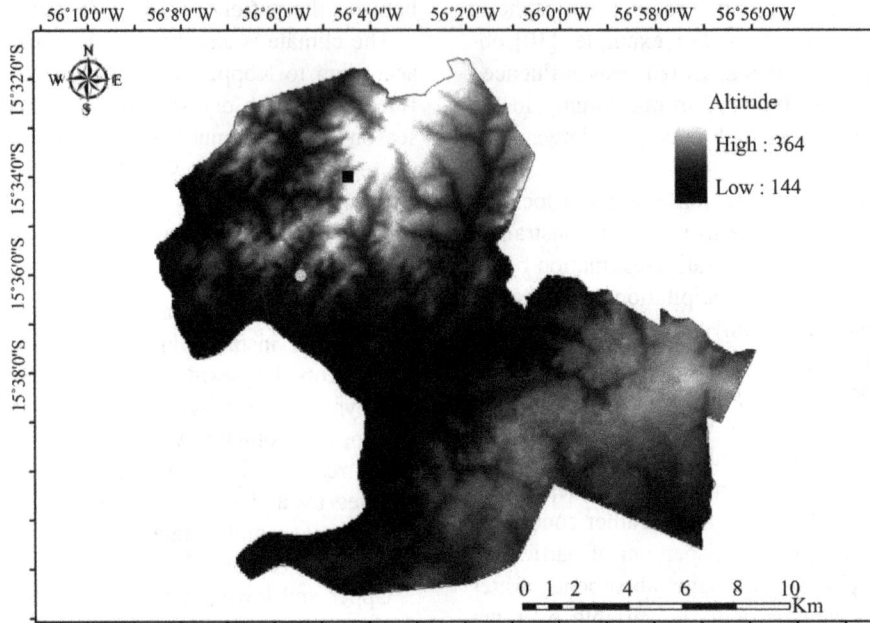

Figure 1. Location of the urban region of Cuiabá city, Mato Grosso state, Brazil and the location of meteorological stations in the Center (gray circle) and CPA station (black square) (SRTM Digital Elevation Model 2000, 90 m spatial resolution, GCSSAD 69 coordinate system). The altitude is in meters.

$$\delta = 23.45 sen\left[\frac{360}{365}(284+dJ)\right] \qquad (5)$$

2.4. Calculated Global Solar Radiation Using the Solar Radiation Spatial Analyst Extension

Solar Radiation is a GIS tool of the GIS software ArcGIS® 9.2 which can be used to describe the spatial distribution of global solar radiation. The influence of latitude, altitude, surface orientation (slope and aspect) was considered as input variables derived from the Shuttle Radar Topography Mission (SRTM) digital elevation model (DEM) v.4.0 (USGS, 2010). The data were reprojected for UTM (Universal Transverse Mercator) South American Datum of 1969 (SAD 69) 21S zone, with a spatial resolution of 3 arc seconds, approximately 90 m of spatial resolution [4]. Equations were used to create the grids of radiation.

The Solar Radiation Area module was used to calculate raster maps and the Points Solar Radiation to calculate vector values of solar radiation in specific points, both expressed in Wh·m⁻². Specific vector points were used to generate the values of radiation related to the Center and CPA stations.

The parameters and input data adopted for the calculation of solar radiation were obtained after the parameterization of the model from the data measured in the field. The parameterization of the model was based on some original parameters of the GISA tool, and adopted for other points in Cuiaba (**Table 1**).

Table 1. Default and modified input parameters of the solar radiation algorithm.

Parameter	Acquisition	Used value
Height offset	Default	0
Z factor (Factor correction units DEM)	Default	1
Calculation directions (Calculating the number of azimuth directions for calculating the Viewshed)	Default	32
Zenith divisions	Default	8
Azimuth divisions	Default	8
Diffuse model type	SRTM	Uniform Sky (uniform diffuse model)
DEM	SRTM	Pixel value
Input points	Modified	Center and CPA station
Mean latitude	Modified	−15.6166116390984
Sky Size/ Resolution (Number of rows and columns of skymap, sunmap and viewshed)	Modified	400
Time configuration	Modified	1 day
Hour interval	Modified	0.5
Slope and aspect	Modified	From DEM
Diffuse proportion	Modified	0.20 - 0.77
Transitivity	Modified	0.68 - 0 - 13

Effects of Sky Conditions Measured by the Clearness Index on the Estimation of Solar Radiation Using a Digital
Elevation Model

75

Solar radiation data from 116 days (collected between 2006 and 2008 at the Centre and CPA stations) were selected considering the classification of the sky cover, of which 40 days were cloudy, 60 partially cloudy and 16 days clear. Global solar radiation was estimated and compared with the developed model using the Solar Analyst algorithm.

The Pearson correlation coefficient r ($p < 0.05$) was used to evaluate the degree of correlation and the direction of this correlation between the global solar radiation estimated from data collected by the sensors in the field and that generated by the solar analyst algorithm.

Another indicator of model performance was the use of statistical indices, such as Root Mean Square Error (*RMSE*) and the Mean Bias Error (*MBE*) recommended by [20]. Finally, the concordance index of [21] Willmott (*d*) (1981) was calculated to evaluate accuracy and to assess the deviation of the estimated values based on the measured values [14].

2.5. Spatial Data Analysis

The representation of the spatial distribution of calculated solar radiation was achieved using raster images in a geographic information system. The spatial resolution of each estimated pixel was 90 m. In total, 18 consecutive days of January (dJ14-31) were selected to represent the wet period and 18 consecutive days in July (dJ191-208) of 2007 represented the dry period. The resulting maps were used to validate the parameters for the sky cover classified as clear. The slope and aspect maps were also calculated based on the DEM and were used to explain the obtained results of solar radiation.

3. Results and Discussion

3.1. Daily Variation of the Clearness Index

The average daily clearness index (K_T) of 0.47 ranged from 0.002 to 0.69 from 2006 to 2008, indicating the percentage of attenuated solar radiation through atmosphere reaching of the surface at the CPA and Center stations (**Figure 2**).

The average value of the clearness index was similar to the values of 0.41 and 0.44 found by [19,22], respectively. These results were observed in Cascavel city, Paraná state, Brazil, and in the rural region of Botucatu city, São Paulo, Brazil, with lowest monthly values of 0.45 and highest values of 0.59 [23].

The highest values K_T occurred most frequently between June and August—these months typically having less cloud cover. The lowest K_T values occurred in January, February and December, these months being characterized by high concentrations of clouds and water vapor. A lower variation in the daily K_T values occurred during conditions of lower cloud cover.

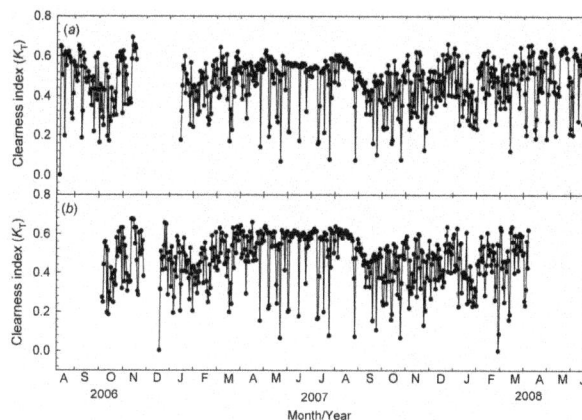

Figure 2. Daily average of clearness index (K_T) in the Centerstation (a) and CPA station (b) from 2006 to 2008.

Similar seasonality was observed by [23,24] in Botucatu city, Brazil, with higher K_T values during dry months (July and August) with less cloud cover. Lower values occurred during the wet months (January, February and November and December).

3.2. Validation of Calculated Global Radiation Considering the Sky Cover

Initially, the model used to estimate the solar radiation was set up using the input parameters suggested by the Solar Radiation Spatial Analyst extension of the GIS software ArcGIS® 9.2. After thoroughly testing the parameters by changing on a parameter at a time, the best statistical adjustments were identified using the highest Pearson correlation coefficient (r), the lower root mean square error (*RMSE*), Mean Bias Error (*MBE*) and a higher index of Willmott (*d*).

The best parameterization for the application of Spatial Analyst over different sky conditions was considered using the following parameters: proportion of diffuse radiation (0.2 for clear skies, 0.32 for partially cloudy sky and 0.77 for cloudy sky conditions) and the fraction of radiation that troughed the atmosphere (0.68 for clear skies, 0.49 for partially cloudy sky and 0.13 for cloudy sky conditions). The other parameters are described in **Table 1**, and did not differ from the default parameters of the Solar Radiation model.

Figure 3 shows the Solar radiation measured versus calculated solar radiation using digital elevation model considering sky conditions.

[25], analyzing the global solar radiation calculated by the Solar Analyst at two radiometric networks stations in Southern Spain, calculated the input parameters of transmissivity and diffuse fraction for all types of sky conditions from different ranges. From this information they were able to generate maps of the global solar radiation of the region.

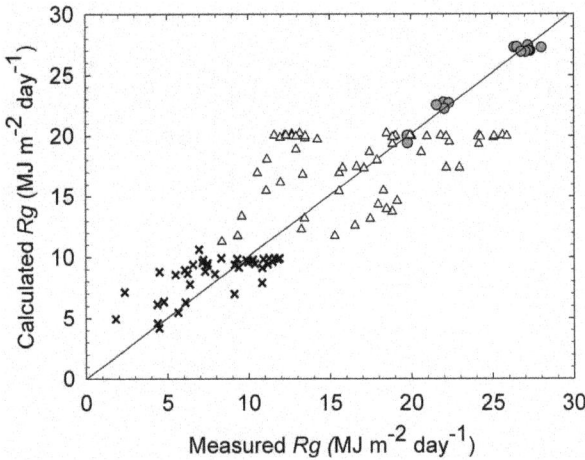

Figure 3. Solar radiation measured versus calculated solar radiation using digital elevation model considering sky conditions (symbol x), partially cloudy (symbol Δ) and clear (symbol •). The solid linere presents a 1:1 ratio.

Table 2. Analysis of adjustment between the calculated and measured of *Rg*, number of data (*n*), linear regression, Pearson's coefficient (*r*) (*p* < 0.05), root mean square error (R*MSE*, MJ·m^{-2}), Mean Bias Error (*MBE*, MJ·m^{-2}) and index of Will mott (*d*) under clearness index (KT). *y* is the calculated value and *x* is the measured value.

K_T interval	*n*	Linear equation	*r*	*RMSE*	*MBE*	*d*
Cloudly	40	$y = 0.43x + 5.17$	0.69	1.23	0.60	0.77
Partially cloudy	60	$y = 0.21x + 14.22$	0.36	2.61	0.75	0.60
Clear	16	$y = 0.94x + 1.57$	0.98	0.51	0.15	0.99
All conditions	116	$y = 0.77x + 4.06$	0.87	3.10	0.61	0.93

In the present study, the input parameters to be used in Solar Radiation (over clear sky and cloudy conditions) were adapted to estimate the *Rg*. However, when calculated for partially cloudy conditions, a greater number of tests were required and the output was not satisfactory. This lack of accuracy probably occurred due to the dry season being less cloudy; consequently more days with clear skies were expected related to the absence of cloud formation in the study area. However, due to frequent outbreaks of fire significant quantities of aerosols were emitted in to the atmosphere. These aerosols absorb solar radiation and reduce incident solar radiation on the surface in relation to their concentration in the atmosphere.

Similar climatic characteristics were observed by the Solar and Wind Energy Resource Assessment (SWERA) project, which mapped solar energy resources in Brazilian territory using a model of radiative transfer from July to December 2005 [14].

The combustion products injected in to the atmosphere during the fire season (July-October) could explain the systematic deviations and be used to validate the calculated global solar radiation by the radiation BRAZIL-SR model, using satellite image and observed surface data [10].

Table 2 shows the relationship between the *Rg* calculated by Solar Radiation and the measured *Rg* considering the three sky conditions.

According to the *MBE* values, calculated *Rg* was overestimated independent of the sky cover condition. As expected, the best correlation between calculated *Rg* and measured *Rg* was obtained during clear sky condition (**Table 2**) with the highest *r*(0.98) and *d*(0.99) and lowest *RMSE* (0.51 MJ·m^{-2}) and *MBE* (0.15 MJ·m^{-2}). There was no satisfactory correlation for partially cloudy and cloudy conditions—over estimating *Rg* by 0.75 and 0.60 MJ·m^{-2},

respectively. Possibly, the parameters used to represent the composition of the atmospheric conditions were responsible for the overestimation of the Solar Radiation model. However, the performance of the adjustment checked by *d*, indicated satisfactory relationship between the calculated and measured *Rg* (values tending to 1).

According to [25] Battles *et al.* (2008) and considering daily data, the best adjustments were observed under clear sky conditions and no satisfactory adjustments were possible for overcast skies. In accordance with the present study, the best results were also generated during clear skies. No satisfactory results were obtained for partially cloudy sky conditions, probably a function of the range of $0.3 \leq K_T \leq 0.65$ adopted in the present study. Cloudiness generated more errors, with an average *RMSE* of about26% when compared to clear sky conditions.

The Spatial Analyst under estimated radiation when applied to very cloudy conditions, confirming the need to estimate solar radiation only over clear skies [26]. Moreover, according to [27] the difficulty of validating the model for all sky conditions was evaluated, confirming the significant influence of aerosols from biomass burning in the atmospheric radiative transfer. Thus, the proximity of fire outbreaks produces a systematic error, similar to the error produced by the presence of clouds in the estimates provided by the BRAZIL-SR model.

[28] used the Spatial Analyst to estimate global radiation using adigital elevation model of 20 m spatial resolution. The authors observed values of *RMSE* of about 18% and the values of transmissivity and diffuse ratio varied depending on the clearness index.

An different model (using the Angstrom equation) was used to estimate daily global solar radiation in the region of Cascavel city, southern Brazil, and had a tendency to overestimate the global radiation in January, February, April, June, August, September, October, November and December and to underestimate during the months of March, May and July [22].

Effects of Sky Conditions Measured by the Clearness Index on the Estimation of Solar Radiation Using a Digital Elevation Model

77

3.3. Monthly Average of Measured and Calculated Global Solar Irradiation for Each Type of Sky Cover

The monthly average of the clearness index (K_T) exhibited seasonal variations, with the highest values occurring during the dry season (June-August) and the lowest values during the wet season, especially in the beginning of the rains (October-November) (**Figure 4(a)**). In a similar study in the region of Maceio city, northeastern Brazil, the sky conditions were also dominated by partially cloudy days with low frequencies of clear or cloudy days, and the clearness index varied from 0.50 (May-August) to 0.61 (November), with an annual average of 0.56 (Souza *et al.*, 2005).

The seasonality of K_T showed the opposite pattern to the seasonality in the monthly average of solar radiation (Rg) (**Figure 4(b)**). The monthly Rg average was 16.41 MJ·m^{-2} considering the entire year and independent of sky conditions. Specifically, for each category of sky condition, the Rg average was 8.15 MJ·m^{-2} (cloudy), 17.97 MJ·m^{-2} (partially cloudy) and 24.35 MJ·m^{-2} (clear), indicating the range of variation in global solar radiation received on the surface due to atmospheric attenuation.

The measured monthly average daily values of the two stations together ranged from 13.23 MJ·m^{-2} (June/2008) up to 20.28 MJ·m^{-2} (November/2006) with an average of 16.41 MJ·m^{-2}. Similar values were reported by [23] for a rural region of Botucatu city, southern Brazil (22.85°S; 48.45°W, altitude of 786 m), with a range of global radiation from 12.6 MJ·m^{-2} in June to 21.0 MJ·m^{-2} in November and an average of 17.62 MJ·m^{-2}. This seasonality of solar radiation in the Southern Hemisphere is expected at Earth's surface, with higher values in the summer and lower values in the winter. This tendency follows the

characteristics of radiation at the top of the atmosphere, decreasing from January through June and July, and increasing in December.

As anticipated, the values for clear sky were always higher than those for partially cloudy conditions and overcast skies.

3.4. Spatial Distribution of Calculated Solar Radiation

The topography factors of altitude, slope and aspect strongly affected the spatial distribution of solar radiation (**Figures 5(a)** and **(b)**).

The distribution of strands orientation (aspect) showed delimited areas with predominance of the slopes facing North and the South. The observed distribution in radiation is consistent with that described for an area of dense vegetation in the South of the state of Goiás, central Brazil, in which for most of the year the solar radiation was focused toward areas facing North, while areas facing to the South were exposed only in January [29]. [30] also observed that Juiz de For a city, southeast Brazil, was characterized by strands oriented to the North and this pattern lead to them receiving a greater amount of energy from the sun over the year.

The representation of the spatial distribution of the daily average of calculated Rg using Solar Analyst during the wet and dry seasons is shown in **Figure 6**. Urban region of Cuiabá city varied in altitude between 146 to 250 m, exhibiting strong trends with the lowest values of Rg at the lowest altitudes with different slopes (0 to 18.4%) (**Figure 6(a)**). [28] reported higher levels of solar radiation (calculated Rg) at both lower and higher altitudesites, with insignificant effects of horizon.

Solar radiation is typically more closely related to the topographic characteristics of the study area than to altitude [25]. According to the authors, differences in altitude were less important than other topographic variables, such as the shadows produced by surrounding mountains. Moreover, no dependence was observed between global solar radiation, altitude and slope. According to the results of the present study, not only the topographic characteristics must be considered for global solar radiation estimation, but also the sky conditions and the occurrence of burning events.

[8] reported that only a small part of the annual solar radiation can be attributed to solar declination, but the cloudiness between dry and wet seasons is a major cause of the high variation in solar radiation.

Variation in solar radiation depends on the slope orientation and was different in places facing South, Southeast and Southwest in January, while in July the orientation of the slopes were more favorable when facing North, Northeast and Northwest. In the Southern hemisphere, North facing slopes receive more sunlight than

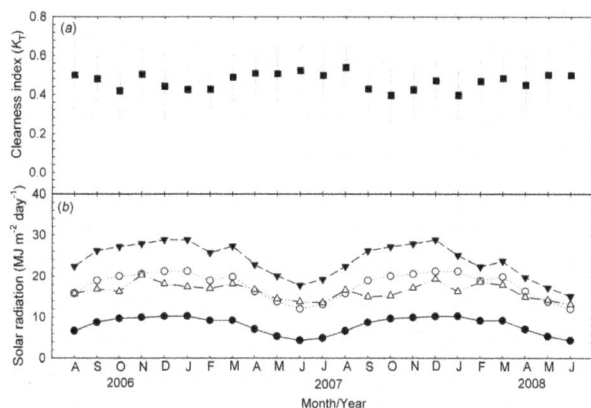

Figure 4. (a) Average monthly clearness index (± SD), (b) Measured solar radiation (MJ·m^{-2}·day^{-1}) and calculated solar radiation (MJ·m^{-2}·day^{-1}) using digital elevation model for the sky conditions cloudy, partially cloudy and clear in the urban region of Cuiabá city, from August 2006 to June 2008.

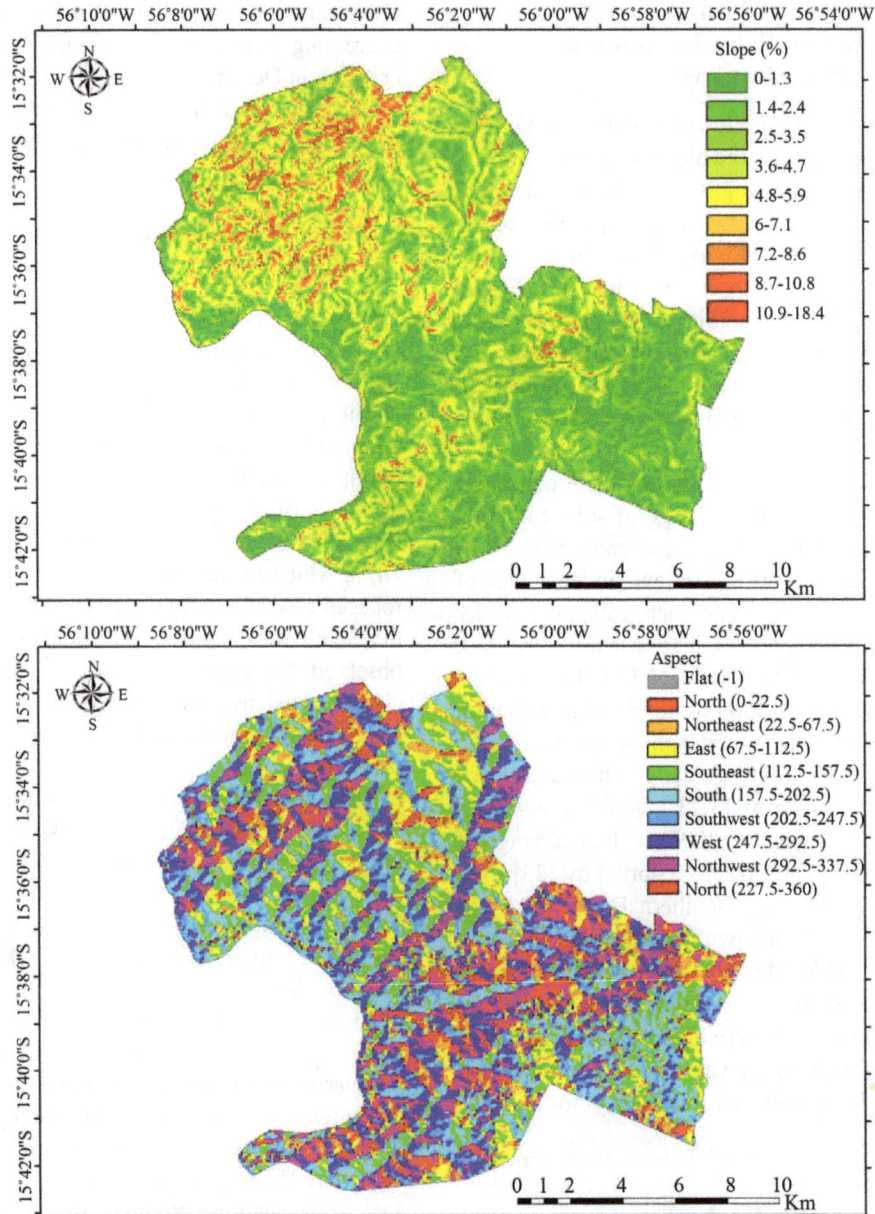

Figure 5. (a) Slope, (b) Strands orientation (aspect) for the urban region of Cuiabá city.

those facing to the South [31].

The spatial distribution of Rg varied according to latitude. The highest daily averages ranged from 7.09 to 7.44 $KWh \cdot m^{-2}$ (wet season) and from 4.24 to 5.26 $KWh \cdot m^{-2}$ (dry season). A similar distribution of Rg maps were observed by [32], with variation from 6.0 to 7.5 $KWh \cdot m^{-2}$ in January and from 4.0 to 5.0 $KWh \cdot m^{-2}$ in July, for the Midwest region of Brazil.

4. Summary and Conclusions

This study evaluated the effects of sky conditions (as measured by the clearness index, K_T) on the estimation of solar radiation and its components, based on SRTM digital elevation model.

The estimation of solar radiation by the Solar Radiation algorithm was more robust under clear and cloudy sky conditions. However, under partially cloudy conditions, the model showed poor adjustment and cannot be satisfactorily applied.

The generated maps of solar radiation enable the visualization of solar radiation in the different topographic regions inside the urban perimeter of Cuiabá city. Through the maps, it was possible to observe that the highest values of Rg occurred in January on strands oriented towards the South, Southeast and Southwest, and in July on strands oriented towards the North, Northeast and Northwest.

Effects of Sky Conditions Measured by the Clearness Index on the Estimation of Solar Radiation Using a Digital Elevation Model

79

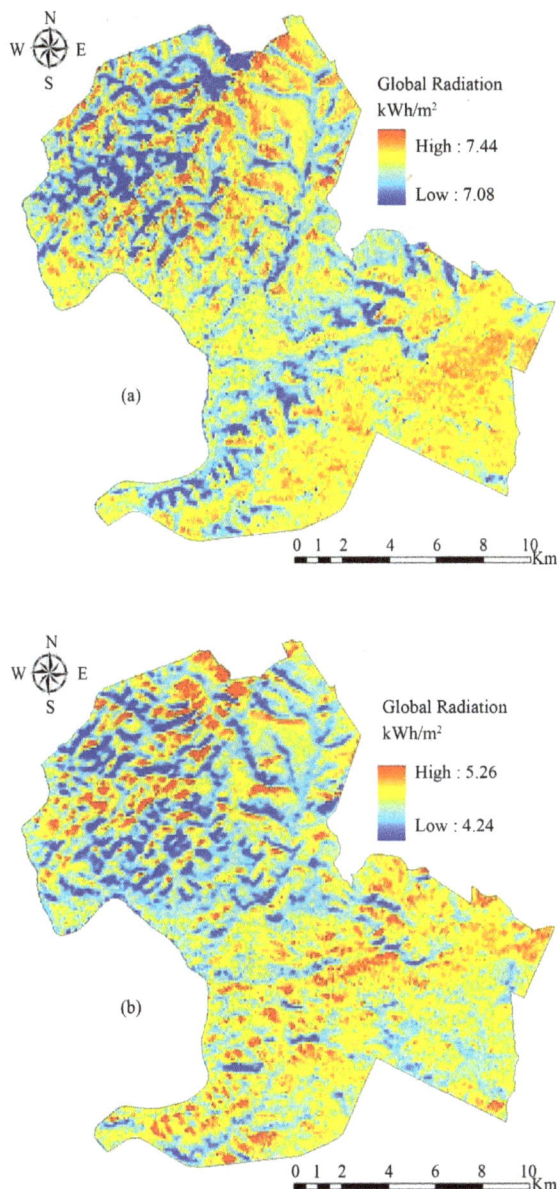

Figure 6. Spatial distribution of solar radiation (KWh·m^{-2}) under clear sky condition in (a) January and; (b) in July 2007 for the urban region of Cuiabá city.

5. Acknowledgements

Partial support was provided by the *Programa de Apoio a Núcleos de Excelência* (PRONEX) provided by Fundação de Amparo a Pesquisa do Estado de Mato Grosso (FAPEMAT), Process N. 823971/2009.

REFERENCES

[1] N. Ilten and T. Selici, "Investigating the Impacts of Some Meteorological Parameters on Air Pollution in Balikesir, Turkey," *Environmental Monitoring and Assessment*, Vol. 140, No. 1-3, 2008, pp. 267-277.

[2] M. Bilgili and M. Ozgoren, "Daily Total Global Solar Radiation Modeling from Several Meteorological Data," *Meteorology and Atmospheric Physics*, Vol. 112, No. 3-4, 2011, pp. 125-138.

[3] K. Zakšek, T. Podobnikar and K. Oštir, "Solar Radiation Modeling," *Computers & Geosciences*, Vol. 31, No. 2, 2005, pp. 233-240.

[4] P. Fu and P. M. A. Rich, "A Geometric Solar Radiation Model with Applications in Agriculture and Forestry," *Computers and Electronics in Agriculture*, Vol. 37, No. 1-3, 2002, pp. 25-35.

[5] M. Martínez-Chico, F. J. Batles and J. L. Bosch, "Cloud Classification in a Mediterranean Location Using Radiation Data and Sky Images," *Energy*, Vol. 36, No. 7, 2011, pp. 4055-4062.

[6] A. Sözen and E. Arcaklıoğlu, "Prospects for Future Projections of the Basic Energy Sources in Turkey," *Energy Sources*, Vol. 2, No. 2, 2007, pp. 183-201.

[7] G. Montero, J. M. Escobar, E. Rodríguez and R. Montenegro, "Solar Radiation and Shadow Modelling with Adaptive Triangular Meshes," *Solar Energy*, Vol. 83, No. 7, 2009, pp. 998-1012.

[8] J. L. de Souza, M. R. Nicácio and M. A. L. Moura, "Global Solar Radiation Measurements in Maceió, Brazil," *Renewable Energy*, Vol. 30, No. 8, 2005, pp. 1203-1220.

[9] D. H. W. Li and J. C. Lam, "An Analysis of Climatic Parameters and Sky Condition Classification," *Building and Environment*, Vol. 36, No. 4, 2001, pp. 435-445.

[10] E. B. Pereira, F. R. Martins, S. L. Abreu, P. Couto, S. Colle and R. Stuhlmann, "Biomass Burning Controlled Modulation of the Solar Radiation in Brazil," *Advances in Space Research*, Vol. 24, No. 7, 1999, pp. 971-975.

[11] A. Ferronato, J. H. Campelo Júnior, E. L. Bezerra and M. M. D. D. Mendonça, "Estimativa da Radiação Solar Global Baseada em Medidas de Temperatura do ar," *Proceedings of XIII Congresso Brasileiro de Agrometeorologia*, Santa Maria, July, 2003, pp. 781-782.

[12] F. F. Blanco and P. C. Sentelhas, "Coeficientes da Equação de Angströn-Prescott para Estimativa da Insolação para Piracicaba-SP," *Revista Brasileira de Agrometeorologia*, Vol. 10, 2002, pp. 295-300.

[13] J. A. Ruiz-Arias, T. Cebecauer, J. Tovar-Pescador and M. Súri, "Spatial Disaggregation of Satellite-Derived Irradiance Using a High-Resolution Digital Elevation Model," *Solar Energy*, Vol. 84, No. 9, 2010, pp. 1644-1657.

[14] F. R. Martins, E. B. Pereira, R. A. Guarnieri, C. S. Yamashita and R. C. Chagas, "Mapeamento dos Recursos de Energia Solar no Brasil Utilizando Modelo de Transferência Radiativa BRASIL-SR," *Proceedings of I Con-*

gresso Brasileiro de Energia Solar, Fortaleza, April 2007, pp. 1-10.

[15] G. Stephens, "Radiation Profiles in Extended Water Clouds. II: Parameterization Schemes," *Journal of Atmospheric Science*, Vol. 35, No. 11, 1978, pp. 2123-2132.

[16] E. B. Pereira, A. W. Setzer, F. Gerab, P. E. Artaxo, M. C. Pereira and G. Monroe, "Airborne Measurements of Aerosols from Burning Biomass in Brazil Related to the TRACE A Experiment," *Journal of Geophysical Research*, Vol. 101, No. D19, 1996, pp. 23983-23992.

[17] M. M. A. Sampaio, "Análise do Desempenho Térmico e Lumínico de Habitações Populares em Cuiabá-MT," Dissertation, Federal University of Mato Grosso, 2006.

[18] B. Liu and R. Jordan, "The Interrelationship and Characteristic Distribution of Direct, Diffuse and Total Solar Radiation," *Solar Energy*, Vol. 4, No. 3, 1960, pp. 1-19.

[19] R. Dallacort, R. P. Ricieri, S. L. Silva, S. L. F. Paulo and F. F. Silva, "Análises do Comportamento de um Actinógrafo Bimetálico (R. Fuess-Berlin-Steglitz) em Diferentes Tipos de Cobertura do céu," *Acta Scentiarum Agronomy*, Vol. 26, 2004, pp. 413-419.

[20] M. Iqbal, "An Introduction to Solar Radiation," Academic Press, Toronto, 1983.

[21] C. J. Willmott, "On the Validation of Models," *Physical Geography*, Vol. 2, 1981, pp. 184-194.

[22] M. I. Valiati, "Estimativa da Irradiação Solar Global com Diferentes Partições para a Região de Cascavel," Dissertation, State University of West of Paraná, Cascavel, 2001.

[23] T. Inácio, "Potencial Solar das Radiações Global, Difusa e Direta em Botucatu," Dissertation, State University of Paulista, Botucatu, 2009.

[24] A. F. Santos, J. G. Z. de Mattos and S. V. de Assis, "Frequência de Dias Claros com Base nos Valores do Índice de Limpidez," *Proceedings of XI Congresso Brasileiro de Meteorologia*, Rio de Janeiro, 2000, pp. 54-58.

[25] F. J. Batlles, J. L. Bosch, J. Tovar-Pescador, M. Martínez-Durbán, R. Ortega and I. Miralles, "Determination of Atmospheric Parameters to Estimate Global Radiation in Areas of Complex Topography: Generation of Global Irradiation Map," *Energy Conversion and Management*, Vol. 49, No. 2, 2008, pp. 336-345.

[26] P. Fu and P. M. A. Rich, "A Geometric Solar Radiation Model and Its Applications in Agriculture and Forestry," *Proceedings of the 2nd International Conference on Geospatial Information in Agriculture and Forestry*, Lake Buena Vista, 10-12 January 2000, pp. I357-364.

[27] F. R. Martins, E. B. Pereira and M. P. S. Echer, "Levantamento dos Recursos de Energia Solar no Brasil com o Emprego de Satélite Geoestacionário: O Projeto Swera," *Revista Brasileira de Ensino de Física*, Vol. 26, No. 2, 2004, pp. 145-159.

[28] M. Martínez-Durbán, L. F. Zarzalejo, J. L. Bosch, S. Rosiek, J. Polo and F. J. Battlles, "Estimation of Global Daily Irradiation in Complex Topography Zones Using Digital Elevation Models and Meteosat Images: Comparison of the Results," *Energy Conversion and Management*, Vol. 50, No. 9, 2009, pp. 2233-2238.

[29] L. E. G. Machado, E. D. Nunes and P. A. Romão, "Análise da Influência da Topografia na Variação Sazonal de Fitofisionomias na Bacia do Rio Veríssimo-GO," XIV Simpósio Brasileiro de Sensoriamento Remoto, Natal, 2009.

[30] F. T. P. Torres, L. N. Moreira, E. A. Soares, J. U. Pierre and G. A. Ribeiro, "Exposição das Vertentes e Ocorrências de Incêndios em Vegetação no Município de Juiz de Fora-MG," *Proceedings of XIII Simpósio Brasileiro de Geografia Física Aplicada*, Viçosa, 2009.

[31] R. J. Hugget, "Geoecology: An Evaluation Approach," London, 1995.

[32] E. R. Martins, E. B. Pereira, S. L. Abreu and S. Colle, "Mapas de Irradiação Solar para o Brasil—Resultados do Projeto SWERA," *Proceedings of the Simpósio Brasileiro de Sensoriamento Remoto*, Goiânia, April 2005, pp. 16-21.

10

What Controls Recent Changes in the Circulation of the Southern Hemisphere: Polar Stratospheric or Equatorial Surface Temperatures?

Isidoro Orlanski

Atmospheric and Ocean Science Program, Princeton University, Princeton, USA

ABSTRACT

Recent research suggests that both tropical ocean warming and stratospheric temperature anomalies due to ozone depletion have led to a poleward displacement of the mid- and high-latitude circulation of the Southern Hemisphere over the past century. In this study, we attempt to distinguish the influences of ocean warming and stratospheric cooling trends on seasonal changes of both the zonally symmetric and asymmetric components of the southern hemisphere circulation. Our analysis makes use of three data sets-the ERA40 reanalysis and results from two different runs of the GFDL global atmosphere and land model (AM2.1) for the period 1870 to 2004. A regression analysis was applied to two variables in each of the three data sets-the zonal component of the surface wind U(10 m) and the height at 300 hPa—to determine their correlation with zonally averaged polar stratospheric temperatures (T_polar—at 150 hPa, averaged over a band from 70S - 80S) and low-level equatorial temperatures (T_equator—at 850 hPa averaged over a band at 5S - 5N). Our analysis shows that the zonally *symmetric* surface winds have a considerably enhanced intensity in high latitudes of the southern hemisphere over the summer period, and that the stratospheric temperature trend, and thus ozone depletion, is the dominant contributor to that change. However, the climatic change of the *asymmetric* component of zonal wind component at z = 10 m (U10) as well as of 300hPa heights has been found to be large for both summer and winter periods. Our regression results show that correlation with T_equator (our proxy for global warming) explains most of the climatic changes for the asymmetric component of U10 and 300 hPa heights for summer and winter periods, suggesting the influence of warming of the global oceans on anticyclones south of the Indian Ocean and south-eastern Pacific Ocean.

Keywords: Southern Hemisphere Changes; Ozone Depletion; Ocean Warming; Poleward Stormtrack

1. Introduction

The polar displacement of the mid- and high-latitude circulation of the Southern Hemisphere over the last decades of the past century has been reported in a large number of articles. An observed polar shift of the surface westerlies derived from reanalysis [1,2], among others) has been verified with radiosonde observations [3] as well as satellite observations [4]. This trend of the positive phase of the Southern Hemisphere Annular Mode (SAM) has been also identified in simulations of the last century as well as projections of future climate change. The trend of the positive phase of the SAM index implies a poleward shift of many different components of the Southern Hemisphere middle and higher latitude circulation, including storm tracks [2] and the southern edge of the Hadley Circulation [5,6].

A number of regional climate changes over the middle and high latitudes of the SH have also been observed over the past century that seem to show the impact of the *asymmetric* component of the westerlies. Regional studies have shown a strong seasonal impact in precipitation over the middle latitudes of the SH [7,8]). For example, as the average surface air temperature of Australia increased by 0.7°C over the past century, there have been marked declines in regional precipitation, particularly along the east and west coasts of the continent ([9,10]. Considerable regional changes which have also been observed over Antarctica-strong warming trends were reported over the west Antarctic region, but there was no significant change over the rest of the continent [3,11-14]). Although there is some evidence of a correlation of trends of the Antarctic Peninsula and the zonally symmetric component of the SAM [1,3], recent results indicate that recent warming has not been restricted to the Antarctic Peninsula, but has been significant across the entire west Antarctica ice sheet [15], (hereafter HR2010); [16]. Also the seasonal pattern of warming is not consistent with a response to SAM trends [14]. Moreover, HR2010 and [16] concluded that the month-to-month variability pattern of the zonally asymmetric SH troposphere is rather dominated by two-quasi-stationary anticyclones in the western section of the Southern Ocean. They stress that the importance of these anticyclones has a profound effect on the climate of the sub-polar regions of the SH by affecting the sea-ice variability and blocking the eddy activity of the storm track.

Much research concerning the SH zonally asymmetric circulation has focused on the Pacific-South American mode (PSA, e.g, [17]) or the major zonal waves. However, these large-scale decompositions may mask important local variability. In HR2010, the month-to-month variability explained by the zonal waves 1 and 3 was examined, and an alternative representation of the SH circulation was suggested based on two quasi-stationary anticyclones in the sub-Antarctic western hemisphere. These anticyclones are related to the zonal waves, but as HR2010 stresses, features of their variability are masked by the zonal wave decomposition; in particular, the anticyclones' strengths are not positively covariant. HR2010 also shows that they capture variance independent of the Southern Annular Mode and explain a generally greater fraction of the variability than the PSA.

The importance of stratospheric ozone depletion on the atmospheric circulation of the troposphere has previously been studied with an atmospheric general circulation model (e.g., [18]). Their focus was the relative importance of ozone depletion contrasted with that of increased greenhouse gases and accompanying sea surface temperature changes. By specifying ozone and greenhouse gases forcing independently, and performing long, time-slice integrations, they concluded that the impacts of ozone depletion are roughly 2 - 3 times larger than those associated with increased greenhouse gases, for the Southern Hemisphere tropospheric summer circulation. However, the [18] study mainly focuses on the zonally symmetric circulation.

By recognizing the importance of both the symmetric and asymmetric components for climate change, the purpose of this work is to identify the respective roles of tropical ocean warming and the cooling trend of stratospheric temperatures anomalies due to ozone depletion on the seasonal changes in both symmetrical and asymmetrical components of the southern hemisphere circulation, with particular emphasis on the lower and upper tropospheric circulation. The limitations of observed and

model data are a major challenge to determine how climate warming and ozone depletion have affected the South hemisphere circulation (the issue of the quality of data used in this analysis will be discussed below in "Data and methods"). However, a relevant and related question that can be answered is how polar stratospheric temperature trends and surface equatorial temperature trends could affect circulation changes in the southern hemisphere.

Our simple approach is to consider the zonally symmetric component as the zonal average of the variable and the asymmetric part as the anomaly of the zonal average. To that end we will consider two variables related to SAM; the zonal component of the surface wind u (10 m) (from hereafter U10) and the height at 300 hPa. The data used for this analysis will be discussed in Section 2. In Section 3, we present the surface wind changes over a period of 36 years, from 1964-1999, and a similar analysis will be shown for the heights (300 hPa) in Section 4. Finally, conclusions and discussions are in Section 5.

2. Data and Methods

We are using three data sets: the ERA40 reanalysis [19], which has been described by Marshall3] as providing a reliable representation of the Southern Hemisphere high latitude atmospheric circulation variability, and two runs of the GFDL global atmosphere and land model, "AM2.1" [20]. The AM2.1 simulations are 135 year runs (1870-2004) consisting of 10 ensemble members using the same changes in forcing functions, but with sea surface temperatures (SSTs) and sea ice prescribed at observed values. "GFDL_A" runs include only observed SST variability while gas concentrations are fixed to their pre-industrial levels. In the "GFDL_B" runs, all the radiative gases and ozone variability are included. The differences between the GFDL all-forcing (GFDL_B) and the GFDL no forcing (GFDL_A) simulations are mainly the effect of the stratospheric ozone variability that was included in GFDL_B but not in GFDL_A scenarios. Greenhouse gases trends that also were included in B, but not in A, do not significantly impact results because both cases incorporate the same SST variability that reflects most of the changes due to greenhouses gases. The difference between the two atmospheric models' results, GFDL_AM2.1 B (10 member ensemble) and GFDL_AM2.1 A (10 member ensemble), will therefore illustrate the effects of ozone variability and change produced by greenhouse gases other than the changes in SST.

Most of the temperature trends in T_polar and T_equator, are probably due to the effects of global warming and ozone depletion. However to prove it from observations requires some assumptions about the data used. First, the ERA40 data to be used before the late seventies are quite unreliable concerning the quality of data over the Southern Oceans. However we will use the reanalysis since the early sixties in order to have enough data (close to forty years) for the polar stratospheric and equatorial surface temperature trends to be clear. As previously stated, we try to estimate how much of the change in circulation was affected by polar stratospheric or equatorial temperatures changes, and leave the cause of the temperature changes for the discussion. We assume here that the ERA40 reanalysis is a self-consistent system regardless what process that produced the observed trends in the data sets.

Second, although both GFDL simulations: GFDL_B and GFDL_A used the same atmospheric model and the same time varying SST's, the runs of GFDL_B have variability of greenhouse gases and ozone concentrations whereas the GFDL_A runs have all the concentrations fixed to climatological values. Since one of the models has a trend in the stratospheric temperatures (GFDL_B) and the other one does not (GFDL_A), the comparison of both solutions is appropriate to answer the question of how much T_polar and T_equator could control the changes of the south hemisphere circulation. Each simulation is analyzed with its own changes in temperature and resulting self-consistent changes in circulation. However it cannot be concluded that changes observed on GFDL_B are only due to Ozone depletion, absent in the GFDL_A runs. Because GFDL_B has also changes in greenhouse gases and may have some effects in the stratospheric temperatures as well.

The period used is 36 years from 1964 to 1999 (similar to that used in [3]). The method used is very simple:

1) We take the difference of monthly climatology between the last 18-year period (1982-1999) and the period from 1964-1981 as the measure of the climate change that should be explained, for summer months (NDJFMA) and winter months (MJJASO).

2) We carry out a regression of the variables (1964-1999) U10 and the 300hPa heights (Z300) with two zonally averaged temperatures; one stratospheric at polar latitudes and the other at low levels in the equatorial region. The time series are divided into summer months and winter months. The temperatures are defined as follows: the zonally averaged polar stratospheric temperature at 150 hPa averaged from 70S - 80S (hereafter T_polar), and the zonally averaged equatorial temperature from 5S - 5N at 850hPa (hereafter T_equator). These two temperatures are proxies for the effect of ozone depletion (T_polar) and global ocean warming (T_equator). The zonally averaged temperature minimizes decadal and interannual variabilities, but does not completely remove them. Note that T_equator is not SST but the atmospheric temperature at 850hPa. Although T_equator has much of the information from the SST, it allows more degrees

of freedom between the two model simulations.

3) The monthly climatological mean has been removed from both temperature proxies, T_polar and T_equator. Both time series have a linear trend for the last part of the century, which is more obvious for T_polar in the summer season. An effort has been made to decorrelate these time series (as described in the appendix) to separate their effects in the atmospheric circulation.

4) To maintain their identity, a linear regression was performed with each individual time series on the U10 wind velocity and the height Z300, rather than using a multiple regression (since in this case for the decorrelated variables it would render the same result).

5) To complete the methodology, we make a simple estimate of how well the regression explains the difference in the monthly climatology by calculating the correlation between both patterns. Obviously, the correlation could be positive (in phase) between the climate change and the regression or negative (out of phase). To create a measure of how well the regression matches the climate change, we define the positive correlation as a measure of "control" (if the correlation is negative we consider it zero or "no control"). By this method we quantify how much control T_polar or T_equator has in explaining changes in climatology.

Figure 1 shows the normalized (by one standard deviation) T_polar and T_equator for the ERA40 reanalysis and both GFDL ensembles runs. It is easy to identify a negative trend in T_polar and a positive trend in T_equator. Also note the difference between T_polar in GFDL_B, which shows a negative temperature trend, and GFDL_A with large variability but no apparent trend. (It should be pointed out that, although the figures for T_polar ERA40 and GFDL_B look very similar, the actual amplitude of the anomaly for the ERA40 is more than double that simulated by GFDL runs. Probably because the GFDL runs have very few stratospheric levels). In contrast, in the right column, T_equator values for the three data sets are very similar, because they are forced with very similar SSTs.

3. Symmetric and Asymmetric Components of U10

3.1. Changes in the Symmetric Component

Large anomalies in the strength of the stratospheric polar jet are followed by persistent anomalies in the tropospheric annular mode [21]. Several idealized models have also shown a poleward shift of surface westerlies in direct response to stratospheric winds [22]. More recently, [23]) suggested that a possible link between the stratospheric and tropospheric changes is the fact that

Zonal Temperature: Polar (150hPa, 70S–80S) and Equatorial (850hPa, 5S-5N)
Anomalies from the monthly climatological mean normalized by their STD

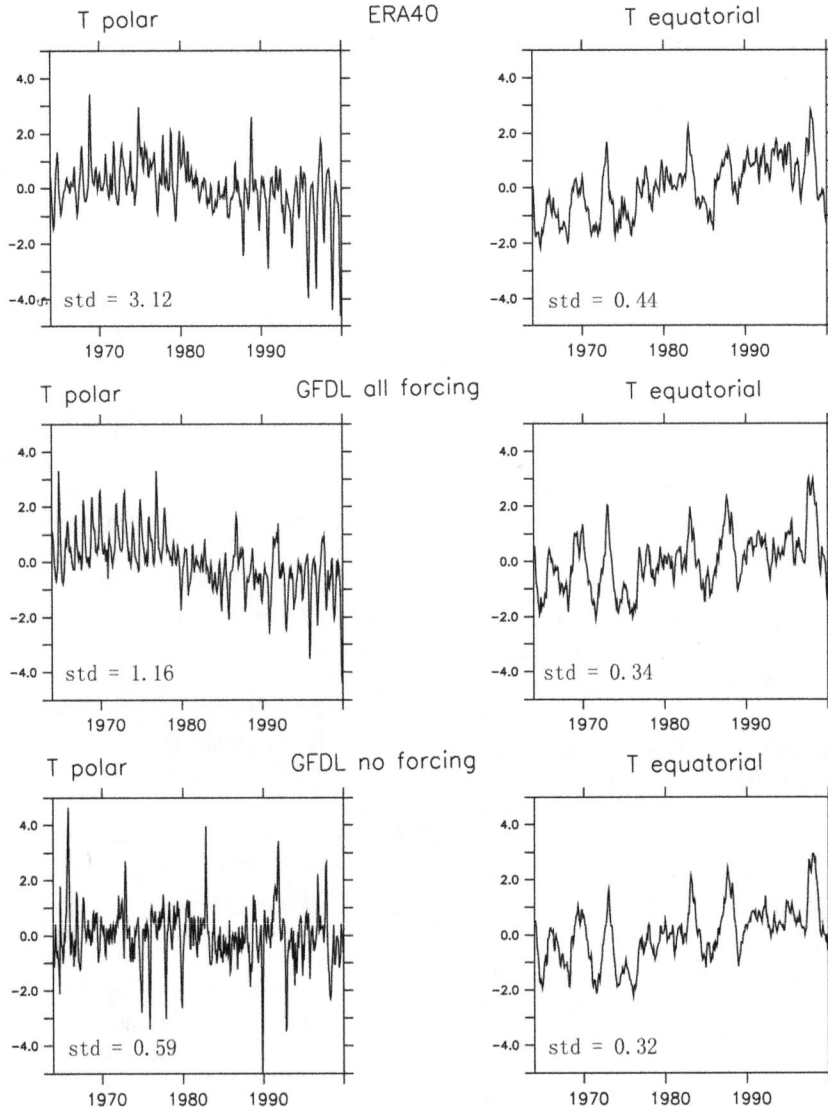

Figure 1. The time series of zonal averaged temperatures, normalized by one standard deviation (STD). Left panels: the lower stratosphere temperature averaged from 70S - 80S at 150 hPa (T_polar). Right panels: the zonal averaged lower level atmosphere equatorial temperature averaged from 5S - 5N at 850 hPa. The first row is for the reanalysis ERA40, the middle row for GFDL ensemble runs with all forcing, and the lower row is for the GFDL no forcing run.

increasing the stratospheric/upper tropospheric winds increases the eastward phase speed of tropospheric eddies and displaces the region of subtropical wave breaking poleward. This shifts the eddy momentum fluxes poleward, as well as the surface westerlies that are maintained by these momentum fluxes [7] (It should be note that the poleward displacement observed for the surface winds in summer months is only a couple of degrees latitude).

Our definition of the zonally symmetric component of the variables treated here follows the commonly used definitions:

The zonally symmetric of the variable Z is

$$<Z(y,z,t)>_{sym} = \text{zonal average of } (Z(x,y,z,t))$$

And the asymmetric part of Z is:

$$Z_{asy} = Z(x,y,z,t)- <Z(y,z,t)>_{sym}$$

It is well known that the climatic change of the zonally symmetric component of the U10 wind has a large seasonal variability-the variability is very large in late spring and summer and very small the rest of the year [23,24]. The summer difference of the climatic mean over the late period (1982-1999) and the earlier period (1964-1981) is shown in **Figure 2**. The two panels in the upper row

What Controls Recent Changes in the Circulation of the Southern Hemisphere: Polar Stratospheric or Equatorial Surface Temperatures?

85

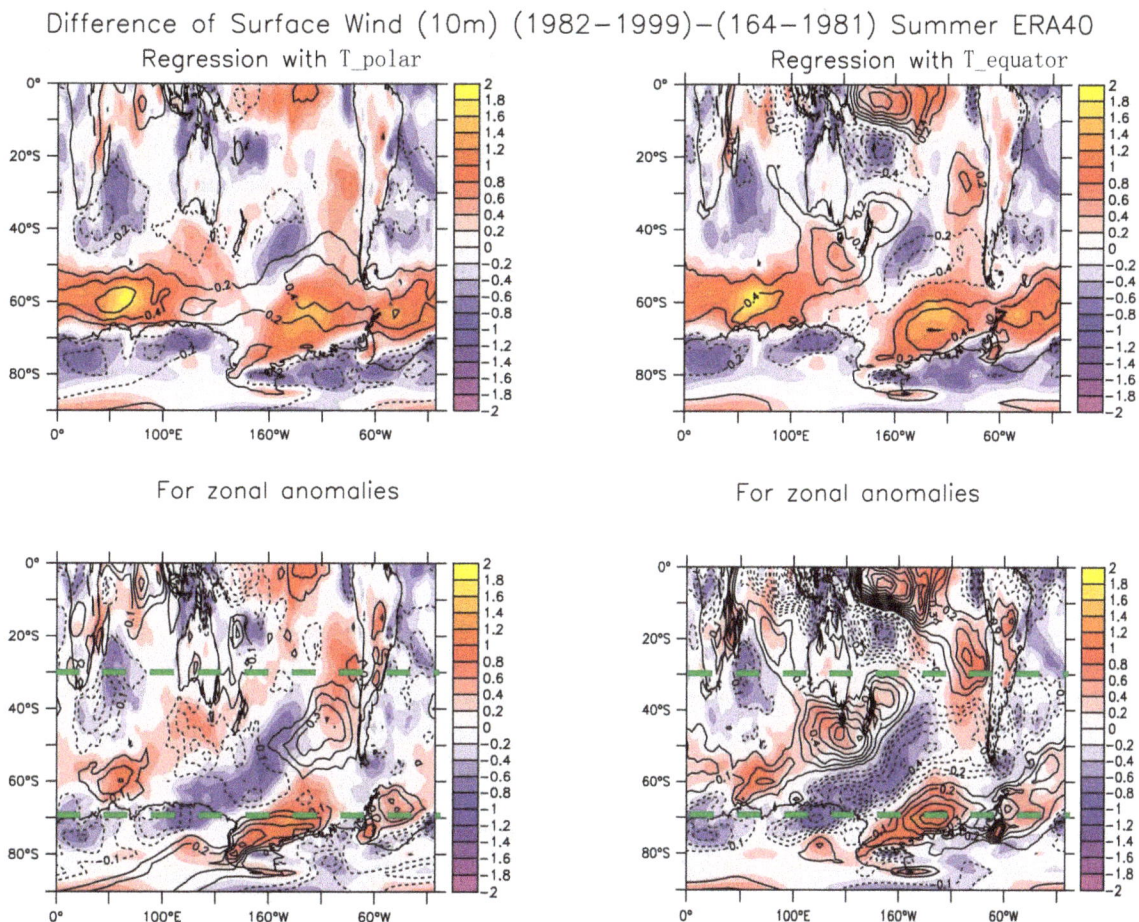

Figure 2. ERA40, Surface wind U10 differences between 1982-1999 and 1964-1981 for the 6 months centered in the austral summer NDJFMA (color). Superposed are regressions (in black contours) of U10 with T_polar (left panels), and with T_equatorial, (right panels). The upper row shows total U10 whereas the zonal anomalies of U10 are shown on the lower row. The green dashed lines show the area in which a similarity test will be performed.

show the climatic difference in color (the same in both panels), with the regressions superposed as black contours for T_polar on the left and with the T_equator on the right. The wind at 10m shows a large zonal change around 60S. The regression with T_polar (left panel) shows a similar behavior, whereas the regression with T_equator (right) shows a significant superposition with the climate change with the asymmetric component of U10 (as can be seen in the lower graphs), but also suggests the zonal mean to be smaller. In contrast, for the zonal anomaly only (lower panels), the climatic difference of the asymmetric component of the wind is clearly better represented by the regression with T_equator.

By splitting the surface westerly wind into its zonally symmetric component and the asymmetric part, we can evaluate the effects of both the upper stratospheric temperature anomaly and the surface equatorial temperature on each component. **Figure 3** shows the zonally symmetric component of the surface westerlies for summer, its climatic change, and the regression of this component

with T_polar and T_equator. The climate change of U10 is much larger in the ERA40 reanalysis than the model simulation (GFDL_B), and the response is quite linear with the stratospheric temperature forcing (as previously mentioned, the ERA40 T_polar is close to double the temperature in the GFDL_B runs (e.g., [18]). For both ERA40 and GFDL_B, regression with T_polar better describes the climatic change. Moreover, the fact that GFDL_A does not show any significant climate change confirms the fact that the zonal mean surface westerlies are controlled mainly by ozone depletion that mostly produces the stratospheric temperature anomalies [18]. It should be noted that the responses to T_equator for GFDL_B and GFDL_A are similar-although small, they tend to be out of phase with observed climatic change of U10. [18] find a similar, but smaller, response in runs with greenhouse gas changes that do not include ozone variability. Changes in upper tropospheric effects from CO_2 and other gases may also influence the circulation in a manner similar to the ozone depletion, increasing the

Symmetric Component of U10, Summer (NDJFMA)

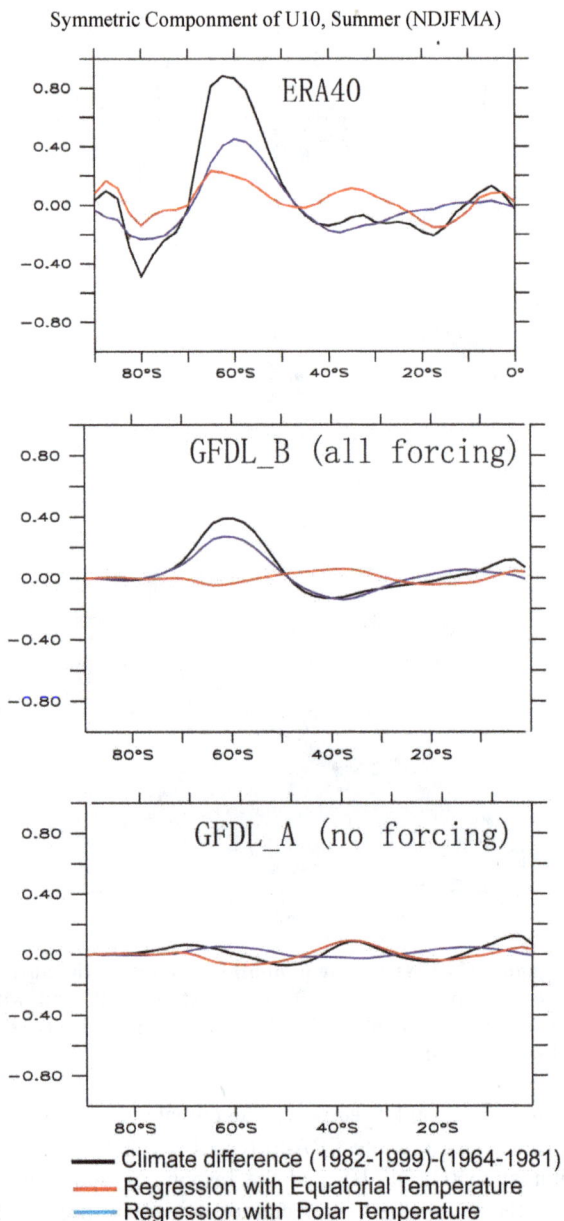

Figure 3. Zonal average of U10, for the summer period. The eighteen year difference (black) and both the regressions: with T_polar (in blue) and T_equator (in red). The upper panel is for ERA40, the middle panel for GFDL_B, and the lower panel for GFDL_A.

surface zonal winds in sub-polar latitudes (as pointed out by [18], to a much weaker degree than the effect of ozone trends).

3.2. Changes in the Asymmetric Component of U10

The asymmetric component of U10 for summer, shown in the lower row of **Figure 2**, shows a very strong similarity between the climatic change and the regression with T_equator. It seems that, in the summer season, the

zonally symmetric component is controlled more by variability in T_polar, whereas the asymmetric component is explained by the regression with T_equator. In order to quantify the role of each regression in explaining the climatic difference patterns, we define a control parameter. This control parameter is defined as the positive correlation between the difference pattern and the regression pattern over a specific region (in particular from 30S to 70S shown as green dashed lines in **Figure 2**). Before showing the correlation, let us review the patterns for the winter season shown in **Figure 4**. As previously mentioned the total U10 in winter, exhibits large positive and negative regions, but a small zonally symmetric component. The asymmetric anomaly seems as large as in the summer months. Again the most relevant feature is the coherence between the climatic difference anomaly and the regression with T_equator.

We can summarize the results by calculating the correlation between the two regressions and the climatic difference over the area shown between the green lines of **Figure 2**. The control parameters for summer and winter are shown in **Figure 5** for the three data sets. There is great consistency between the reanalysis and the two model solutions. The outlier seems to be GFDL_B for the summer season, but in general there is a very good agreement among the figures, particularly for the winter season. The averaged calculated correlation for each member that composed the ensemble (**Figure 5**) of GFDL_B runs is shown in **Table 1**. Values for GFDL_A are not shown since only one major forcing was acting for that case Note that each individual correlation between the climate change and the regression is a nonlinear function for each independent member and its average may not coincide with the ensemble mean correlation. The results are fairly consistent among the members. Which is remarkable given that the forcing for stationary waves in the South Hemisphere is very weak; as a consequence the variability between each member is considerable. Because these asymmetries in the surface westerly wind should also be manifested in height anomalies, we next apply the same analysis to that field.

4. Asymmetric Component of the Geo-Potential Heights 300 hPa

The climatic change of the zonal anomalies of the 300 hPa height field for the ERA40 and both GFDL simulations are shown in **Figure 6** for the summer season. As in **Figure 2**, we show the climatic difference in both panels for the same row and the two regressions of that field, with T_polar on the left and T_equator on the right. The main characteristic of the response seems to be a wave train of Rossby waves that emanate from the Indian Ocean and propagate south east of the Pacific Ocean that

What Controls Recent Changes in the Circulation of the Southern Hemisphere: Polar Stratospheric or Equatorial Surface Temperatures?

87

Difference of Surface Wind (10m) (1982−1999)−(164−1981) Winter ERA40

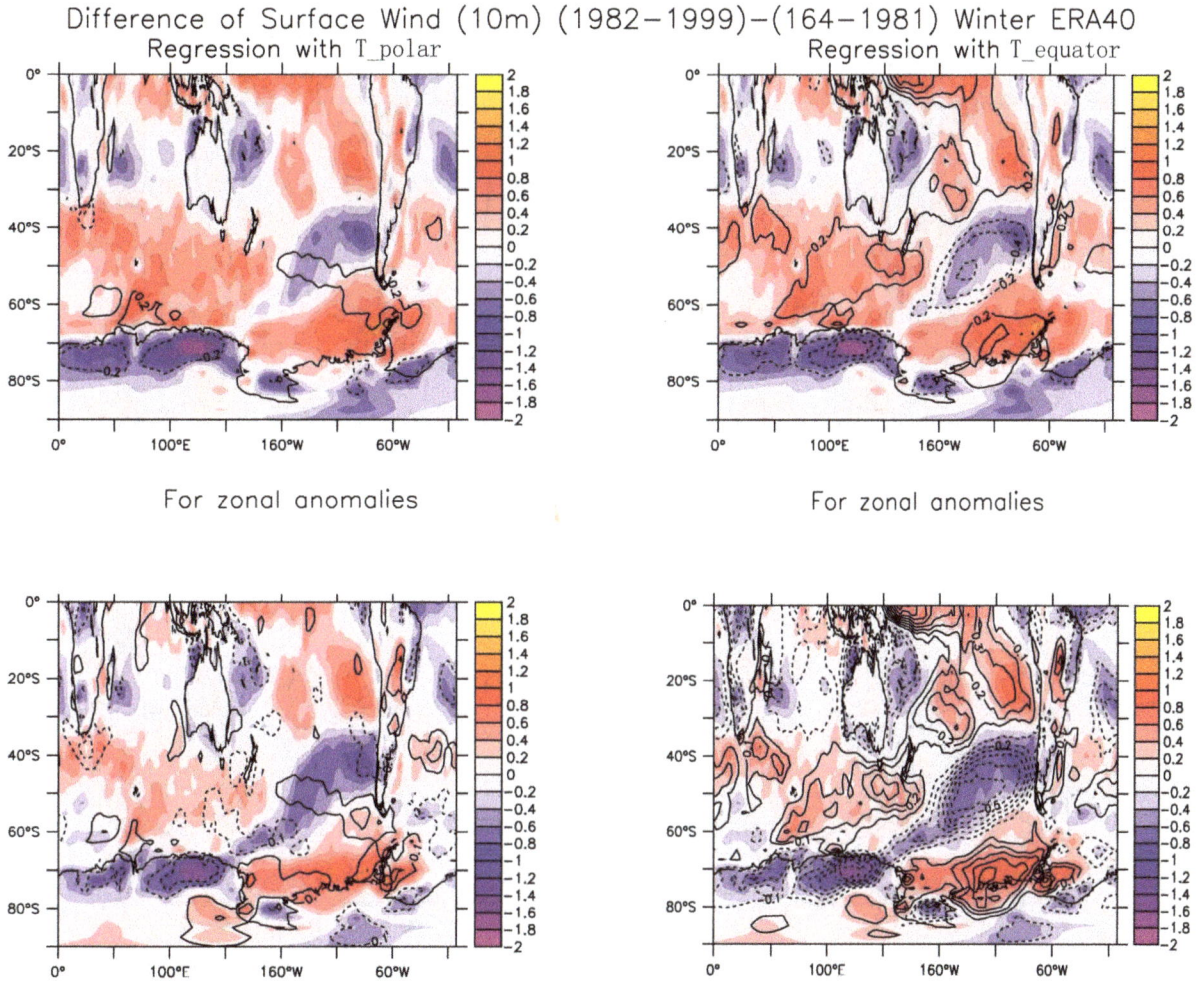

Figure 4. The same as in Figure 2, but for the 6 months centered in the austral winter.

Control Diagram of Polar and Equatorial Temperatures on Heights.

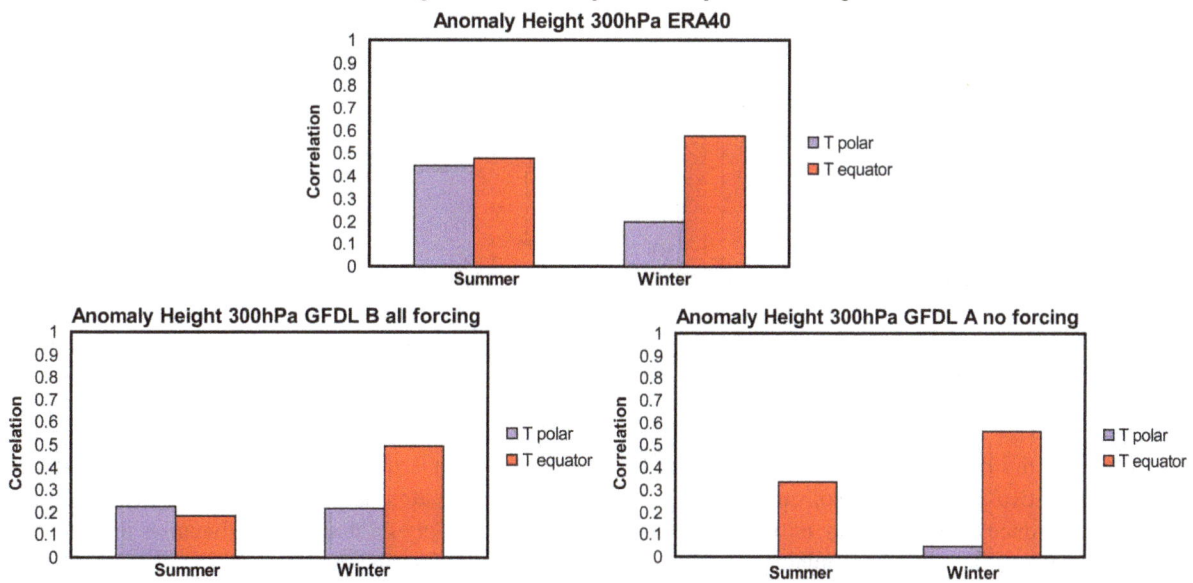

Figure 5. The control diagram summarizing how much the regressions with T_polar (blue) and T_equator (red) explain the climate differences for each season. Each dataset is shown as in Figure 3.

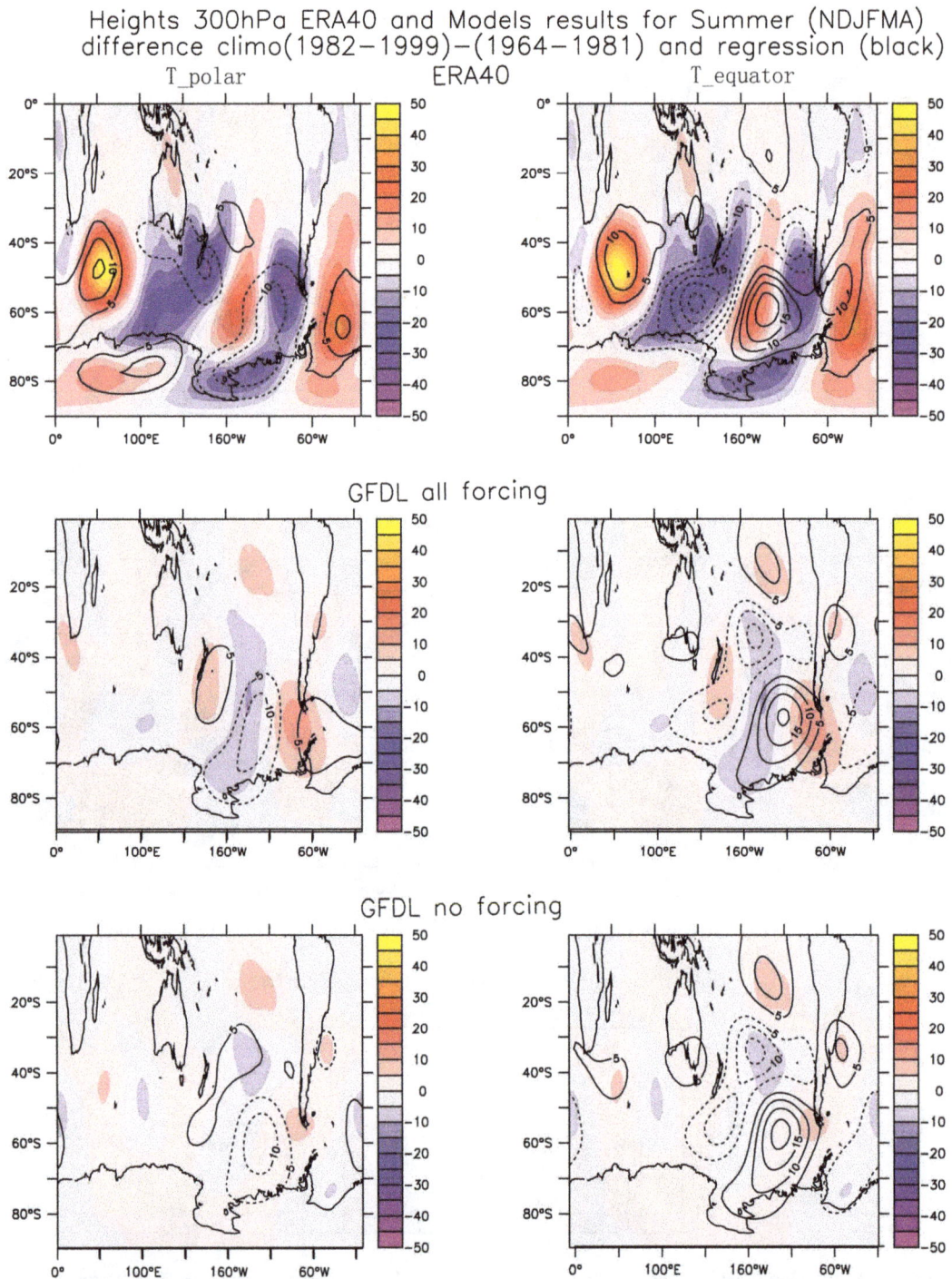

Figure 6. The zonal anomalies for the 300 hPa heights are shown for the summer period. The climate differences are shown in color. As in Figure 2, the regressions of T_polar and T_equator are shown in black contour. The upper row is for ERA40, the middle row for GFDL_B and the lower row for GFDL_A.

is clearly seen in the ERA40, but is more weakly exhibited in the other two simulations. Note that while both regressions reproduce some features of the difference fields for the ERA40, the GFDL simulations tend to be consistent with ERA40 for the regression with T_equator, in particular a height anomaly center around 110W and

60S. This pattern also seems to be characteristic of an interannual pattern known as the PSA [17]. A superficial inspection suggests that the combination of both regressions could describe the summer difference pattern well. However, for the winter patterns shown in **Figure 7**, the T_equator regression practically reproduces the entire-

What Controls Recent Changes in the Circulation of the Southern Hemisphere: Polar Stratospheric or
Equatorial Surface Temperatures?

89

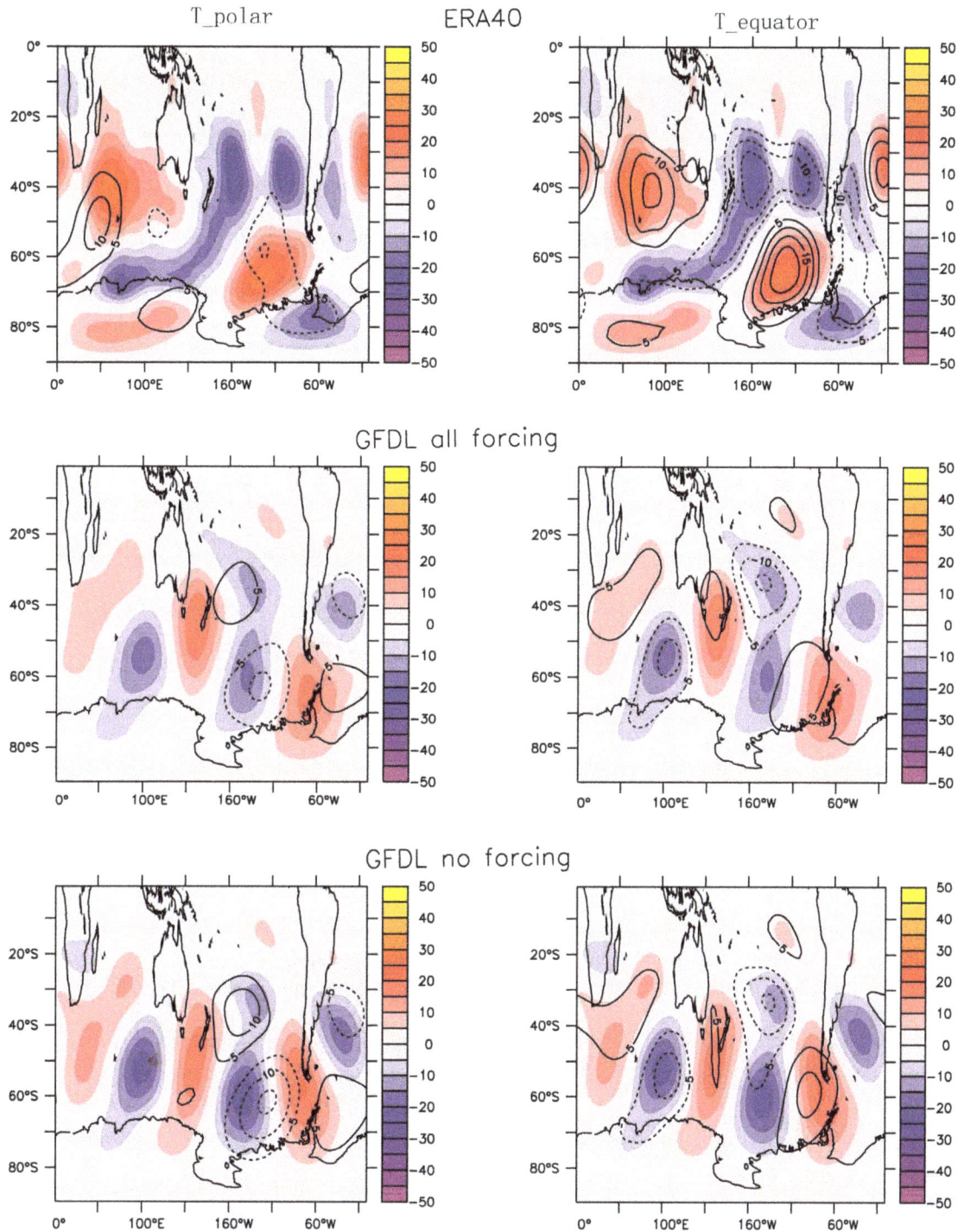

Figure 7. The same as in Figure 6, but for the winter period.

difference pattern in the ERA40 reanalysis, and shows weaker but still good correlation with the differences in the two GFDL simulations. In contrast, the T_polar regression tends to be quite out of phase with all three difference patterns.

The results of **Figures 6** and **7** can be summarized by calculating the correlation pattern as has been shown for U10 (**Figure 5**). The control parameter shown in **Figure 8** contains clearly what we have seen in the previous figures. The control is shared between T_polar and

Control Diagram of Polar and Equatorial Temperatures on Heights.

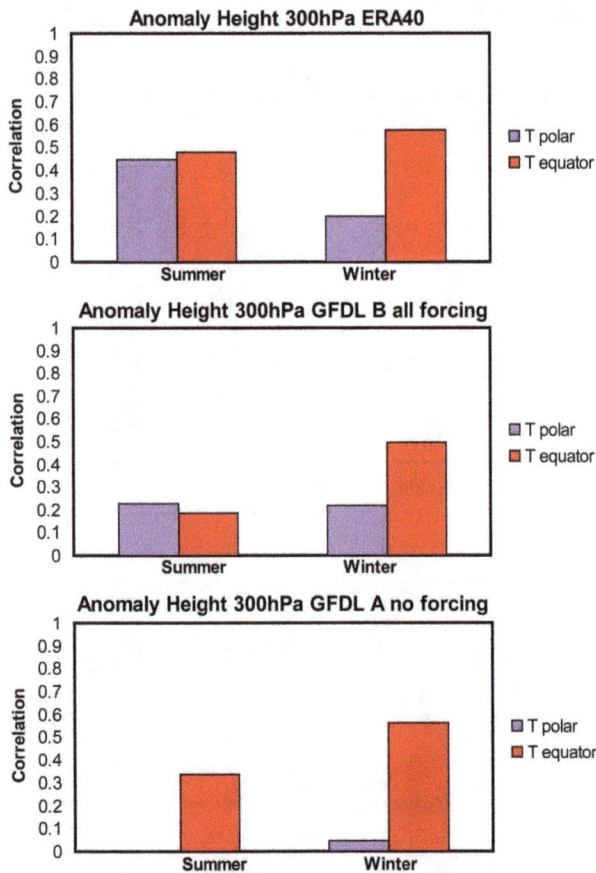

Figure 8. The control diagram for the asymmetric component of 300 hPa Height (see Figure 5).

Table 1. Values of the individual correlations for members of GFDL All forcing. Individual correlation of ΔZ300 and ΔU10 for regressions of T_polar and T_equator.

ΔZ300	Member	Summer		Winter
#	T_polar	T_equator	T_polar	T_equator
1	0.3096	0.2741	0.176	0.364
2	0.090	0.1065	−0.0354	0.180
3	−0.001	0.248	−0.057	0.305
4	0.2859	0.061	0.421	0.414
5	0.062	0.107	0.095	0.151
6	0.069	0.278	0.138	0.167
7	0.148	0.327	−0.022	0.293
8	0.299	−0.153	0.161	0.458
9	0.142	0.259	−0.053	0.280
10	0.225	0.094	−0.275	0.426

ΔU1	Member	Summer		Winter
#	T_polar	T_equator	T_polar	T_equator
1	0.619	0.372	0.331	0.520
2	0.325	0.395	0.179	0.576
3	0.072	0.229	−0.164	0.635
4	0.460	0.107	0.616	0.586
5	0.360	0.259	0.560	0.196
6	0.170	0.546	0.601	0.133
7	0.213	0.402	0.133	0.487
8	0.646	−0.176	0.538	0.658
9	0.394	0.366	−0.067	0.638
10	0.247	0.304	−0.273	0.883

T_equatorial for the summer period and T_equator exerts more control for the winter season. These results are corroborated for the ERA40 as well as both GFDL simulations. The averaged controls for the individual members for GFDL_B also exhibit a consistent pattern (see **Table 1**). It should be pointed out that the climate difference response for the ensemble mean seems rather weak compared with the ERA40 response. An inspection of the height response for each individual member shows that, although they consistently reproduce a height anomaly at the South-East Pacific Ocean (some of them as large as the ERA40 response), the individual centers vary as much as 50 degrees longitude from each other, producing a weaker ensemble averaged.

One feature that clearly stands out in all three datasets is a Rossby wave train from the Indian Ocean to the southeast Pacific Ocean that ends in a blocking high similar to the summer pattern of **Figure 6**. The zonally asymmetric climate change signal in the SH troposphere is dominated by two quasi-stationary anticyclones south of the Indian Ocean and south-eastern Pacific Ocean. These patterns are very similar to those found by HR2010. As has been discussed by HR2010 and others,

these patterns are a response to the Indian Ocean warming and the interannual-decadal variability of the equatorial Pacific Ocean.

Although the work presented here focuses on seasonal data, the trend in the anticyclones can be linked to blocking of the high frequency eddies. [25] analyzed SH blocking events lasting 5 days or longer, and found that persistent geopotential height anomalies occurred throughout the year south of New Zealand, in a region east of the Indian Ocean. A second region was also evident in the south-east Pacific Ocean, corresponding to the anticyclone location which was much more significant in winter season and very similar to that shown in **Figure 7**.

5. Summary and Discussion

The consistency of the results obtained with the ERA40 and both GFDL solutions makes the conclusions rather

What Controls Recent Changes in the Circulation of the Southern Hemisphere: Polar Stratospheric or Equatorial Surface Temperatures?

91

robust. Our results show that the zonally symmetric surface winds in the summer period, over the last decades of the past century, have a considerably enhanced intensity in high latitudes of the SH (around 60S as shown in **Figure 3**), consistent with previously reported results [1,2], among others). Our analysis also suggests that the stratospheric temperature trend (T_polar) is the major contributor to that change. Moreover, the symmetric component of U10 displays a similar, although weaker, maximum for the GFDL_B solutions, but a very weak maximum for the GFDL_A runs. This difference in model response reaffirms that the trend of T_polar (ozone depletion) is the main cause of the drifting of the zonal winds to sub polar latitudes.

The response expected from the surface equatorial temperatures (T_equator), although being consistent between the two GFDL runs, disagrees with the ERA40 analysis over the sub-polar regions. From both GFDL runs, it appears that the global ocean warming solution can not by itself produce a sustainable sub-polar drift. It should be remembered that the trend in ocean temperature, although it is the largest effect of the increase of greenhouses gases, is not the only one. Changes in upper tropospheric effects from CO_2 and other gases may also influence the circulation in a matter similar to the ozone depletion, increasing the surface zonal winds in sub polar latitudes (as pointed out by [18], to a much weaker degree than the effect of ozone trends).

The climatic change of the *asymmetric* component of U10 has been found to be large for summer and winter periods. To quantify the response of both regressions, T_polar and T_equator, with the climatic differences, we compute the correlation between both patterns and refer to the positive correlation as "control". In contrast with the findings for the zonally symmetric component, it was found that T_equator exerts a strong control over climatic changes for the asymmetric component of U10 for summer and winter periods. These results are consistent between ERA40, and both GFDL ensembles run. However the stratospheric temperature T_polar has some effect, in particular for the summer period. The analysis of the asymmetric component for the 300hPa Heights has a similar response with respect to the control exerted by T_polar and T_equator as for the surface winds, U10. These results are corroborated for the ERA40 as well as both GFDL simulations.

The zonally asymmetric climate change signal in the SH troposphere is dominated by two quasi-stationary anticyclones south of the Indian Ocean and south-eastern Pacific Ocean. These patterns are very similar to those found by HR2010. This simple analysis tries to suggest that the asymmetric component changes are considerable over the high latitudes of the SH and are not as much a product of the ozone depletion, which in future climates may recuperate, but are mainly forced by the warming of the global oceans (Indian and Pacific Ocean). This regional ocean forcing tends to project better on the asymmetric component of the high latitude SH circulation. This is mainly because of the response from Rossby waves rays directed to the sub-polar regions [26], in contrast with the more zonally symmetric forcing that stratospheric temperatures produce. These asymmetries, through deflection of storm tracks, modify the sea-ice distribution and change sub-polar ocean currents, with significant implications for future climate.

6. Acknowledgements

The author is deeply indebted to Dr. Isaac Held for his valuable comments along this research, also appreciates his comments on the paper as well as Dr. Thomas Delworth and Dr. Silvina Solman that helped clarify the manuscript. Special appreciation to Dr. Roberta M. Hotinski for editing the manuscript. Acknowledgments to ERA-interim project of ECMWF, Dr Fanrong. Zeng and Dr. Andrew Wittenberg for providing data from GFDL models and ERA40 data base and advise on Ferret analysis.

This report was prepared by Isidoro Orlanski under award NA08OAR4320752 from the National Oceanic and Atmospheric Administration, US Department of Commerce. The statements, findings, conclusions, and recommendations are those of the author(s) and do not necessarily reflect the views of the National Oceanic and Atmospheric Administration, or the US Department of Commerce.

REFERENCES

[1] D. W. J. Thompsonand S. Solomon, "Interpretation of Recent Southern Hemisphere Climate Change," *Science*, Vol. 296, No. 5569, 2002, pp. 895-899.

[2] C. Archer and K. Caldeira, "Historical Trends in the Jet Streams," *Geophysical Research Letters*, Vol. 35, No. 8, 2008, Article ID: L08803,

[3] G. J. Marshall, "Trends in the Southern Annular Mode from Observations and Reanalysis," *Journal of Climate*, Vol. 16, No. 24, 2003, pp. 4134-4143.

[4] Q. Fu, J. C. Wallace and J. Reitchler, "Enhanced Mid-Latitude Tropospheric Warming in Satellite Measurements," *Science*, Vol. 312, No. 5777, 2006, p. 1179.

[5] Y. Hu and Q. Fu, "Observed Poleward Expansion of the Hadley Circulation Since 1979," *Atmospheric Chemistry and Physics*, Vol. 7, 2007, pp. 5229-5236.

[6] J. Lu, G. A. Vecchi and T. Reichler, "Expansion of the Hadley Cell under Global Warming," *Geophysical Research Letters*, Vol. 34, 2007, Article ID: L06805.

[7] C. Vera, G. Silvestri, B. Liebmann and P. Gonzalez, "Climate Change Scenarios for Seasonal Precipitacion in South America from IPCC-AR4 Models," *Geophysical Research Letters*, Vol. 33, 2006, Article ID: L13707.

[8] S. Solman and I. Orlanski, "Subpolar High Anomaly Preconditioning Precipitation over South America," *Journal of the Atmospheric Sciences*, Vol. 67, No. 5, 2010, pp. 1526-1542.

[9] B. L. Preston and R. N. Jones, "Climate Change Impacts on Australia and the Benefits of Early Action to Reduce Global Greenhouse Gas Emissions," Report CSIRO, 2006.

[10] H. D. Pritchard, *et al.*, "Antarctic Ice-Sheet Loss Driven by Basal Melting of Ice Shelves," *Nature*, Vol. 484, 2012, pp. 502-505.

[11] D. G. Vaughn, G. J. Marshall, W. M. Connolley, C. Parkinson, R. Mulvaney, D. A. Hodgson, J. C. King, J. C. Pudsey and J. Turner, "Recent Rapid Regional Climate Warming in the Antarctic Peninsula," *Climate Change*, Vol. 60, 2003, pp. 243-274.

[12] J. Turner and J. Overland, "Contrasting Climate Change in the Two Polar Regions," *Polar Research*, Vol. 28, No. 2, 2009, pp. 146-164.

[13] J. Turner, T. A. Lachlan-Cope, S. Colwell, G. J. Marshall and W. M. Connolley, "Significant Warming of the Antarctic Winter Troposphere," *Science*, Vol. 311, No. 5769, 2006, pp. 1914-1917.

[14] E. J. Steig, D. P. Schneider, S. D. Rutherford, M. E. Mann, J. C. Comiso and D. T. Shindell, "Warming of the Antarctic Ice-Sheet Surface Since the 1957 International Geophysical Year," *Nature*, Vol. 1457, No. 7228, 2009, pp. 459-463.

[15] R. W. Hobbs and M. N. Raphael, "Characterizing the Zonally Asymmetric Component of the SH Circulation," *Climate Dynamics*, Vol. 35, 2010, pp. 859-873.

[16] N. Mathewman and G. Magnusdottir, "Clarifying Ambiguity in Intraseasonal Sothern Hemisphere Climate Modes during Austral Winter," *Journal of Geophysical Research*, Vol. 113, No. D7, 2012, Article ID: D03105.

[17] K. C. Mo and J. N. Paegle, "The Pacific-South American Modes and Their Downstream Effects," *International Journal of Climatology*, Vol. 21, No. 10, 2001, pp. 1211-1229.

[18] L.-M. Polvani, D. W. Waugh, G. J. P. Correa and S.-W. Son, "Stratospheric Ozone Depletion: The Main Driver of Twentieth-Century Atmospheric Ciruclation Changes in the Southern Hemisphere," *Journal of Climate*, Vol. 24, 2011, pp. 795-812.

[19] S. M. Uppala, *et al.*, "The ERA-40 Re-Analysis," *Quarterly Journal of the Royal Meteorological Society*, Vol. 131, 2005, pp. 2961-3012.

[20] J. Anderson, *et al.*, "The New GFDL Global Atmosphere and Landmodel AM2/LM2: Evaluation with Prescribed SST Simulations," *Journal of Climate*, Vol. 17, 2004, pp. 4641-4673.

[21] D. W. J. Thompson, J. M. Wallace and G. C. Hegerl, "Annular Modes in the Extra-Tropical Circulation, Part II: Trends," *Journal of Climate*, Vol. 13, 2000, p. 1018.

[22] L. Polvani and P. J. Kushner, "Tropospheric Response to Stratospheric Perturbations in a Relatively Simple General Circulation Model," *Geophysical Research Letters*, Vol. 29, No. 7, 2002, 18 p.

[23] G. Chen and I. M. Held, "Phase Speed Spectra and the Recent Poleward Shift of Southern Hemisphere Surface Westerlies," *Geophysical Research Letters*, Vol. 34, No. 21, 2007, Article ID: L21805.

[24] P. Kushner, I. M. Held and T. L. Delworth, "Southern Hemisphere Atmospheric Circulation Response to Global Warming," *Journal of Climate*, Vol. 14, No. 10, 2001, pp. 2238-2249.

[25] K. E. Trenberth and K. C. Mo, "Blocking in the Southern Hemisphere," *Monthly Weather Review*, Vol. 113, No. 1, 1985, pp. 3-21.

[26] I. Orlanski and S. Solman, "The Mutual Interaction between External Rossby Waves and Thermal Forcing: The Sub-Polar Region," *Journal of the Atmospheric Sciences*, Vol. 67, 2010, pp. 2018-2038.

What Controls Recent Changes in the Circulation of the Southern Hemisphere: Polar Stratospheric or Equatorial Surface Temperatures?

93

Appendix

To Partially Decorrelate the Time Series

Assume two time series that are slightly correlated, say:

$$Y = Y(t) \quad \text{and} \quad T = T(t) \tag{A1}$$

The cross-correlation between them is:

$$C_{YT} = \sum \left(Y' * T' / \left((N-1)\ \sigma_Y \sigma_T \right) \right) \tag{A2}$$

where σ_Y, σ_T are the standard deviations of each function.

$$\sigma_U = \left(\sum (U - \bar{U})^2 / (N-1) \right)^{0.5} \tag{A3}$$

where $\bar{U}$ is the mean of the variable U.

To decorrelate Y with T, it is enough to calculate the regression of Y on T as follows:

$$R_Y = \sum \left(Y' * T' / \left((N-1)\sigma_T \right) \right) \tag{A4}$$

Then the decorrelated Y anomaly can be written as:

$$Y'_{dc} = Y' - R_Y T' / \sigma_T \tag{A6}$$

It is easy to see the correlation between T' and Y'_{dc} is equal to zero. However it seems asymmetric: one time series has been modified and the other not. We could apply the same algorithm to both time series to partially decorrelate them. In this case, the correlation between Y'_{dc} T'_{dc} is not zero, but it is very small.

$$Y'_{dc} = Y' - 0.5 * R_Y * T' / \sigma_T \tag{A7}$$

$$T'_{dc} = T' - 0.5 * R_T * T' / \sigma_T \tag{A8}$$

$$\begin{aligned}
C_{YdcTdc} &= \sum \left(Y'_{dc} * T'_{dc} / (N-1) \sigma_{Ydc} \sigma_{Tdc} \right) \\
&= \left[\sum \left\{ Y' * T' + 0.25 * (R_T R_Y / \sigma_Y \sigma_T) * Y' * T' - 0.5 * (R_T Y'^2 / \sigma_Y + R_Y Y'^2 / \sigma_T) \right\} \right] / \left((N-1)\sigma_{Ydc} \sigma_{Tdc} \right)
\end{aligned} \tag{A9}$$

Replacing R_T and R_Y for their definition A4, it is easy to show that the last term of A9 can be written as follows:

$$\sum 0.5 * \left(R_T Y'^2 / \sigma_Y + R_T Y'^2 / \sigma_T \right) = \sum (Y' * T') \tag{A10}$$

And the expression in A9 can be reduced to:

$$C_{YdcTdc} = 0.25 * (R_T R_Y / \sigma_Y \sigma_T) \sum Y'T' / \left((N-1)\sigma_{Ydc} \sigma_{Tdc} \right) \tag{A11}$$

Recognizing that the Standard deviation σ_{Ydc} and σ_{Tdc} can be rewritten as:

$$\sigma_{Ydc} = \left\{ \sum Y'^2 - R_Y Y'T' / \sigma_T + 0.25 * (R_Y / \sigma_T) \sum T'^2 \right\} / (N-1) \tag{A12}$$

Using the definitions A2, A3 and A4, A12 can be written as follows:

$$\sigma_{Ydc} = \sigma_Y \left(1 - 0.75 C_{YT}^2 \right)^{0.5}$$

And similarly for: A13

$$\sigma_{Tdc} = \sigma_T \left(1 - 0.75 C_{YT}^2 \right)^{0.5} \tag{A13}$$

Finally, the correlation of the partially decorrelated time series A7-A8

$$C_{YdcTdc} = 0.25 * (C_{YT})^3 / \left(1 - 0.75 C_{YT}^2 \right) \tag{A14}$$

shows that for cases in which $C_{YT} \sim -0.3$, as in our case,

$$C_{YdcTdc} = -0.0072$$

the time series is very well decorrelated.

11

Characteristics of Central Southwest Asian Water Budgets and Their Impacts on Regional Climate

Khalid M. Malik[1*], Peter A. Taylor[2], Kit Szeto[3], Azmat Hayat Khan[4]

[1]National Agromet Center, Pakistan Meteorological Department, Islamabad, Pakistan
[2]York University, Toronto, Canada
[3]Climate Research Division, Environment Canada, Toronto, Canada
[4]National Drought Monitoring Center, Pakistan Meteorological Department, Islamabad, Pakistan

ABSTRACT

Water budgets terms, evapotranspiration (E), precipitation (P), runoff (N), moisture convergence (MC) and both surface as well as atmospheric residual terms have been computed with National Centers for Environmental Prediction (NCEP) (1948-2007) and European Centre for Medium-Range Weather Forecasts (ECMWF) ERA-40 (1958-2001) reanalysis data sets for Central Southwest Asia (CSWA).The domain of the study is 45° - 75°E & 25° - 40°N. Only the land area has been used in these calculations. It is noted in the comparison of both reanalysis data sets with Global Precipitation Climatology Centre (GPCC) that all three data sets record different precipitation before 1970. The maximum is from NCEP and the minimum with ERA-40. However, after 1970 all the data sets record almost the same precipitation. ERA-40 computes two phases of MC. Before 1975, the domain acts as a moisture source, whereas after 1975 it behaves as a moisture sink. The region CSWA is divided into six sub areas with rotational principle factor analysis and we distinguish them by different approached weather systems acting on each area. Finally, NCEP yearly precipitation is further divided into seasons; winter (November to April) and summer (May to October) and two phases have been noted. The variation in winter precipitation is more than summer during last 60-year analysis.

Keywords: Water Budgets; Drought over Central Southwest Asia; Moisture Flux Convergence; Principle Component Analysis; Climate Change

1. Introduction

Central Southwest Asia, which mainly includes, Iran, Afghanistan, Pakistan, Tajikistan, Iraq and Saudi Arabia, is located between 45° - 75°E and 25° - 40°N, from mid-latitudes to the tropics, with the Arabian Sea to the south, Great Himalayas and Karakoram to the east, the Caspian Sea and Russian states in north and the Black Sea and Tigris river basin in west. Southwest Asia is a region of diverse climates and is generally divided into three main climate types. There are arid, semiarid, and temperate as per the Köppen-Geiger classification [1], which factors in seasonal distribution of precipitation and the degree of dryness/coldness of the season. Upland and mountain parts of Pakistan with adjacent areas of India have a dry continental and subtropical climate with a main feature being the southwest monsoon, which lasts for 4 months, from June to September. In this southwest

monsoon period, the precipitation is one order of magnitude more than during the rest of the year. The region lies at the boundary of three climate regimes. These are: 1) the cold Siberian High in winter over Central Asia; 2) the monsoon Asian Low in summer over India; and 3) eastward propagating secondary low-pressure systems traveling through the Mediterranean and adjacent areas during non-summer seasons. They are called western disturbances WDs [2].

The region is badly affected during 1999-2001 drought, categorized as severe in the history of drought persisted over the globe in 20th century. To find out the causes of drought, it is vital to understand water budget terms of the region. In southwest Asia precipitation primarily falls from winter storms moving eastward from the Mediterranean, with the high mountains of the region intercepting most of the water. This wintertime precipitation generally occurs between the months of November and April, with the peak between January and March. Much of the

---

*Corresponding author.

precipitation falls as snow at higher elevations and the timing and amount of snowmelt is an important factor in the irrigated agriculture prevalent in the region. Very little precipitation falls in most of southwest Asia during the summer season. However, in eastern parts of Pakistan the primary precipitation season is summer. This is associated with the northernmost advance in the Asia monsoon, which results in a summertime maximum in precipitation in the northern mountain regions of Pakistan but generally suppresses precipitation over Iran and Afghanistan. Our study area is the central southwest Asia covers an area of approximately 4.67×10^{12} m^2, (**Figure 1**).

2. Water Budget Equation

Two dimensional (vertically integrated) horizontal variations of a key water process in the atmosphere-land surface system adopted by [3] are used in the study.

The atmospheric water equation can be written as

$$\frac{\partial Q}{\partial t} = E - P + MC + RESQ' \qquad (1)$$

and the surface water equation is

$$\frac{\partial W}{\partial t} = P - E - N + RESW' \qquad (2)$$

The temporal change in surface water W including soil moisture, snow and liquid water and in vertically integrated specific humidity or on atmospheric precipitable water Q, are represented by $\partial W/\partial t$ and $\partial Q/\partial t$ in above water conservation equations. Solid and liquid water evaporates (E) into the atmosphere from ocean and land surfaces which includes snow and vegetation. Water vapor is transported into the region from the surroundings by atmospheric winds and its convergence flux (MC), will increase atmospheric water vapor over the region while decreasing water vapor over the surroundings. Clouds formed through nucleation, grow by condensation (or diffusion if solid) and convert into large liquid and solid drops by accumulation, which fall as precipitation to the surface, P. Eventually, surface water is increased by precipitation and decreased by evapotranspiration. Rivers and canals transport surface fresh water to other locations and the net divergence of this transport, N (runoff), will increase surface water in low lying regions before discharging it into the oceans. $RESQ'$ and $RESW'$ are the residuals or errors in the budgets. They are implicitly included to force the analyses state variables close to observations. After Roads *et al.* [3], all vertically integrated reanalysis water budget terms (kg/m^{2}s) are multiplied by 8.64×10^4 s/day to provide individual values in kg/m^{-2} day or mm/day after dividing by the density of water (1000 kg/m^3).

3. Datasets and Methodology

The Global Precipitation Climatological Centre (GPCC) has provided gauge-gridded precipitation data for the whole globe with resolutions of 0.5°, 1.0°, and 2.5° from 1900 to the present. In this study, we used GPCC-version 4 precipitation data with a resolution of 1° × 1° for the period 1950-2007 [4]. This data set was formed by gridding monthly precipitation anomalies using the complete set of gauge measurements held at the time in the GPCC station database. The global reanalysis data sets used in this study include the European Center for Medium-Range Weather Forecasts [5], 40-years reanalysis (ERA-40) [6], and the National Centers for Environmental Prediction-National Centre for Atmospheric Research (NCEP-

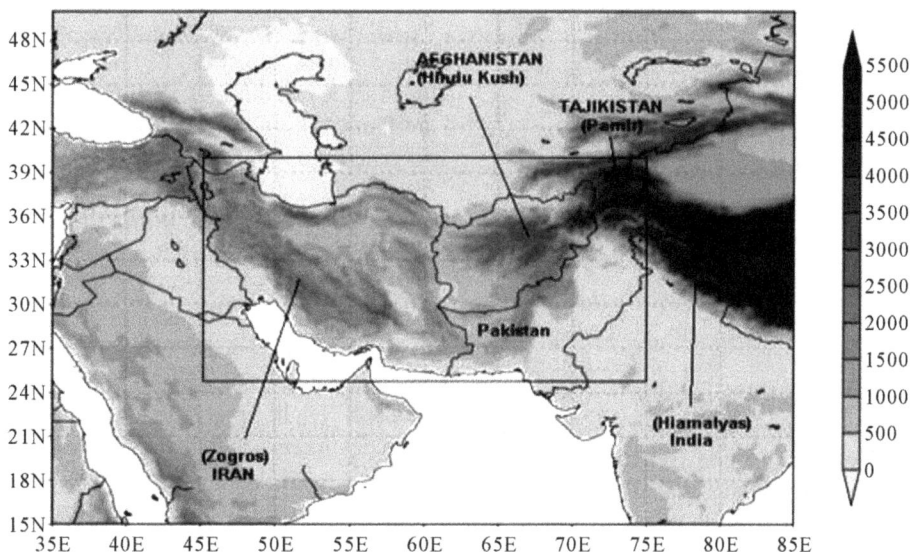

Figure 1. Model domain and topography (m) with a marked boundary with black line used in the study.

NCAR) reanalysis [7]. Data taken from each of these reanalysis data set NCEP for the period (1948-2007) and ERA-40 for (1958-2001) have been used to evaluate the individual terms in both surface and atmospheric water budget equations over a wide range of spatio-temporal scales. To compute the MFC, meridonal wind (u), zonal-wind (v) and specific humidity (q) at 2.5° resolution with six hourly data from surface to 200 hPa level were used. Evapotranspiration in the NCEP dataset are computed from surface latent heat flux (4x/day) by using,

$E = L/\lambda\rho_m$, where L is the latent heat flux at the surface in W·m^{-2}, λ is latent heat of evaporation (2270 kJ·kg^{-1}) and ρ_m is density of water (1000 kg·m^{-3}). Then evapotranspiration was multiplied by 8.64×10^4 s·day^{-1} to convert into mm day^{-1}. Other surface variables such as runoff, precipitation and surface pressure (six hourly, 2.5° resolution in ERA-40 and 1.875° resolution in NCEP) were used for the study. Areal average budgets were computed from land areas only.

Principal Component analysis is used for data reduction and can identify different groups with similar characteristics via variance correlation. There are two main techniques used in applications of factor analysis, 1) to reduce the number of variables; and 2) to detect structure in the relationships between variables. The factor analysis is applied as a data reduction or structure detection method (the term *factor analysis* was first introduced by [8]. The same technique is used on time series of precipitation data of each grid point and forms the different groups of the region. Each group has same characteristics of precipitation data.

4. Analysis, Comparison and Discussion

4.1. Water Budget Comparison-Yearly

Figure 2 illustrates a comparison of the annual averages of water budget terms, Evapotranspiration (E), Runoff (R), Precipitation (P) and Moisture convergence (MC). Inter-comparison of water budget terms in the ERA-40 reanalysis suggests that the region is behaving as a source of moisture during 1958-2001 because no moisture is transported in ($MC < 0$) and precipitation occurs due to local recycling (**Figure 2**).

Generally on land it could happen if most of the area in the region is covered with a water surface that causes more evaporation but in this study only land areas have been used. In case of NCEP, $MC > 0$, implies that moisture is transported into the region because of precipitation. Higher average values of MC indicate that regional precipitation is dependent on moisture advection into the region. The NCEP analysis shows that water is transported in ($MC > 0$) whereas ERA-40 shows water is transported out from the region. Comparisons of water budget terms of NCEP, ERA-40 and GPCC version 4 (for

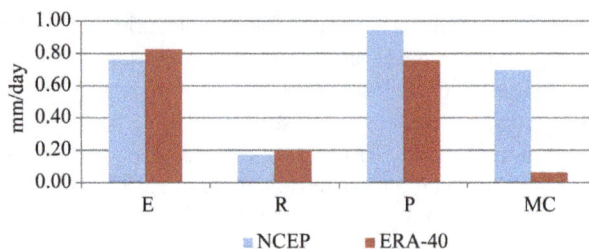

Figure 2. Comparison of annual averages (1948-2007 for NCEP and 1958-2001 for ERA-40) of Evapotranspiration (E), runoff (R), precipitation (P) and Moisture flux convergence (MC) of ERA-40 (dark red) and NCEP (light blue).

precipitation) are given in **Figures 3(a)-(d)**.

The patterns of runoff in both reanalysis datasets are almost the same except during the period of 1965-1969. Despite the use of sophisticated surface modules in NCEP and ERA-40 models, runoff processes are crudely represented [9]. ERA-40 records maximum peak in 1969 whereas the NCEP maximum is in 1966. The former may be associated with continuous persistence of the same amount of precipitation for the previous three consecutive years and the latter was extremely wet year throughout the study period.

ERA-40 showed two phases of moisture flux, divergence during (1958-1976) and convergence during (1977-2001). This can be confirmed with the CSWA precipitation during their respective phases; moisture converges with less precipitation and diverges with more precipitation. Moisture flux convergence drops after 1998 with a start to the drought period in the region (1998-2001). In the NCEP analysis [10], two different averages of moisture flux convergence can be observed, one from 1948-1966 with a higher value and other during 1967-2006 with a lower value.

All the time ERA-40 computes more evaporation than NCEP reanalysis except during 1963-1967. During this period, more moisture had advected in whereas less moisture is lost though runoff from the region. Two peaks are found during 1972 and 1976 with high moisture convergence and runoff from previous years. It implies that a rise in temperature increases the water in rivers because of more water melted from snow.

In panel (d) of **Figure 3**, a comparison of precipitation of ERA-40, NCEP and GPCC (v4) has been given. All data sets show similar values from 1978 onwards but significant differences before 1968. During 1963-1967, precipitation predicted by NCEP is higher than other two data sets. Higher precipitation relates to higher moisture flux convergence and lower evaporation.

4.2. Water Budget Comparison-Monthly

Comparison of monthly averages over 60 years for NCEP and 44 years for ERA-40 of runoff, MC and

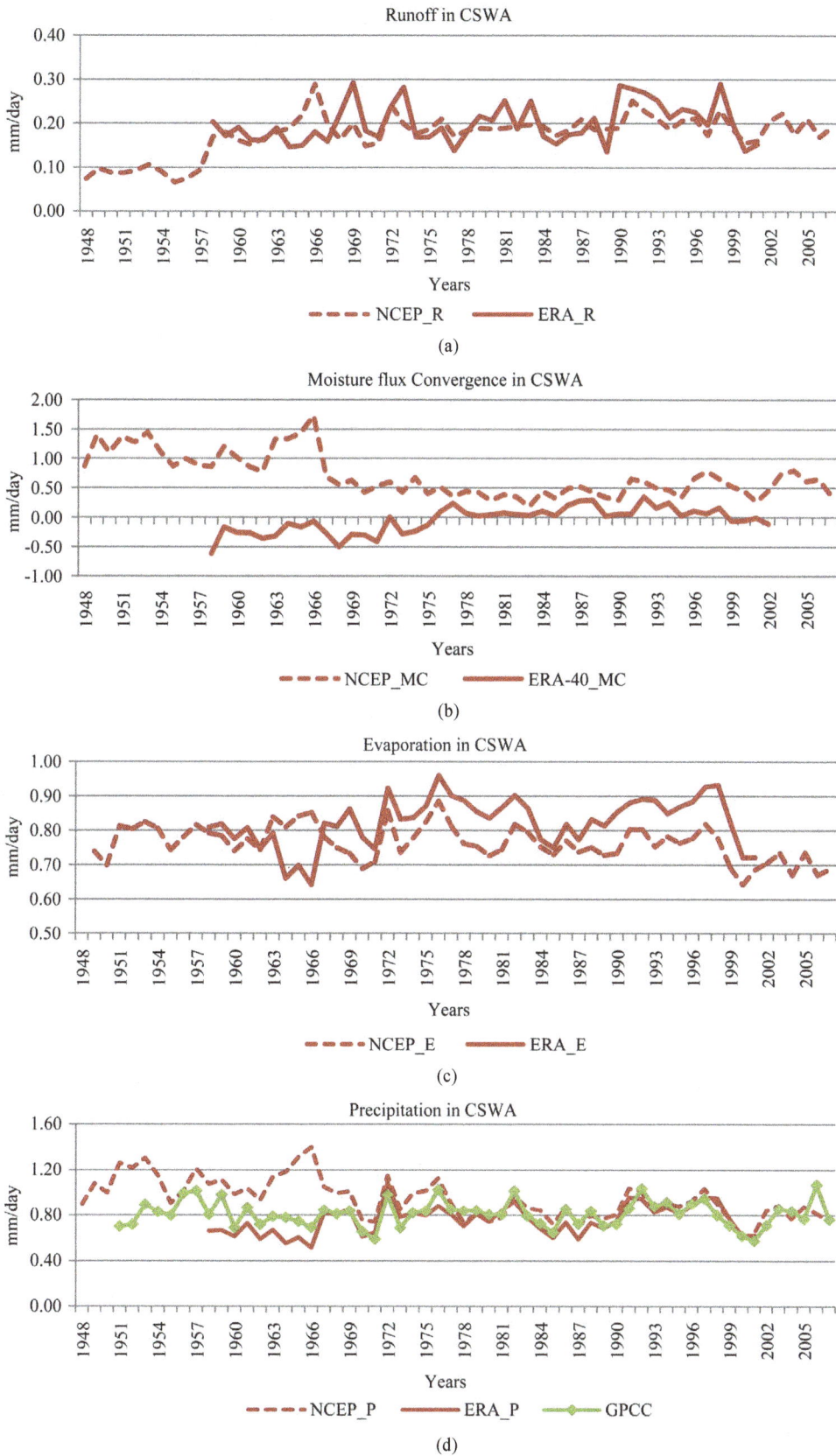

Figure 3. Comparison of annual runoff, evaporation, moisture flux convergence and precipitation from CSWA of NCEP (dotted) ERA-40 (solid) and GPCC (solid with cross rectangle mark).

evaporation in CSWA are shown in **Figures 4 (a)-(c)** and in **Figure 4(d)** with GPCC of precipitation in addition. The computational details of moisture flux convergence are not mentioned here and can be found in Liu *et al.* [11]. Both data sets shows maximum values of runoff in March and minimum in October related with the amount of precipitation in the respective months in the region. In August runoff increase from previous months because of two factors, one is due to precipitation in the eastern portion of the region and the second is due to the melting of snow from glaciers which increases inflow in the rivers of the eastern side of the domain.

The moisture convergence occurs from November to April with a maximum in February. Divergence is from May to October with a maximum in June from the ERA-40 analysis. The convergence is related with weather systems approaching from the west in winter and divergence is related to continuous heating and less precipitation in the region. In February, more water is transported into the region from the west and in June, net water transport in to the region is negative due to heating. In July, moisture enters from southeast in the form of monsoon currents causes decreases in moisture flux divergence in the region. In August and September, with the offset of monsoon currents causes increases the divergence again in the region and persists until the onset of winter weather systems.

Figure 4(c) represents the comparison of average monthly evaporation of both data sets. In winter, the region is under the influence of humid air and lower surface temperatures causes less evaporation. In summer dry air and high surface temperatures in the region causes high evaporation. Two peaks of high evaporation can be seen in April and June. The former is due to discontinuous western disturbances caused convert moist air into dry and the latter is due to high heating of the region. In July moist air returns because of the eastern monsoon system and causes a decrease in evaporation. It is observed that in winter NCEP records higher values of runoff, *MC* and evaporation than ERA-40 whereas in summer ERA-40 computes higher values than NCEP (**Figures 4(a)-(c)**). These variations are related to the variations of Sea Level Pressure (SLP) in both data sets. It is noted that NCEP predict higher SLP during summer and lower during winter than ERA-40.

Comparison of the average monthly precipitation of CSWA is shown in **Figure 4(d)**. All the data sets have recorded more than 1 mm/day precipitation in January to April and December. In July and August, precipitation increases because the eastern monsoon enters the region from the southeast to cover a portion of the region. The graph shows that winter behaves as a wet period and summer as a dry. The region receives maximum precipitation in March and minimum in October with an average

of 0.8 mm·day^{-1}. NCEP computes more precipitation than ERA-40 almost throughout the year and the difference of computed precipitation between both data sets increases in winter and decreases in summer suggesting that temperature plays an important role in the computation of all the water budget terms in both data sets.

4.3. Sub Division of CSWA

Principal component analysis technique is applied on time series of all grid points (variables) (8×17) of precipitation of CSWA and formed six factors (region).

Each group has similar precipitation characteristics based on NCEP reanalysis data sets from 1948-2007. In **Figure 5**, the precipitation at each grid point is divided into six factors and apparently, the first factor (A-1) is generally more highly correlated with the grid point's precipitation than the second factor (A-2) and so on. This is to be expected because, these factors are extracted successively and will account for less and less variance overall.

A-1 covers most of the western portion of the domain and some hilly portion of Pakistan. A-2 includes most of the central parts of the domain which is covered with rocks and uneven surfaces. A-3 includes the northeast portion with high mountain ranges and glaciers. A-4 includes south eastern portion over Pakistan with plains and forests. A-5 includes the central northern portion with a lower altitude rocky surface. A-6 splits into two portions one is the along extreme North West and second coves some of the boundary area of Pakistan and Afghanistan.

Monthly and yearly average precipitation values of each area [12] and the whole domain are given in **Figures 6** and **7**. Winter precipitation of all sub areas except A-4 is higher than summer, with maxima in March, because of strong western disturbance (WDs) inputs. Precipitation in A-4 is higher in summer (July to September) due to the strong Southeast monsoon. In winter, strong WDs enter from west and give maximum precipitation in the northern half of the domain which includes A-6, A-5 & A-3. The intensity of precipitation reduces as WDs move to the southeast. As a result, WDs enter in A-4 with weak intensity causing less precipitation in winter as compared to the other areas. Shifting of the Inter Tropical Convergence Zone (ITCZ) plays an important role in the prediction of the maximum precipitation month in the winter of each area. Maximum precipitation belts start from north of the Zagros Mountains in the west and turn towards the northeast after crossing the Hindu Kush in Afghanistan. This implies that maximum precipitation would occur in A-1 and A-3 when the ITCZ is in the extreme southern location as in March. As it starts shifting towards the no th in April, the maximum

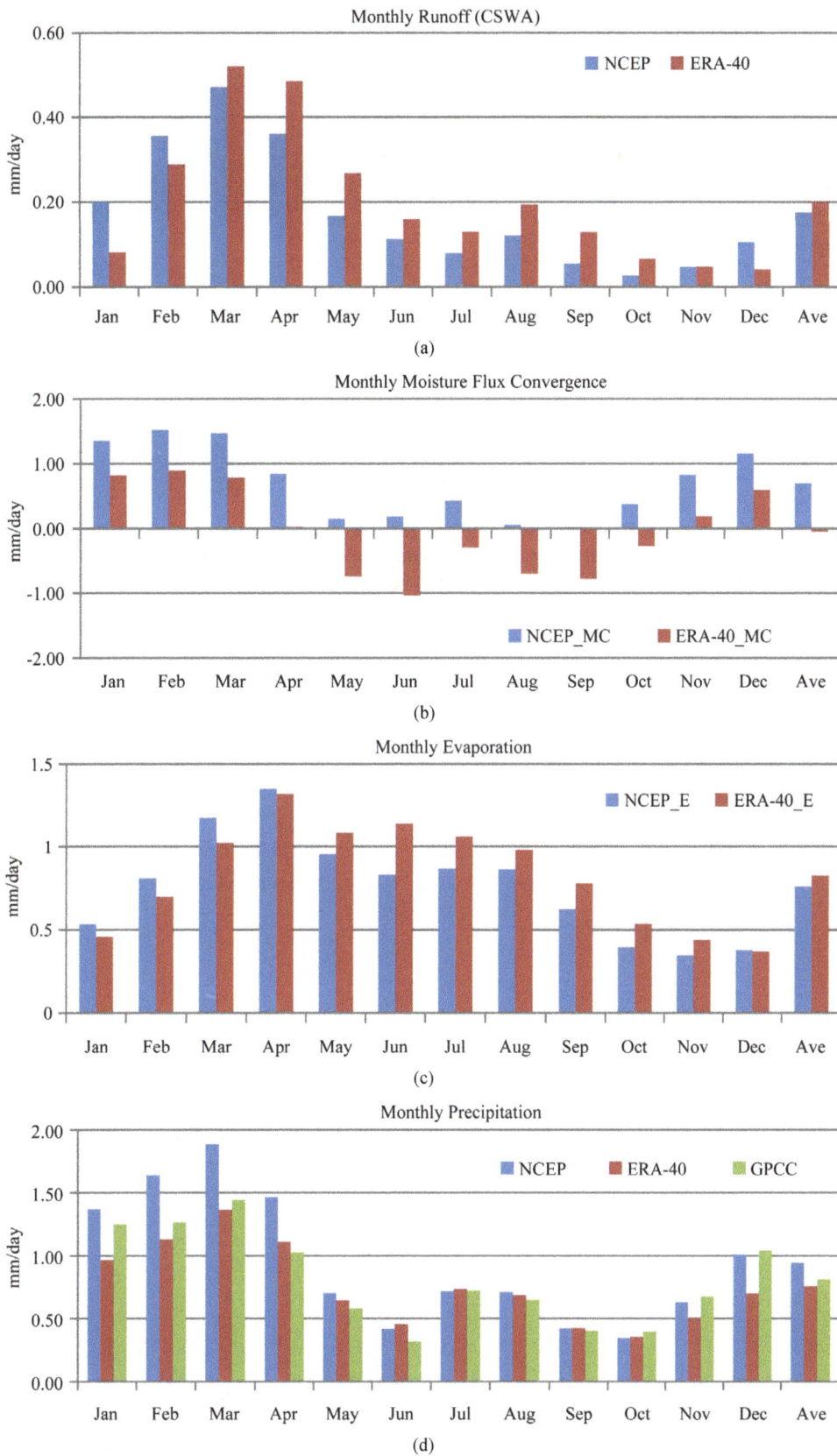

Figure 4. Comparison of monthly moisture budget terms in CSWA (a) Runoff; (b) Evaporation; (c) Moisture flux convergence with both reanalysis and (d) Precipitation of NCEP (blue), ERA-40 (red) and GPCC (green).

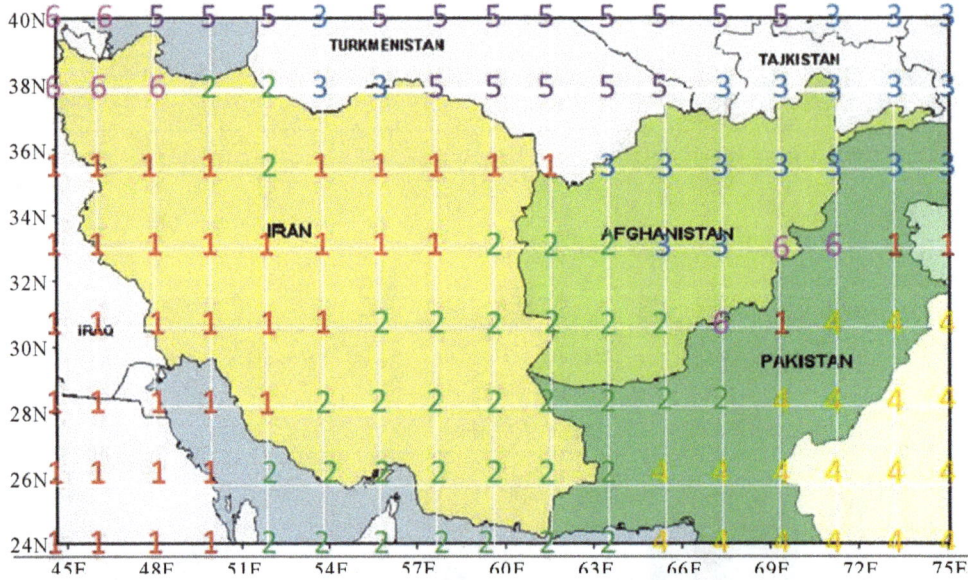

Figure 5. Sub division of CSWA, (A-1 to A-6) with different color & numbers from 1 to 6. The position of each number represents the grid points (1.875° × 1.875°) by longitude and latitude of the NCEP reanalysis-1 data sets [10]. Number 1 (red) shows sub region A-1, 2 (green) is A-2, 3 (blue) is A-3, 4 (yellow) is A-4, 5 (Dark purple) is A-5 and 6 (light purple) is A-6.

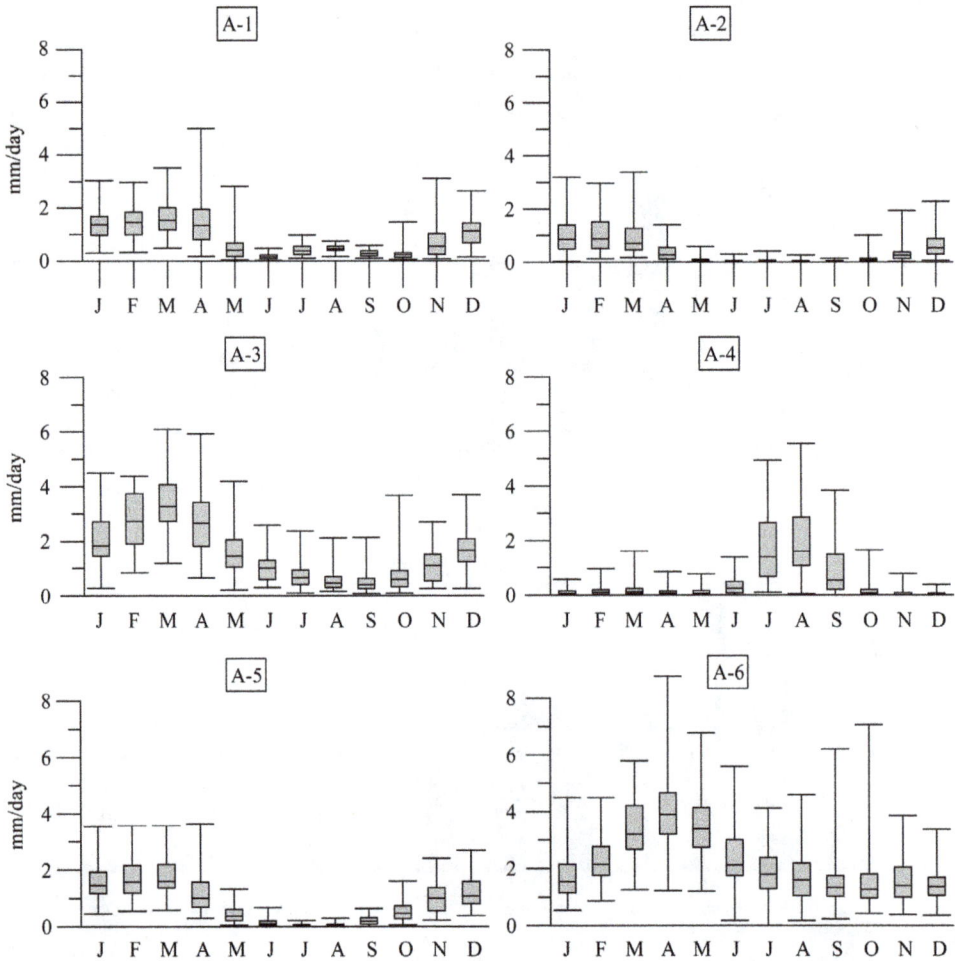

Figure 6. Monthly average precipitation with variance, in mm/day, of each (A-1 to A-6) sub region of CSWA. Box represents the range of the data, 25% as bottom, 75% as top and 50% with centre line.

precipitation area will also shift northwards towards A-6 and A-5.

A-1 shows slightly higher precipitation in July and August during summer, which is not due to monsoon current in the region. This relates to adjacent areas of extreme northwest of the domain. Precipitation is higher on higher latitudes and decreases on lower latitudes during summer. As the area is close to A-6 in the Northwest, we expect more precipitation and as the area goes towards south, it records less in summer (June to August).

In **Figure 7**, yearly average precipitation of each area is given. A-6 and A-3 cover the northwest and northeast portions of the domain and show higher yearly average precipitation than CSWA with maximum more than 2 mm/day in A-6. A-6 records higher precipitation because of intense WD and continuous increasing moisture from Caspian Sea. Higher precipitation in A-3 is the result of its topography. High mountains ranges and glaciers in A-3 produce anticylonic circulation which captures more moisture from water surfaces in the area and causes increased precipitation. Weather systems enter in to the region from west, give more variance in A-1 and from east, give more variance in A-4. The central portion of the domain A-2, compressed in between WDs and southeast monsoon weather systems has less average precipitation with less variance as in A-5. A-1 and A-5 show the same yearly average precipitation as CSWA in the range of 0.75 - 0.85 mm·day^{-1}. A-2 and A-4 show less precipitation in the region with a minimum of 0.39 mm·day^{-1} in A-2.

Figure 8, shows how the decadal precipitation has decreased in CSWA and its sub areas. All the areas except A-4 and A-5 show a decreasing trend whereas A-4 shows increasing and A-5 is steady. Precipitation increasing in A-4 means that monsoon precipitation increases in the region. In the last two decades, precipitation in A-5 also increases suggesting a higher intensity of WDs in the

extreme north of the region. Between 1989-1998 and 1999-2007, all the areas except A-6 showed an increasing trend and over all precipitation increases in the region. During 1999-2007, precipitation in the entire region except A-5 decreases due to drought (1999-2001).

4.4. Drought in CSWA

Figure 9 shows annual precipitation anomalies during (1997-2002) based on NCEP reanalysis (1948-2007). CSWA received less precipitation then normal and this pushed the area into severe drought. Drought starts in the region from 1999 and persisted for three years. During 1999, it started from west in A-1 & A-6 and moved towards the southeast. The region got severe drought during 2000 which affected the whole domain except for A-5 & A-6 (extreme north portion of the region) with maximum impart in A-4. During 2001, severity of drought has been decreased in A-4 because of more than average monsoon precipitation. The most affected areas were A-4 and A-2 during 2000 and 2001 respectively. Drought in the CSWA occurred as a result of weak WDs during this period. This implies that recent drought in

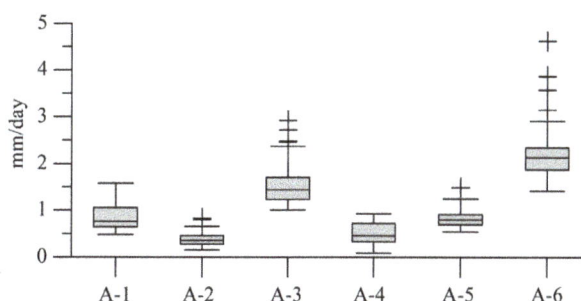

Figure 7. It represents yearly average precipitation of each area (A-1 to A-6) and + sign indicates outliers. Box represents the range of the data, 25% as bottom, 75% as top and 50% with centre line.

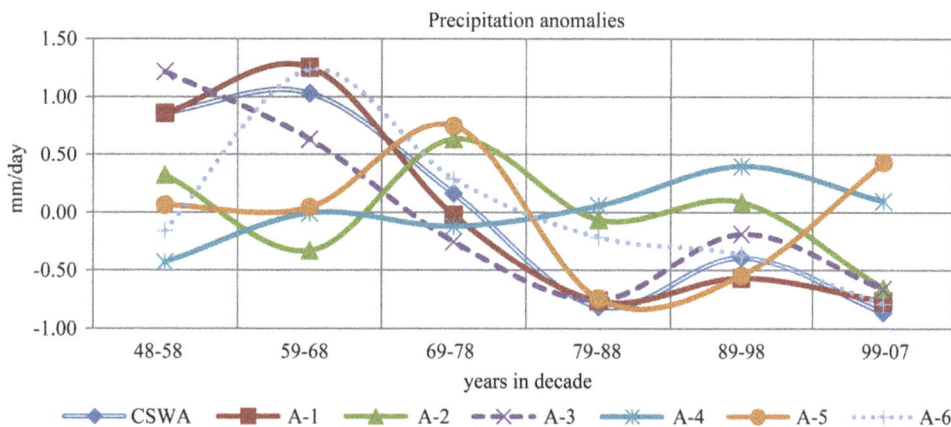

Figure 8. Decadal variation of precipitation anomalies of CSWA (line with square mark), A-1 (double solid line with rectangle mark), A-2 (line with triangle), A-3 (dashed line with cross), A-4 (line with star), A-5 (line with solid circle) and A-6 (dotted line with cut vertically) is given. The grid point on x axis represents one decade.

CSWA was a global phenomenon and occurred with a variation of the weather systems that approached the region.

in summer is 0.55 mm/day which is 30% of total yearly precipitation and the remaining 70% falls in winter.

4.5. Variation of Seasonal Precipitation

Two peaks can be seen in two phases, in late sixties (phase I) and late nineties (phase II). In phase-I, precipitation increases from 1966 up till 1969 and then start decrease up till mid eighties (**Figure 10(b)**). In next 6 - 7 years it persists at the same average and then in phase-II, it increases from the early nineties up till late nineties

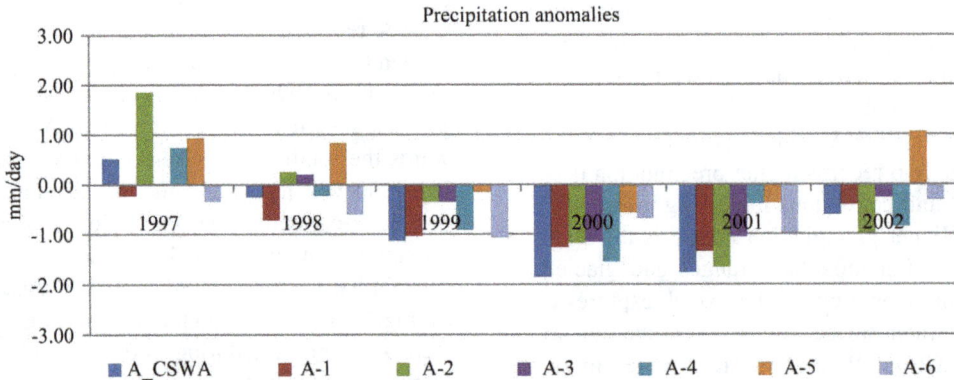

Seasonal precipitation: 1) winter from November to April and 2) summer from May to October, is illustrated in **Figure 10(a)**. Average seasonal precipitation of CSWA

Figure 9. Anomalies of CSWA and its sub areas precipitation during drought period (1997-2002) A_CSWA stands anomalies of CSWA and A_GP1 to A_GP6 is anomalies for each area from A-1 to A-6.

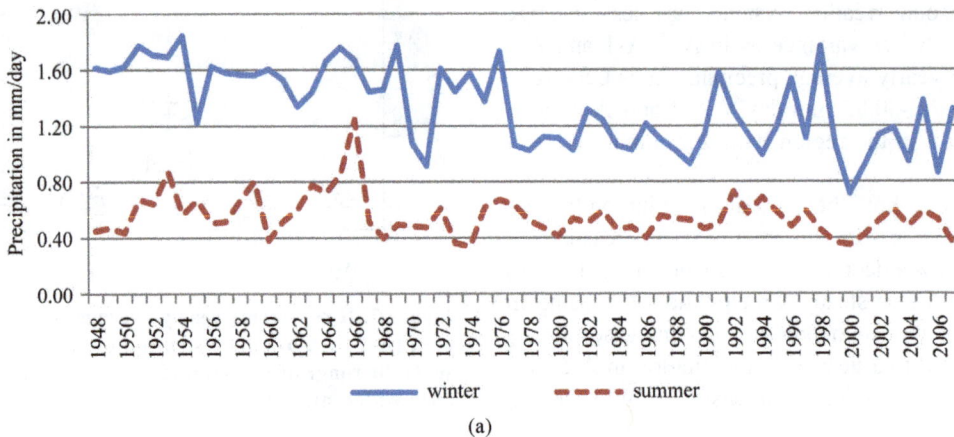

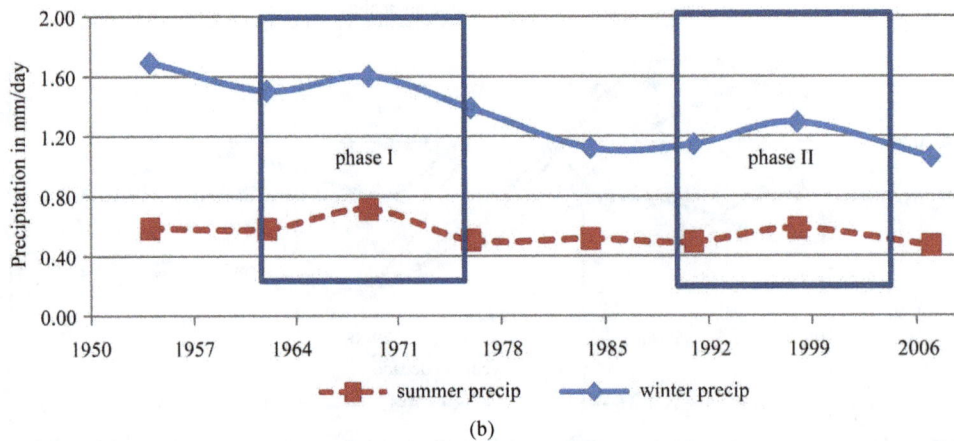

Figure 10. (a), (b) show yearly seasonal and 6 - 7 years average precipitation respectively. Summer includes May to Oct with dotted line and winter is from Nov to Apr with solid line.

and start decreases from 1998 up till 2007. It suggests that a decreasing trend in precipitation may be continued in future if it follows the same pattern. The larger decrease in average winter season precipitation than summer season indicates that precipitation changes in the region through western disturbance has more influence than monsoon effects.

5. Summary and Conclusions

The study has been conducted to compute the water budgets in Central Southwest Asia region and will be used to identify the drought areas. Water budget terms including evapotranspiration, runoff, moisture convergence and precipitation over CSWA have been summarized and examined in a comprehensive manner.

Precipitation data over CSWA of all three data sets (NCEP, ERA-40 and GPCC) gave similar results after 1978 whereas from 1948 to 1977, NCEP computes more precipitation than other data sets. During this period (1958-2001) ERA-40 computes more evaporation than precipitation in the area which in general is not realistic. The NCEP data set computes more precipitation than evaporation and gave confidence that it would be a better option to compute water budget terms from NCEP, especially for CSWA area. Two weather systems frequently approach the region, one from the west in winter and other is from the east in form of southeast monsoon in summer.

Maximum runoff occurs in April because of winter precipitation and one peak is observed in August because of snow melting in the Northern areas of the region. In winter, moisture converges with maximum magnitude in February and in summer moisture diverges in the region with a maximum in June because of high temperatures. Evaporation has a maximum in April due to winter precipitation and in June it increases again from May because of the rise in temperature. All the data sets record maximum precipitation in March during winter and in July during summer. It is noted that meteorological drought occurred in the areas of CSWA which are under the influence of WDs as well as the southwest monsoon and implies that it is a global phenomenon and not a local one. Weak approaching weather system decreased precipitation in the region.

Precipitation occurs more during winter on the western side and more during summer on the southeastern side of the region. The areas towards the extreme northwest had maximum precipitation throughout the year. Precipitation

showed a decreasing trend in decadal anomalies in all areas except A-4. The increasing precipitation trend in A-4 indicates that monsoon precipitation in the region is increasing with time. Furthermore, It is concluded that winter precipitation is decreasing continuously whereas summer is receiving essentially same for the whole study period.

REFERENCES

[1] Köppen-Geiger, "Climate zones," Centre for International Development at Hervard University, 1999.

[2] D. Martyn, "Climates of the World," Elsevier, Amsterdam, 1992, 436 p.

[3] J. Roads, et al., "CSE Water and Energy Budgets in the NCEP-DOE Reanalysis II," Journal of Hydrometeorology, Vol. 3, No. 3, 2002, pp. 227-248.

[4] T. Dinku, et al., "Comparison of Global Gridded Precipitation Products over a Mountainous Region of Africa," International Journal of Climatology, Vol. 28, No. 12, 2008, pp. 1627-1638.

[5] ECMWF, ECMWF Reanalysis Data. http://data-portal.ecmwf.int/data/d/era40_daily/

[6] S. M. Uppala, et al., "The ERA-40 Re-Analysis," Quarterly Journal of the Royal Meteorological Society, Vol. 131, No. 612, 2005, pp. 2961-3012.

[7] R. Kistler, et al., "The NCEP-NCAR 50-Year Reanalysis: Monthly Means CD-ROM and Documentation," Bulletin of the American Meteorological Society, Vol. 82, No. 2, 2001, pp. 247-267.

[8] L. L. Thurstone, "Multiple Factor Analysis," Psychological Review, Vol. 38, No. 5, 1931, pp. 406-427.

[9] K. K. Szeto, et al., "The MAGS Water and Energy Budget Study," Journal of Hydrometeorology, Vol. 9, No. 1, 2008, pp. 96-115.

[10] NCEP, NECP Reanalysis Data. http://www.cdc.noaa.gov/cdc/reanalysis/reanalysis.shtml

[11] J. L. Liu, et al., "Characteristics of the Water Vapour Transport over the Mackenzie River Basin during the 1994/95 Water Year," Atmosphere-Ocean, Vol. 40, No. 2, 2002, pp. 101-111.

[12] N. Panigrahy, et al., "Algorithms for Computerized Estimation of Thiessen Weights," Journal of Computing in Civil Engineering, Vol. 23, No. 4, 2009, pp. 239-247.

12

Evaluation of Spatial-Temporal Variability of Drought Events in Iran Using Palmer Drought Severity Index and Its Principal Factors (through 1951-2005)

Mojtaba Zoljoodi[*], Ali Didevarasl[#]

Atmospheric Sciences and Meteorological Research Centre (ASMERC), Tehran, Iran

ABSTRACT

Intensity and variability of droughts are considered in Iran during the period 1951 to 2005. Four variables are considered: the Palmer Drought Severity Index (PDSI), the soil moisture, the temperature and the precipitation (products used for the analysis are downloaded from the NCAR website). Link with the climatic index La Nina is also considered (NOAA downloadable products is used). The analysis is based on basic statistical approaches (correlation, linear regressions and Principal Component Analysis). The analysis shows that PDSI is highly correlated to the soil moisture and poorly correlated to the other variables—although the temperature in the warm season shows high correlation to the PDSI and that a severe drought was experienced during 1999-2002 in the country.

Keywords: Intensity and Variability of Droughts; Palmer Drought Severity Index (PDSI); Basic Statistical Approaches; La Nina; Iran

1. Introduction

Drought incidences, regardless their severity, have became more common in recent years in parallel with global climate changes. Drought is a gradual phenomenon, slowly taking hold of an area and tightening its grip with time. Sometimes, in severe cases, drought can last for many years and can have devastating effects on the socioeconomic, agricultural, and environmental conditions that may result from one or more of the water-scarcity factors by insufficient precipitation, high evapotranspiration, and over-exploitation of water resources [1-3].

Severe drought over this spell of three years (1998-2001), in combination with the effects of protracted socio-political disruption, has led to widespread faming affecting over 60 million people in central and southwest Asia [4].

Regarding physical geography, Iran has arid and semi-arid climates mostly characterized by low rainfall and high potential evapotranspiration [5]. The annual precipitation varies from about 1800 mm over the west-

ern Caspian Sea coast and western highlands to less than 50 mm over the uninhabitable eastern and central deserts.

The average annual precipitation over the country is estimated to be around 250 mm, occurring mostly from October to March. Annual precipitation is lower in the eastern half of Iran compared with the western half. Drought events and the rainfall shortage result in many natural difficulties, and characterize the climatic behaviour throughout this country. Drought annually hits most Iranian provinces. This was particularly the case during the recent spell of 1999-2002 which was the worst drought event since 1950 to the present [6]. Drought, naturally, is a recurring phenomenon whose duration and intensity are unpredictable. Drought can occur in any place with precipitation. Keep in mind that, dry and semi-dry places, or places with little precipitation (climatically dry-lands) are not considered as places with permanent drought. Droughts occur when the needed water for a site is basically less than a specific amount. Droughts are long-term hydrological events affecting vast regions and causing significant non-structural damage. Droughts are the costliest natural disaster in the world and affect more people than any other natural disaster [3]. The Middle East is a region of

[*]Associate professor and chairman of ASMERC.

[#]Corresponding author: Ali Didevarasl (Senior expert of ASMERC), ASMERC, Islamic republic of Iran meteorological organization, Tehran, 1497716385, Iran. E-mail: ali_didehvar714@yahoo.com.

extremes. It is almost one of the driest and most water scarce areas of the world [7].

Iran, located in the south-western part of Asia and the Middle East receives one-third of the world's average precipitation [8].

In order to mitigate the destructive drought impacts on local ecosystems, economy and society, it is necessary to generate various studies. Prior to any research on drought impact assessment, this natural disaster ought to be defined in detail.

Generally, droughts can be classified into agricultural, hydrological or meteorological in which avoiding meteorological drought is impossible; however, they can be predicted and monitored to alleviate their adverse impacts [9-13]. To quantify drought and monitor its development, many drought indices have been developed and applied [14-19]. A large number of drought indices have been suggested to date, including Palmer Drought Severity Index [12], Crop Moisture index [20], Agro-hydro Potential [21], Surface Water Supply Index [22], vegetative drought index of Normalized Difference Vegetations Index [23], Standardized Precipitation Index [24], Deciles [25], and multiple indices of low river flow [26].

Among them, the Palmer Drought Severity Index (PDSI) is the most prominent index of meteorological drought used in the United States for drought monitoring and research [15]. Besides PDSI's routine use for monitoring droughts in the United States, the PDSI has been used to study drought climatology and variability in the United States [27,28], Europe [29,30], Africa [31], Brazil [32], and other areas. The PDSI was also used in tree ring-based reconstructions of droughts in the United States [33-35]. Most of these studies are regional and focus on a particular location or nation. One exception is Dai *et al.* (1998) who calculated the PDSI for global land areas for 1900-1995 and analyzed the influence of El Niño-Southern Oscillation (ENSO) on dry and wet areas around the globe [36]. This study updates the global PDSI dataset of Dai *et al.* (1998), provides a detailed evaluation of the PDSI against available soil moisture and stream flow data, examines the trends and leading modes of variability in the twentieth-century PDSI fields, and investigates the impact of surface warming in the latter half of the twentieth century on global drought and wet areas [36].

Since Iran is located in an area of arid and semi-arid climates and is frequently affected by droughts, a great deal of research on drought monitoring and analysis has already been carried out. For example: Rahimzadeh P. *et al.* in 2008 [37], developed a research study "Using AVHRR-based vegetation indices for drought monitoring in the Northwest of Iran", the results indicated that NOAA-AVHRR derived NDVI well reflects precipitation fluctuations in the study area, promising a possibility for the early drought awareness necessary for drought risk

management. Raziei T. *et al.* in 2008 studied [38] "A precipitation-based regionalization for Western Iran and regional drought variability"; Results show that the northern and southern regions of western Iran are characterized by different climatic variability. Shiau J. T. and Modarres R. in 2009 [39], have studied "Copula-based drought severity-duration-frequency analysis in Iran"; this research implies that the drought severity in humid regions might be more severe if high rainfall fluctuations exist in that region.

Morid S. *et al.* in 2006 [40], developed a research study "Comparison of seven meteorological indices for drought monitoring in Iran", in which they found the SPI and EDI were able to consistently detect the onset of drought, as well as its spatial and temporal variation, and may be recommended for operational drought monitoring in the Tehran province. However, the EDI was found to be more responsive to the emerging drought and performed better.

Rahimzadeh F. *et al.* in 2008 [41], studied "Variability of extreme temperature and precipitation in Iran during recent decades", and observed a negative trend for about two-thirds of the country for annual total wet days precipitation.

2. Study Area

Iran is one of the large semi-arid countries of the world with an area of 1,648,000 km^2, respectively with elevations ranging from –28 m (Caspian sea) to 5671 m (Damavand), and a mean rainfall of 250 mm yearly. Iran is located in the southwest of Asia and it borders the Gulf of Oman, the Persian Gulf (in the south), and the Caspian Sea (in the north) with a geographical position of 25°N - 40°N and 44°E - 64°E. The topography of the country features two main mountain chains: the Alborz Mountains (from northwest towards northeast of the country) and the Zagros Mountains, a series of parallel ridges interspersed with plains that bisect the country from northwest to southeast. The center of Iran consists of several closed basins that collectively are referred to as the Central Plateau. The eastern part of the plateau is covered by two salt deserts, the Dasht-e Kavir (Great Salt Desert) and the Dasht-e Lut (**Figure 1**).

3. Data

3.1. Palmer Drought Severity Index

Palmer (1965) developed a soil moisture algorithm, which uses precipitation, temperature data and local Available Water Content (AWC) of the soil. AWC is effectively a "model parameter", which has to be set at the start of calculations. Calculations result in an index (PDSI), which indicates standardized moisture conditions and

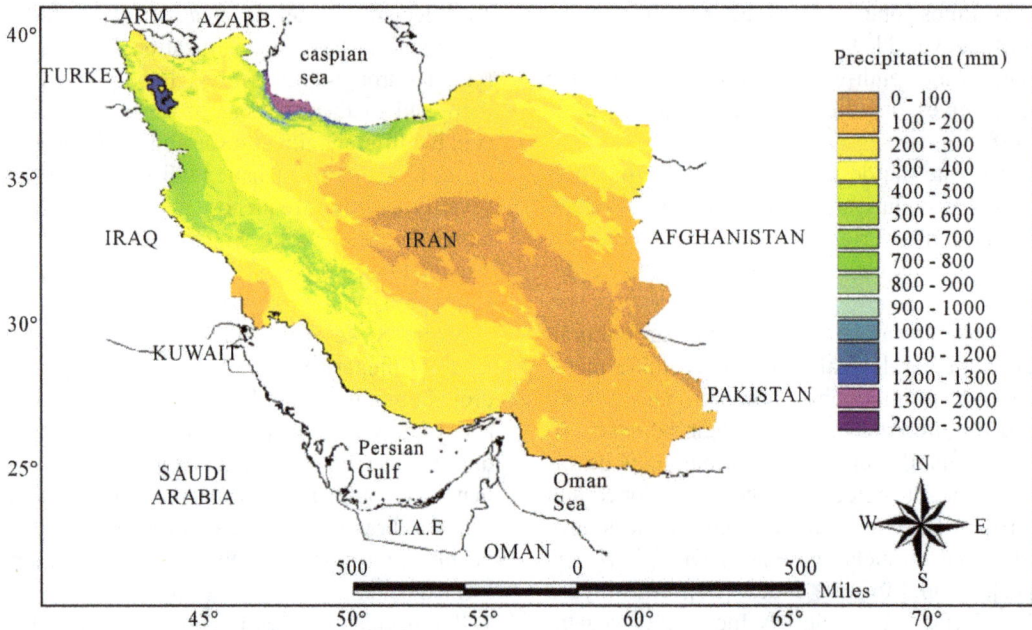

Figure 1. Geographic position and average annual precipitation of the study area (Iran).

allows comparisons to be made between locations and between months. PDSI varies roughly between –6.0 and +6.0. wetter conditions are indicated by positive values of PDSI, drier by its negative values. Thresholds for classification of different wetness are arbitrary. PDSI values between −2 and +2 would normally indicate normal conditions, although the sub-range of −1 to −2 could also be treated as mild drought. PDSI in values in the range of −2 to −3 are indicative of moderate drought, −3 to −4 points to severe drought and values less than −4 would be associated with extreme drought.

The computation of the PDSI begins with a climatic water balance using historic records of monthly precipitation and temperature. Soil moisture storage is considered by dividing the soil into two layers. The upper layer is assumed to contain 1 inch (25.4 mm) of available moisture at field capacity. The underlying layer has an available capacity that depends on the soil characteristics of the site. Palmer used an available water capacity (AWC) of 9 inches for central Iowa and 5 inches for western Kansas. The AWC value should be representative of the area soils in general. Moisture cannot be removed from the lower layer until the top layer is dry. Runoff (RO) is assumed to occur when both layers reach their combined moisture capacity (AWC).

Four potential values are computed:

1) Potential evapotranspiration (PE, e.g. by Hargreaves equation or other),

2) Potential recharge (PR)—the amount of moisture required to bring the soil to field capacity.

3) Potential loss (PL)—the amount of moisture that could be lost from the soil to evapotranspiration provided

precipitation during the period was zero.

4) Potential runoff (PRO)—the difference between the potential precipitation and the PR.

The climate coefficients are computed as a proportion between averages of actual versus potential values for each of 12 months. These climate coefficients are used to compute the amount of precipitation required for the Climatically Appropriate for Existing Conditions (CAFEC). The difference, d, between the actual (P) and CAFEC precipitation ($\hat{p}$) is an indicator of water deficiency for each month.

$$d = P - \hat{P} = P - \left(\alpha PE + \beta PR + \gamma PRO + \delta PL\right) \quad (1)$$

where $\alpha = \overline{ET}/\overline{PE}$, $\beta = \overline{R}/\overline{PR}$, $\gamma = \overline{RO}/\overline{PRO}$, and $\delta = \overline{L}/\overline{PL}$ for 12 months. The value of d is regarded as a moisture departure from normal because the CAFEC precipitation is an adjusted normal precipitation.

A Palmer Moisture Anomaly Index (PMAI), Z, is then defined as

$$Z = Kd \quad (2)$$

where K is a weighting factor. The value of K is determined from the climate record before the actual model calculation. Palmer suggested empirical relationships for K such that

$$K_i = \left(\frac{17.6}{\sum_{i=1}^{12} \overline{D}_i K_i'}\right) K_i' \quad (3)$$

where $\overline{D}_i$ is the average of the absolute values of d, and K_i' is dependent on the average water supply and demand, given by

$$K'_i = 1.5 \log_{10} \left[\left(\frac{\overline{PE} + \overline{R} + \overline{RO}}{\overline{P} + \overline{L}} + 2.8 \right) \overline{D} - 1 \right] + 0.5 \quad (4)$$

where PE is the potential evapotranspiration, R is the recharge, RO is the runoff, P is the precipitation, and L is the loss. The PDSI is now given by

$$PDSI_i = 0.897 \, PDSI_{i-1} + \frac{1}{3} Z_i \quad (5)$$

where the PDSI of the initial month in a dry or wet spell is equal to $\frac{1}{3} z_i$.

The basic spatial calculation of the PDSI values was based on the station's datasets, which consider the measure-points. The values obtained through these measure-points are used for calculation of the PDSI average values in each grid-point (a square of $2.5° \times 2.5°$). This calculation is done through the statistic analyses (filtrage, Krigeage, etc.) which eliminate essentially the under network irregularities

This research is developed based on the calculated PDSI data-set which is available on the NCAR website (http://iridl.ldeo.columbia.edu/SOURCES/.NCAR/.CGD/.CAS/.Indices/.PDSI2004/.PDSI). This data set has been calculated on a worldwide scale for a period extending more than 130 years (1870-2005). The PDSI dataset is arranged into geo-points with the dimension of $2.5° \times 2.5°$, and is useable in monthly series.

We have obtained the PDSI dataset during the study period (1951-2005) in a monthly series (55 years = 660 months) over an area more vast than Iran's precise territorial extent of 40° to 65° longitude and 25° to 40° latitude. As mentioned above the dataset is in geo-points of $2.5° \times 2.5°$ that cover the study area as a network (surface) with 60 grid-points with a dimension of 6 grid-points (15 degrees of latitude) to 10 grid-points (25 degrees of longitude).

3.2. Soil Moisture

The study of the surface hydrology invariably starts with the equation below

$$dw = dt = P \_ E \_ R \_ G$$

where w: soil moisture in a single column of depth 1.6 meter, mm; P: precipitation, mm/month; E: evaporation, mm/month; R: runoff, mm/month; G: loss to groundwater, mm/month.

Equation (1) is applied locally. All quantities are positive, and P is taken to be the input source, while E, R and G are the loss terms. H96 designed a water balance model; that is, E is calculated (adjusted Thornthwaite) via observed T, and R (surface and base runoff separately) and G are parameterized, such that we have 5 tunable parameters in the expressions for R and G. P is observed. The depth

of 1.6 meter came about as follows. Tuning the model (see H96) to runoff of several small river basins in eastern Oklahoma resulted in a maximum holding capacity of 760 mm of water. Along with a common porosity of 0.47 this implies a soil column of 1.6 meter. This depth seems reasonable for our goals since evaporation of moisture from deeper levels must be small [42].

The soil moisture dataset is provided through the NCAR website for a period during 1951-2005 in a monthly series (55 years = 660 months) over an area from 40° to 65° longitude and 25° to 40° latitude. This global dataset has a high spatial resolution in geo-points of $0.5° \times 0.5°$ that cover the study area as a network (surface) with 1500 grid-points with a dimension of 30 gridpoints (15 degrees of latitude) to 50 grid-points (25 degrees of longitude).

3.3. Temperature

The temperature anomaly dataset is provided from NOAA NCEP CPC CAMS: Climate Anomaly Monitoring System monthly gridded and station precipitation and temperature data. Spatial resolution of data is $2° \times 2°$; longitude and latitude are global; Time from Jan 1950 to present in monthly series. We used the dataset over an area from 40° to 65° longitude and 25° to 40° latitude. The details of data-production are presented in **Appendix 1.**

3.4. Precipitation

The "CAMS_OPI" (Climate Anomaly Monitoring System ("CAMS") and OLR Precipitation Index ("OPI") is a precipitation estimation technique which produces real-time monthly analyses of global precipitation. To do this, observations from raingauges ("CAMS" data) are merged with precipitation estimates from a satellite algorithm ("OPI"). The analyses are on a 2.5×2.5 degree latitude/longitude grid, are updated each month, and extend back to 1979. This data set is intended primarily for real-time monitoring. For research purposes, we refer users to the GPCP and CMAP products which are more quality-controlled and use both IR and microwave-based satellite estimates of precipitation.

The CAMS_OPI data files contain, for each month:
- raingauge/satellite merged analysis
- gauge-only precipitation analyses
- the number of gauge reports in each gridbox
- OPI-only precipitation estimates
- gauge/satellite merged analysis anomalies (1979-1995 base period)
- anomalies expressed as a percentage of the Gamma distribution

The merging technique is very similar to that described in Xie and Arkin (1997), and the CAMS_OPI technique has also been published recently [43]. Briefly, the merging methodology is a two-step process. First, the random error is reduced by linearly combining the satellite estimates using the maximum likelihood method, in which case the linear combination coefficients are inversely propostional to the square of the local random error of the individual data sources. Over global land areas the random error is defined for each time period and grid location by comparing the data source with the raingauge analysis over the surrounding area. Over oceans, the random error is defined by comparing the data sources with the raingauge observations over the Pacific atolls. Bias is reduced when the data sources are blended in the second step using the blending technique of Reynolds, 1988 [44]. Here the data output from Step 1 is used to define the "shape" of the precipitation field and the rain gauge data are used to constrain the amplitude.

3.5. ENSO Values

The ENSO values in monthly series have been provided from climate prediction centre website (www.cpc.ncep.noaa.gov) for the period of 1950-2005. Warm and cold episodes based on a threshold of $+/- = 0.5C$ for the Oceanic Niño Index (ONI) [3 month running mean of ERSST.v3b SST anomalies in the Niño 3.4 region (5oN-5oS, 120o-170oW)], based on centered 30-year base periods updated every 5 years. For historical purposes cold and warm episodes are defined when the threshold is met for a minimum of 5 consecutive over-lapping seasons.

4. Methodology

In order to analyze the aforementioned datasets (PDSI, precipitation, soil moisture and temperature) and also to process the spatial-temporal patterns, the Scilab software (http://www.inria.org) was used in the statistical analysis. The applied techniques are based on the statistical methods for analyzing the spatial and temporal variability of the drought events. All the statistical methods have been applied through the codes, which were added to the Scilab software for data processing. The Scilab-codes are written by V. Moron in CEREGE (Centre Européen de Recherche et d'Enseignement des Géosciences de l'Environnement).

For beginning, after the opening of the datasets through a matrix definition in Scilab the missing values (−99999) in all datasets (PDSI, precipitation, soil moisture and temperature datasets) have been removed before the practical analyses.

Practically, the analysis has begun through the yearly and monthly data series in order to develop the general spatial-temporal patterns of the original values of the PDSI dataset during the period of 1951-2005. This overview allows us to consider in general the variations of

drought severity over the period. And, to find a climatic teleconnection linkage with droughts in Iran the ENSO phases (cold and warm episodes of El Niño/La Niña-Southern Oscillation) as the episodes of large-scale climate variability have been considered.

In order to characterize the spatial-temporal variations of droughts, we have applied a statistical method using the Principal Component Analysis (PCA) (traditionally known as Empirical Orthogonal Functions (EOFs) in studies of the atmospheric sciences). PCA and the closely related principal factor analysis (PFA) of multivariate techniques have been widely used in meteorology and climatology [45,46]. The PCA is a standard tool in modern data analysis—in diverse fields from neuroscience to computer graphics—because it is a simple, non-parametric method for extracting relevant information from confusing data sets. With minimal effort PCA provides a roadmap for how to reduce a complex data set to a lower dimension to reveal the sometimes hidden, simplified structures that often underlie it [47]. In this research the PCA has been used for explaining the temporal variation of the PDSI values and its geographical distribution patterns.

In addition to the PCA, another performed analysis was the calculation of correlation functions for the PDSI data series in: 1) an inter-annual mode (between all months of year) and 2) a yearly correlation in two forms of 2 and 3 consecutive overlapping years. Correlation is a statistical technique that can show whether and how strongly pairs of variables are related. Here through correlation technique we measured the PDSI-resulted drought variability over the mentioned time scales.

Then, the precipitation, soil moisture and temperature datasets were processed through the common statistical approaches, such as correlation, anomaly and regression analyses to determine their consistency with the PDSI behavior over the monthly and yearly time scales. Anomaly analysis in the mentioned data sets allowed us to detect the changes of time-series, and then we compared the detected changes. Also through the regression analysis we tried to ascertain the causal effect of the variables upon the PDSI.

5. Results

5.1. Drought Monitoring during Study Period (1951-2005)

5.1.1. Temporal and Spatial Patterns of the PDSI
Regarding the temporal pattern of the PDSI for 55 years (1951-2005) that is in evidence there are some considerable departures on the PDSI monthly values. The average, maximum and minimum values in this long term temporal pattern of the PDSI dataset are respectively of the order: −0.72, 4.78, and −7.66, and the most extreme

Evaluation of Spatial-Temporal Variability of Drought Events in Iran Using Palmer Drought Severity Index and Its Principal
Factors (through 1951-2005)

109

drought spell lasted 4 years from 1999-2002. On the other hand the most considerable wet spell was in 1954-1957. Generally, in respect of the onset and end points of the PDSI values in this pattern, it is evident the drought severity is increasing, as the PDSI values have a tendency towards the negative over the study period (**Figure 2**). At the same time, this plot reveals four spells respectively 1951-1960 (average of PDSI: 0.46) as a wet spell, and 1978-1988 (average of PDSI: −0.43) that is considered a weak drought spell, while on the other hand 1960-1978 (average of PDSI: −0.6) as well as 1988-2005 (average of PDSI: −1.93) are considered two weak and mild drought spells. Also, as is evident, most of the study months during the period 1950-2005 demonstrate the negative PDSI ranks (droughts), so that about 67% of the study months have negative signals in contrast to the 33% of positive signals or wet months.

This PDSI plot clearly indicates a non-linear trend on the period 1951-1999, whereas a very different behavior seems to have taken place in the 1999-2005.

Also, for comparing the temporal PDSI variations with large-scale climate variability, the ENSO phases (cold and warm episodes of El Niño/La Niña-Southern Oscillation) have been considered during 1950-2005. Regarding the monthly variations of ENSO phases, some drought and wet spells of PDSI in Iran seem to correspond to the ENSO variations (**Figure 3**). For example, the cold episode of ENSO (La Nina) in 1999-2002 links to the extreme drought spell found via PDSI in Iran. Although the other drought and wet spells of PDSI respectively correspond with the La Nina and El Nino phases, we cannot ignore the time delay between their fluctuations and durations, which may be resulted from: 1) the nature of PDSI as a complicated drought severity index, and 2) the severity and duration of the droughts over the study area are related to a combination of the prolonged duration of the La Niña and the unusually warm SSTs in

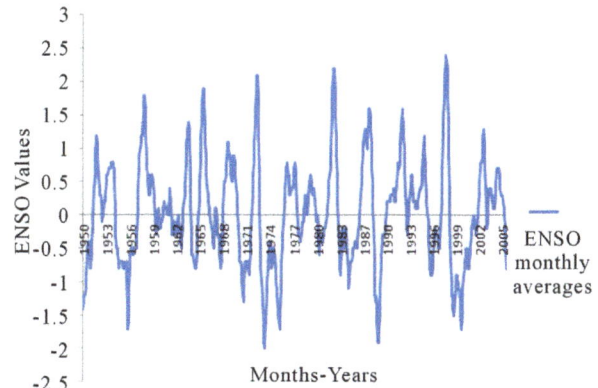

Figure 3. ENSO phases (cold and warm episodes of ENSO) based on 3 months running average during 1950-2005 (data source: climate prediction centre).

the west Pacific, which may enhance the regional dynamics of the warm pool [4], and so only the La Nina does not cause the long-term droughts in our study area.

The subset of ENSOs with a strong warm pool signal are associated with a vigorous extension of positive precipitation anomalies into the Indian ocean and negative anomalies over central and south-west of Asia, wherein Iran is placed. The similarity between this rainfall pattern and the drought period rainfall is striking [4].

The spatial pattern of the PDSI for 55 years (1951-2005) like the temporal one has been developed through the average values of the index.

Based on this spatial pattern (**Figure 4**) it is clear that the spatial variability of drought in Iran generally reveals the existence of three regions countrywide with respect to the intensity of drought: Northern Iran (north western and north eastern parts) which exhibits a high frequency of drought events with more intensity, Central Iran which experiences moderate drought conditions and southern Iran which has a low frequency of droughts.

Thus, during the study period (1951-2005) the northern parts of the country should reveal the precipitation deficit and positive signal of the temperature anomaly relatively more than the other areas. This condition apparently has been caused by global surface warming as a complementary cause to the precipitation deficit particularly after 1980, in certain global regions such as the Middle East [48].

Extreme drought period of 1999-2002; PDSI mean values during this period reveal an extreme condition of drought in both intensity and duration compared with the other drought spells found during 55 years of the study period. Socio-economic and environmental sections were seriously damaged due to this rainfall deficit, and obviously Iran was hit by the most intense drought event in the 3 year period 1999-2002.

For the determination of this intensive drought spell the yearly PDSI value of −4 is considered a threshold

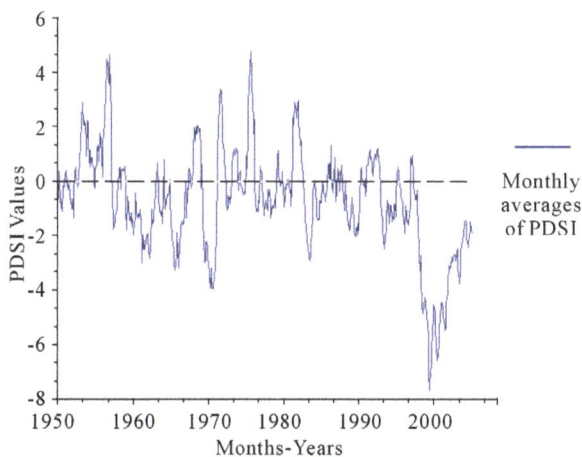

Figure 2. Temporal average of the PDSI mean values based on the monthly series during the study period (1951-2005).

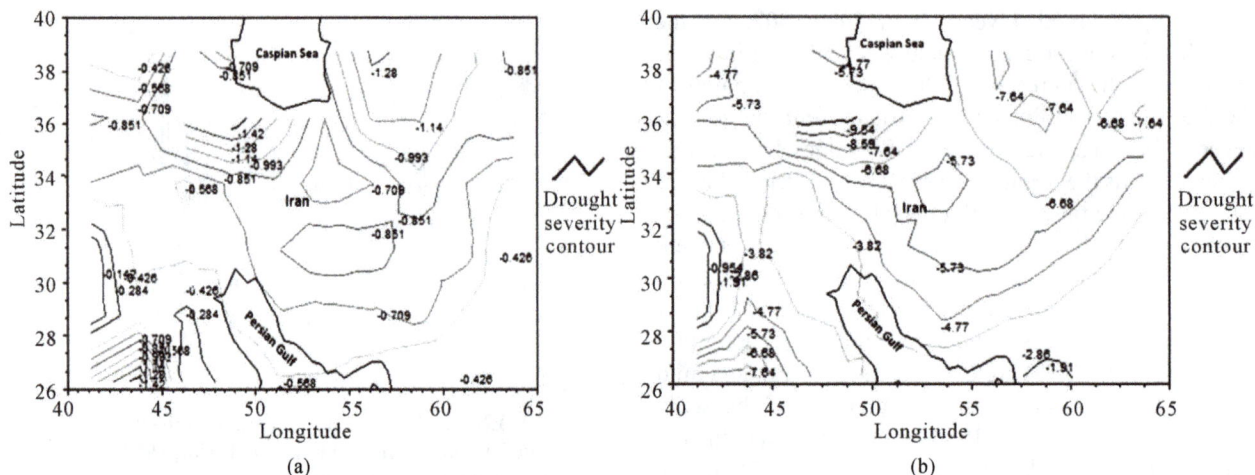

Figure 4. Spatial patterns of the PDSI during 1951-2005 (a) and 1999-2002 (b) in Iran.

point in this classification (extreme drought). Regarding the role of large-scale climatic variabilities in the periodic changes of precipitation and temperature regimes over the global regions, from 1988 to 2005 Iran was hit by progressive dry climatic conditions, which reached its peak in 1999-2002. This drought spell apparently corresponds to La Nino phase (cold episode of ENSO). The similarity between the enhanced warm pool-La Nina composite and the climate anomalies of 1998-2002, suggests that the prolonged, westward-concentrated La-Nina during this spell was one of the important factors in the central and southwest Asia drought [4]. Spatial variability of the drought intensity over this spell in the country (**Figure 4**) also shows that the northwestern and the northeastern regions respectively display the most intensive drought event with a severity around −8. So the severe drought conditions particularly on the northwestern regions would confirm the existence of a high correlation between ENSO teleconnection and the precipitation variations over the northwest of Iran especially in the cold season [49,50].

5.1.2. Monthly Patterns of PDSI

Following the processing of the index for the entire study period that has released the average values of the PDSI through spatial-temporal patterns, a monthly based spatial-temporal processing for the PDSI original dataset has been taken into consideration in order to explain the drought intensity variation over the months of year. In order to identify the monthly difference of the drought intensity in the course of 1951-2005, the mean values of the PDSI for each month is calculated (**Figure 5**). All months demonstrate negative values, although the months of the cold seasons release relatively low negative mean values (weak drought) and on the contrary the warm seasons include higher negative mean values (strong drought). So here we can see the temperature's effect on increasing

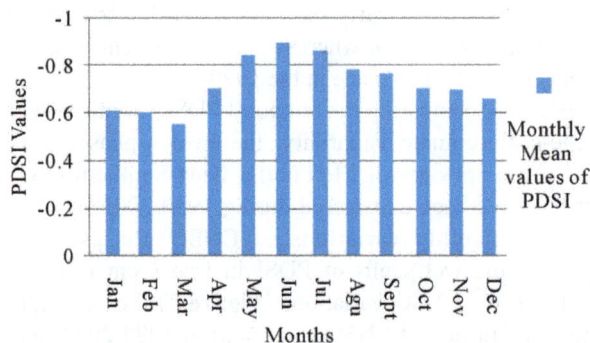

Figure 5. The average values of the PDSI for 12 months of year during 1951-2005.

drought intensity in the warm season, due to the raising of evaporation. The lowest average value of PDSI is in March (−0.55) and the highest in Jun (−0.89).

The geographical distribution of the PDSI values for each of the 12 months over Iran clearly shows 3 regions with intensive and frequent drought events in the Northern Regions, moderate droughts in the central Iran and weak droughts in the South (in particular the South-east). In fact drought intensity gradually decreases from the north of the country toward the central and southern parts. The monthly developed maps exhibit that the intensity contour of −1 covers some regions in central Iran from May to Oct (warm season), whereas this intensity contour mainly during the cold months of year is limited to the northwest and northeast (**Figure 6**). Thus regarding the spatial pattern of drought intensity over the country, it seems that the precipitation is negatively correlated with ENSO teleconnections (by La-Nina phase) over the northern parts of the country [51]. Furthermore, the western climatic systems produced especially by North Atlantic Oscillation may play an important role in precipitation variability (here rainfall deficit) over northwestern, western and northern Iran [52].

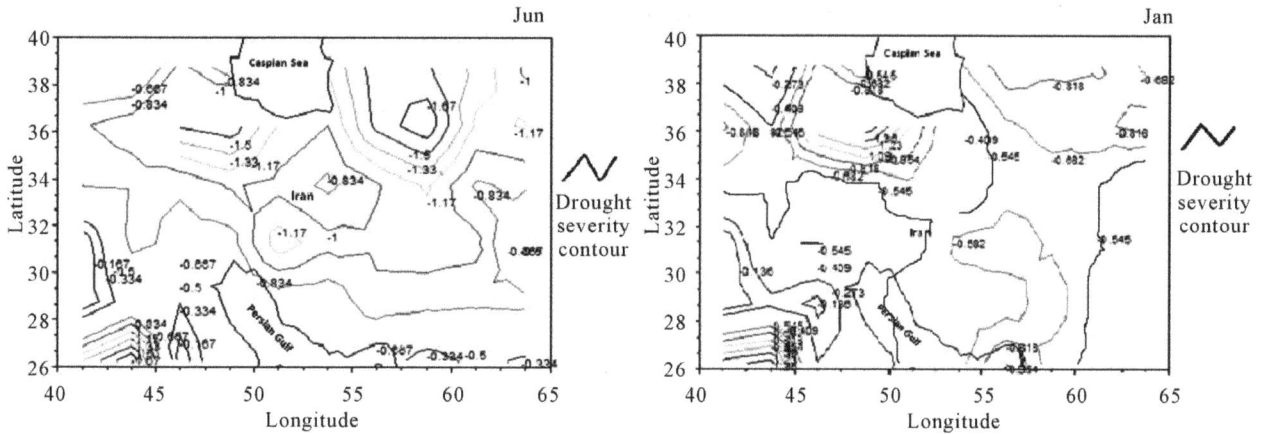

Figure 6. Examples of the spatial monthly patterns of the PDSI during 55 years for Jun (left) and Jan (right), the produced maps for each of the months of year (12 months) are available at Annex.

5.2. Principal Component Analysis (PCA)

In order to find out the independent axes or variances in the PDSI original dataset, the PCA for each of the months has been computed. The explained variances derived from this analysis results 57 independent variances. We have used only two initial variances in our analysis, since from the third variance up the rates found are below 10%. In fact the first two PCs will explain the large patterns of the PDSI, whiles the rest of the PCs are mostly related to the small or local patterns. Through this part of the work, the first (PC1) and second (PC2) modes explain respectively in average the variance rates of about 46.39% (PC1) and 11.45% (PC2) in the PDSI dataset. Thus, the correlation maps for each month have been developed based on these two variance rates which include the majority of variance percentages calculated. Also the cumulative variance of these two rates explains the values above 55% for all months (**Table 1**).

Regarding the geographical description of the correlation calculated through this analysis between temporal variations of the PDSI and its original data via the PC1 (first mode) in the course of 55 years, there is evidently a remarkable coherence in the most areas of the country over all months of the year. In fact this spatial pattern of the PCA as the leading mode explains the fact that the drought events during a given year often demonstrate a high spatial correlation over most Iranian regions especially in the central and eastern parts of the country, where it shows the homogenous geographic and meteorological status (desert plains with low rainfall as well as high temperature). But the southeastern regions with low ranks seem to have different climatic characters, for example; monsoon precipitation regime influences usually these regions.

The results derived through PC2 (second mode) as the residual of the first mode explain a dipolar spatial pattern through the geographical distribution of the correlation

calculated for each of the months. So it is evident that the southern and northern parts of the country normally demonstrate the high correlation around ±0.5, whereas toward central regions the correlation rates are going decrease continuously towards a belt around 33°N where the correlation values are about zero.

Thus as mentioned above, based on the leading mode, which is reckoned as a dominant portion from this principal component analysis; the drought events in Iran may follow a coherent spatial pattern (**Figure 7**). We give more explanations about this drought pattern in the discussion part.

5.3. Monthly and Yearly Correlation Functions in PDSI Values

For this analysis firstly a matrix of 12×12 is designed; this matrix reveals the relative correlations between all months of the year. Through this analysis the highest correlation-ranks usually are obtained between each month to itself, on the other hand the lowest one indicates a correlation between two months in cold and warm seasons (for example: 0.6 between Jan & Aug). The monthly correlation analysis confirms that during a given year, drought event may be persistent over the months, as this matrix gives on average a correlation-rank between all months of the year about 0.83 that is a considerably high monthly correlation.

In order to examine the above hypothesis, the moving correlation between each two consecutive months over a year has been computed. As this performance indicates the behavior of the monthly correlations during the seasons of year (**Figure 8**), the drought events in Iran show a high monthly coherence throughout a year (correlation ranks from 0.88 to 0.97). Also, we have performed a yearly based correlation analysis. So we selected a 20-year period from 1961 to 1981 as a sample period. This yearly based correlation has been performed in two modes:

Table 1. Monthly rates of the explained variances for PC1 & PC2 and their cumulative variance.

PCs \ Months	Jan	Feb	mar	Apr	may	Jun	Jul	Aug	Sep	Oct	Nov	Dec	Entire period
(PC1)	46.5	46.97	49.48	48.30	48.8	47.5	47.1	44.96	44.95	44.88	48.00	45.0	46.39
(PC2)	12.2	12.74	11.79	12.76	11.7	11.8	11.0	10.89	10.60	11.20	11.76	12.7	11.45
Cumulative variance	58.8	59.1	61.27	61.07	60.6	59.3	58.2	55.85	55.55	56.08	59.7	57.8	57.84

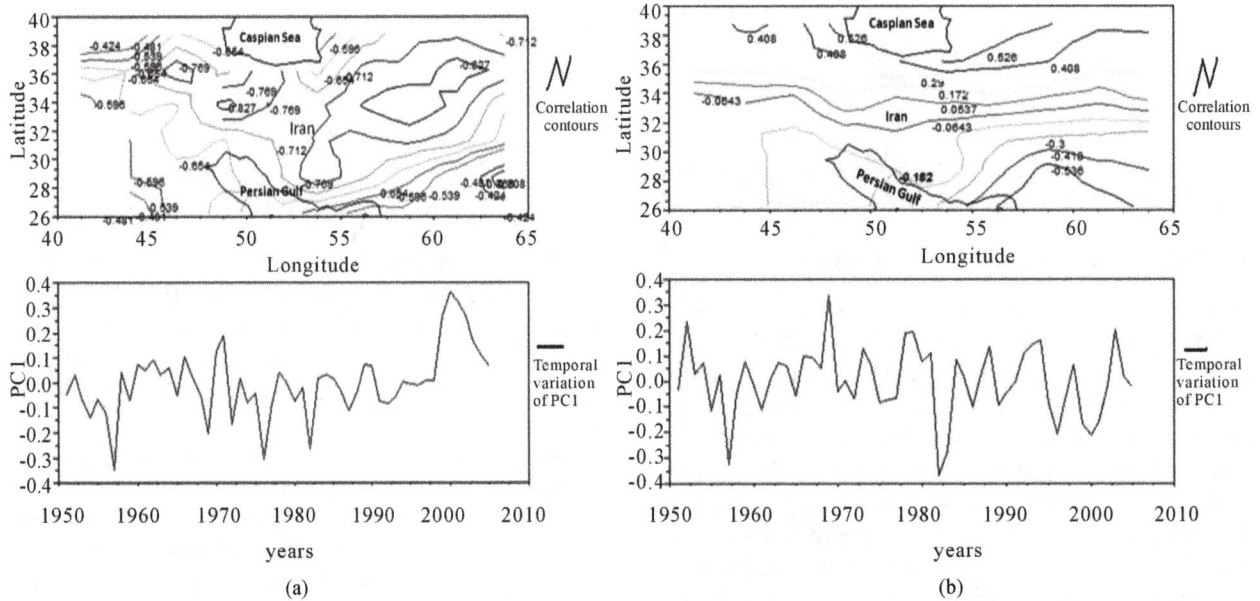

(a) (b)

Figure 7. The yearly average of the spatial-temporal PCA processed through the leading mode (a) and second mode (b) during the study period (1951-2005).

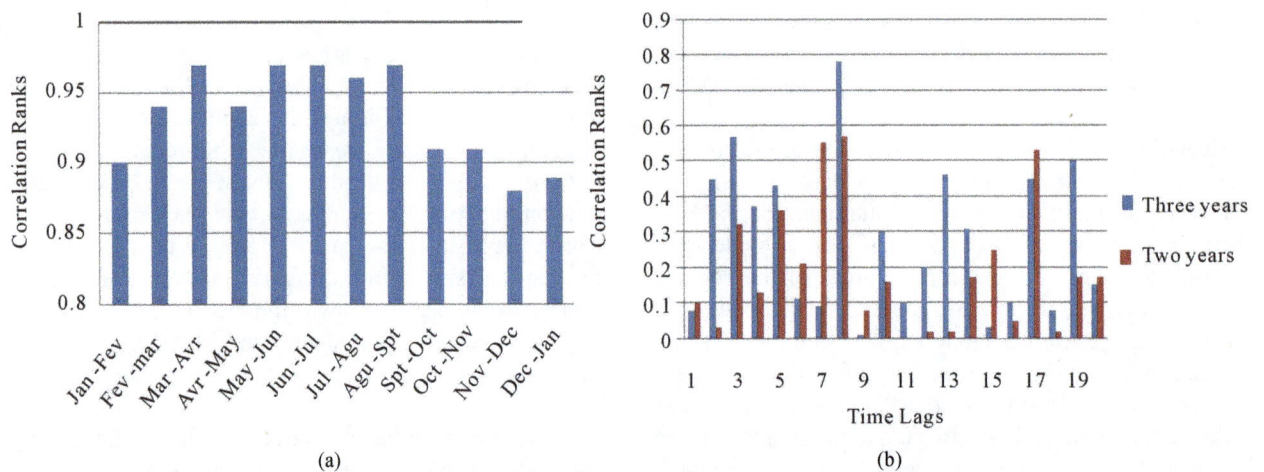

(a) (b)

Figure 8. The correlation functions of the PDSI values through PC1 (leading mode of PCA) between the consecutive months (a) as well as consecutive years in two modes of 3 years and 2 years (b).

the first one for 2 consecutive years and the second one for 3 consecutive years. In fact, the correlation successively between 1 year with the next one (2 year mode), and between 1 year with 2 next years (3 year mode) has been performed through the selected 20 years. Both these two

yearly correlation modes confirm that there is a low coherence between yearly variations of drought (PDSI values). So, the resulted correlation-ranks show on average 0.19 for 2-year mode and 0.27 for 3-year mode, we cannot expect to have a yearly or periodic high regulation for

Evaluation of Spatial-Temporal Variability of Drought Events in Iran Using Palmer Drought Severity Index and Its Principal Factors (through 1951-2005)

113

drought events in Iran.

5.4. Considering of the Precipitation, Temperature and Soil Moisture Datasets

The precipitation, soil moisture and temperature datasets in monthly series also were provided via the same web site of the PDSI dataset (http://iridl.ldeo.columbia.edu). Through this part of the work the contributions of the mentioned factors as the principal factors of the PDSI have been quantified in the PDSI spatial-temporal variability over the study period (1951-2005).

5.4.1. Correlation and Anomaly Analyses between PDSI with the Soil Moisture, Precipitation and Temperature

The correlation analysis has been preformed through the leading principals of the PDSI and the precipitation, temperature as well as soil moisture during 1951-2005. This three-aspect correlation matrix is designed in the form of the monthly moving average correlation (1, 3, 6 and 12 months moving average). The results basically explain that the PDSI variability is more compatible with the soil moisture variation than with precipitation and temperature variations. Also, we found that the correlation-ranks between both precipitation and temperature with PDSI are gradually growing if we increase the length of the monthly moving average (**Table 2**). Thus, this analysis indicates that the precipitation and temperature may affect the PDSI variability by a delay of a few months, although the effect of soil moisture in PDSI seems to be mostly direct. Why is this, the PDSI has a fairly long memory, reflecting the memory of soil moisture, so that temperature and precipitation for example in spring or summer can still affect autumn PDSI, thus this relationship between the PDSI and its principal meteoro-

logical factors result in a delay of few months in PDSI's response to their variations. Furthermore, water holding capacity of soils, and the depth in which available water content has been estimated play an effective role in this temporal lag between the PDSI and its metrological factors.

In addition, an inter-annual analysis in order to find out the correlation variations between PDSI and the three mentioned factors during the months of the year has been performed. With respect to the monthly correlation averages, it is evident that the monthly ranks mostly explain low correlation between the PDSI with temperature and especially with precipitation, although during the warm months PDSI and temperature are relatively more matched. But, the soil moisture and PDSI exhibit the high correlations over all months with the ranks above 0.88 (**Figure 9**). This performance explains that the PDSI values correlate closely to the soil moisture variability during all seasons of a year.

Also, a monthly anomaly analysis was performed separately for all three variables and PDSI, during a 22 year period (1979-2001). Then, the linear correlation function between the anomaly values of each variable and PDSI has been performed for all 12 months of the year (**Figure 10** shows examples of Jul, Nov, Feb, Aug). The highest R^2 rank is found in November between the anomaly values of PDSI and soil moisture with a rank of 0.63. Linear monthly correlations of the anomaly values of precipitation and temperature with PDSI are relatively low and irregular over all months of the year. However, the highest R^2 ranks for PDSI/precipitation and PDSI/temperature respectively are found in February (0.47) and in August (0.46).

The monthly linear correlations as well as monthly anomaly analyses on PDSI and its basic factors demonstrate that the variations of PDSI in a year (over the months of a year) would be closely compatible to the soil moisture changes, since, the out puts of PDSI are as the

Table 2. Correlation analysis between precipitation (P), temperature (T) and soil moisture with the PDSI through the entire study period (1951-2005).

Correlation time-scales / Variables	Soil moisture	Temperature	Precipitation
1-month correlation between PDSI and the variables	0.88	−0.27	0.14
3-month correlation between PDSI and the variables	0.86	−0.51	0.29
6-month correlation between PDSI and the variables	0.82	−0.59	0.38
12-month correlation between PDSI and the variables	0.72	−0.65	0.6

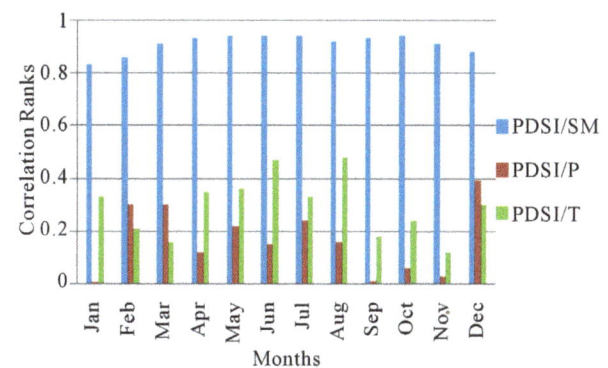

Figure 9. Three-aspect monthly correlation analysis between soil moisture (SM), precipitation (P), temperature (T) with the PDSI for each of the months of year during the study period (1951-2005).

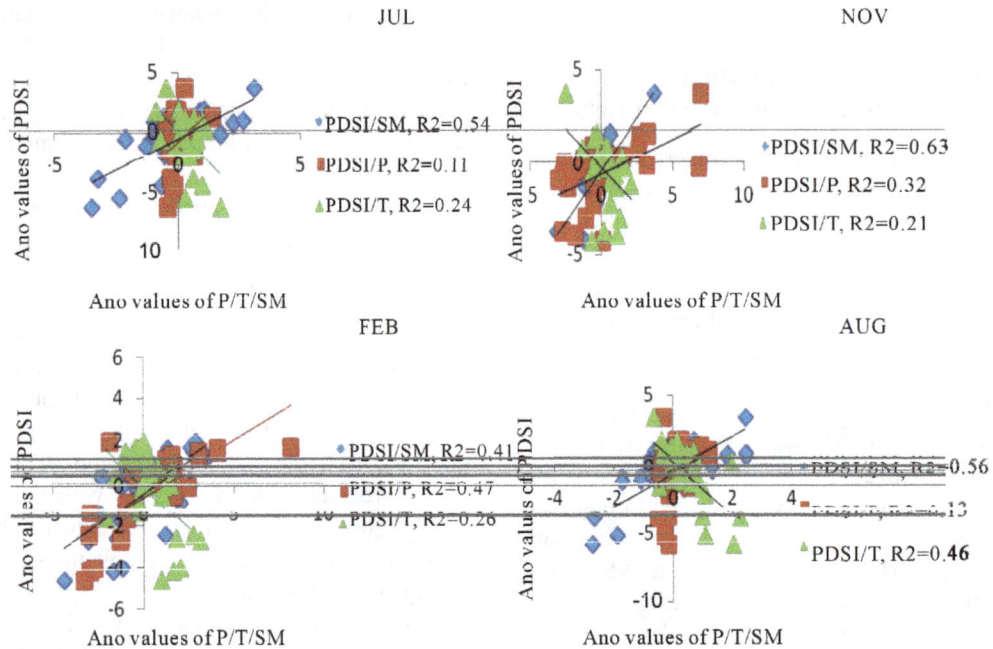

Figure 10. Monthly anomaly analysis for PDSI, soil moisture, precipitation and temperature. And the linear correlation between the monthly anomalies of PDSI and the anomalies of three mentioned variables for each of the 12 months of the year. Examples of Nov., Jul., Feb. and Aug.

soil moisture model.

5.4.2. Regression Analysis for the PDSI via Soil Moisture, Precipitation and Temperature

This regression analysis has considered separately the relations between the PDSI with the soil moisture, precipitation and temperature. So, regarding the coefficients of determination (r2) which are derived through this analysis between the PDSI values and the predictive values of PDSI by the mentioned factors, it is evident that the estimated r2 by the soil moisture (0.77) is more reliable than that of the precipitation (0.037) and temperature (0.15). In fact via prediction of soil moisture values, it seems to be more reliable to estimate PDSI (**Figure 11**), although the derived statistical significances between all variables and PDSI are around zero.

Also, this analysis indicates that there is no linear correlation (or very weak linear correlation) between PDSI with precipitation and temperature, in spite of what is found for PDSI and soil moisture (strong linear correlation). As mentioned already, PDSI is not regularly and directly compatible to the monthly variations of precipitation and temperature.

6. Discussion

The long term and monthly spatial-temporal patterns of the PDSI original dataset through the entire study period (1951-2005) indicate the drought severity during the mentioned study period has been going on increasingly

throughout Iran, and in particular over the northwest and northeast regions. The peak of this increasing severity was found over the period of 1999-2002 as the worst drought period, which is linked to the La Nina phase (cold episode of ENSO). We found, the PDSI is good responsive to the long-term droughts, which may be resulted from the prolonged duration of the La Nina phase in addition to the unusually warm SSTs in the west Pacific.

Regarding the performed analyses (correlation functions and PCA), the PDSI monthly values seem to follow a dominant and persistent pattern in the Iranian regions over the months of a drought year. The derived results through the leading mode of principal component analysis indicate the high ranks over the most parts of the study area, in particular on central and eastern areas with a high spatial coherence, although the south-eastern part where receives the monsoon precipitation shows the weak ranks, and then has been isolated (**Appendix 2**). This spatio-temporal pattern of the PDSI's variations may confirm the influence of a dominant climatic index, such as the enhanced warm pool-la Nina composite, to cause drought conditions over the Iranian regions. Spatial high coherence in the PDSI's variations over some parts of the study area not only depends on a given climatic index, but also homogenous topography and soil type play a significant role. As we found; the PDSI is very responsive in the central and eastern desert plains of the country, where are characterized with the dry climate as

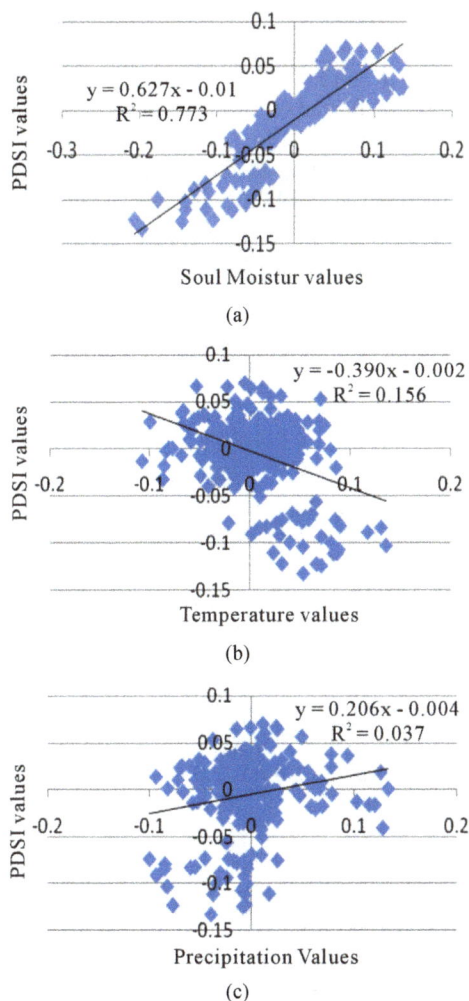

$$y = 0.627x - 0.01$$
$$R^2 = 0.773$$

Soul Moistur values

(a)

$$y = -0.390x - 0.002$$
$$R^2 = 0.156$$

Temperature values

(b)

$$y = 0.206x - 0.004$$
$$R^2 = 0.037$$

Precipitation Values

(c)

Figure 11. Regression analysis on PDSI monthly values based on soil moisture (a), temperature (b) and precipitation (c).

well as the flat lands.

At the same time, the PDSI relationships with its principal factors are very complicated. Although PDSI is closely correlated to the monthly variations of the soil moisture, this is not the case with precipitation and temperature. However, temperature would raise the potential evapotranspiration, which results high intensity of droughts, so the correlation between PDSI and temperature becomes relatively strong in the warm months of the year. Basically, the detected incompatibility between the PDSI and precipitation as well as temperature, seems to result from the fact that, the PDSI is a hydro-climatic index, so its values naturally are too complicated. As, four hydrologic models of potential recharge (PR), potential evapotrasipiration (PE), potential loss (PL) and potential runoff (PR), have been taken into account in the PDSI model. Long memory of the PDSI holds the effect of temperature and precipitation from few months ago. Also, one of the important factors in PDSI calculation is AWC (avai-

lable water content), which is strongly related to the soil types, topography and the depth in which the AWC has been estimated. Incidentally, for calculation of PDSI for a certain month, it's necessary to use the result of preceding month in the formula. Thus, regarding the explanations above, precipitation and temperature variations may correlate to the PDSI variations with a delay around several months.

7. Acknowledgements

We are grateful to the ASMERC (Atmospheric Sciences and Meteorological Research Centre) for the funding of this project. Also we deeply appreciate Mr. Gerard Serviant and Mme. Sue Serviant for revising the English-text.

REFERENCES

[1] C. Bhuiyan, *et al.*, "Monitoring Drought Dynamics in the Aravalli Region (India) Using Different Indices Based on Ground and Remote Sensing Data," *International Journal of Applied Earth Observation and Geoinformation*, Vol. 8, No. 4, 2006, pp. 289-302.

[2] K. Yurekli, A. Kurunc and O. Cevik, "Simulation of Drought Periods Using Stochastic Models," *Turkish Journal of Engineering and Environmental Sciences*, Vol. 28, No. 3, 2004, pp. 181-190.

[3] D. A. Wilhite, "Drought as a Natural Hazard: Concepts and Definitions," In: A. Donald and A. Wilhite, Eds., *Drought: A Global Assessment*, Routledge, New York, 2000, pp. 3-18.

[4] B. Mathew, H. Cullen and B. Lyon, "Drought in Central and Southwest Asia: La Niña, the Warm Pool, and Indian Ocean Precipitation," *Journal of Climate*, Vol. 15, No. B4, 2002, pp. 697-700.

[5] M. J. Nazemosadat, N. Samani, D. A. Barry and N. M. Molaii, "Enso Forcing on Climate Change in IRAN: Precipitation Analysis," *Iranian Journal of Science & Technology, Transaction B, Engineering*, Vol. 30, No. B4, 2006, pp. 47-61.

[6] S. Morid and V. Smakhtin, "Comparison of Seven Meteorological Indices for Drought Monitoring in Iran," *International Journal of Climatology*, Vol. 26, No. 7, 2006, pp. 971-985.

[7] S. Lonergan, "Climate Warming, Water Resources and Geopolitical Conflict: A Study of Nations Dependent on the Nile, Litani and Jordan River Sysems," Defence Research and Development Canada, 1991.

[8] A. Alizadeh and A. Keshavarz, "Status of Agricultural Water Use in Iran, Water Conservation, Reuse and Recycling," The National Academies Press, Paris, 2005.

[9] C. Agnew, "Spatial Aspects of Drought in the Sahel," *Journal of Arid Environment*, Vol. 18, No. 2, 1990, pp. 279-293.

[10] C. Agnew and A. Warren, "A Framework for Tackling

Drought and Land Degradation," *Journal of Arid Environment*, Vol. 33, No. 3, 1996, pp. 309-320.

[11] H. N. Le Houerou, "Climate Change, Drought and Desertification," *Journal of Arid Environment*, Vol. 34, No. 2, 1996, pp. 133-185.

[12] W. C. Palmer, "Meteorological Drought," Research Paper No. 45. US Department of Commerce Weather Bureau, Washington DC, 1965.

[13] V. U. Smakhtin and D. A. Hughes, "Automated Estimation and Analyses of Meteorological Characteristics from Monthly Rainfall Data," Environmental Model Software, Vol. 22, No. 6, 2007, pp. 880-890.

[14] R. R. Heim, "Drought Indices. A Review In: Drought A Global Assessment, Hazards Disaster Series," Routledge, New York, 2000.

[15] R. R. Heim, "A Review of Twentieth-Century Drought Indices Used in the United States," *Bulletin of the American Meteorological Society*, Vol. 84, No. 6, 2002, pp. 1149-1165.

[16] J. A. Keyantash and J. A. Dracup, "An Aggregate Drought Index: Assessing Drought Severity Based on Fluctuations in the Hydrologic Cycle and Surface Water Storage," *Water Resources Research*, Vol. 40, No. 9, 2004.

[17] M. J. Hayes, C. Alvord and J. Lowrey, "Drought Indices," International West Climate Summary, Vol. 3, No. 6, 2007, pp. 2-6.

[18] Vicente-Serrano, M. Sergio, S. Beguería and J. I. López-Moreno, "A Multiscalar Drought Index Sensitive to Global Warming: The Standardized Precipitation Evapotranspiration Index," *Journal of Climate*, Vol. 23, No. 7, 2010, pp. 1696-1718.

[19] A. G. Dai, "Characteristics and Trends in Various Forms of the Palmer Drought Severity Index during 1900-2008," *Journal of Geophysical Research*, Vol. 116, No. D12, 2011, p. D12.

[20] W. C. Palmer, "Keeping Track of Crop Moisture Conditions, Nationwide: The New Crop Moisture Index," *Weatherwise*, Vol. 21, No. 4, 1968, pp. 156-161.

[21] I. Petrasovits, "General Review on Drought Strategies," *Transactions of the 14th Congress on Irrigation and Drainage, Rio de Janeiro*, Vol. 1-C, Internationa Commission on Irrigation and Drainage (ICID), 1990, pp 1-12.

[22] B. A. Shafer and L. E. Dezman, "Development of a Surface Water Supply Index (SWSI) to Assess the Severity of Drought Conditions in Snowpack Runoff Areas," *Proceedings of the Western Snow Conference*, Colorado State University, Fort Collins, 1982, pp. 164-175.

[23] F. N. Kogan, "Remote Sensing of Weather Impacts on Vegetation in Non-Homogeneous Areas," *International Journal of Remote Sensinsing*, Vol. 11, No. 8, 1990, pp. 1405-1419.

[24] T. B. McKee, N. J. Doesken and J. Kleist, "The Relation of Drought Frequency and Duration to Time Scales," *Proceedings of the 8th Conference on Applied Climatol-*

ogy," American Meteorological Society, Boston, 1993, pp 179-184.

[25] W. J. Gibbs and J. V. Maher, "Rainfall Deciles as Drought Indicators," Bureau of Meteorology Bulletin, Melbourne, 1967.

[26] V. U. Smakhtin, "Low-Flow Hydrology: A Review," *Journal of Hydrology*, Vol. 240, No. 3, 2001, pp. 147-186.

[27] T. R. Ka and A. J. Koscielny, "Drought in the United States: 1895-1981," *Journal of Climatology*, Vol. 2, No. 4, 1982, pp. 313-329.

[28] T. R. Karl, "Sensitivity of the Palmer Drought Severity Index and Palmer's Z-index to Their Calibration Coefficients including Potential Evapotranspiration," *Journal of Climate Applied Meteorology*, Vol. 25, No. 1, 1986, pp. 77-86.

[29] P. Domonkos, S. Szalai and J. Zoboki, "Analysis of Drought Severity Using PDSI and SPI Indices," *Idoejaras*, Vol. 105, No. 2, 2001, pp. 93-107.

[30] B. Lloyd-Hughes and M. A. Saunders, "A Drought Climatology for Europe," *International Journal of Climatology*, Vol. 22, No. 13, 2002, pp. 1571-1592.

[31] H. K. Ntale and T. Y. Gan, "Drought Indices and Their Application to East Africa," *International Journal of Climatology*, Vol. 23, No. 11, 2003, pp. 1335-1357.

[32] R. M. N. Dos Santos and A. R. Pereira, "PALMER Drought Severity Index for Western Sao Paulo state, Brazil," *Revista Brasileira de Agrometeorologia*, Vol. 7, No. 1, 1999, pp. 139-145.

[33] J. E. Cole and E. R. Cook, "The Changing Relationship Between ENSO Variability and Moisture Balance in the Continental United States," *Geophysical Research Letters*, Vol. 25, No. 24, 1998, pp. 4529-4532.

[34] E. R. Cook, D. M. Meko, D. W. Stahle and M. K. Cleaveland, "Drought Reconstructions for the Continental United States," *Journal of Climate*, Vol. 12, No. 4, 1999, pp. 1145-1162.

[35] F. K. Fye, D. W. Stahle and E. R. Cook, "Paleoclimatic Analogs to Twentieth-Century Moisture Regimes Across the United States," *Bulletin of the American Meteorological Society*, Vol. 84, No. 7, 2003, pp. 901-909.

[36] A. G. Dai, K. E. Trenberth and T. R. Karl, "Global Variations in Droughts and Wet Spells: 1900-1995," *Geophysical Research Letters*, Vol. 25, No. 17, 1998, pp. 3367-3370.

[37] P. Rahimzadeh, A. Darvishsefat, A. Khalili and M. Makhdoum, "Using AVHRR-Based Vegetation Indices for Drought Monitoring in the Northwest of Iran," *Journal of Arid Environments*, Vol. 72, No. 6, 2008, pp. 1086-1096.

[38] T. Razei, I. Bordi and L. S. Pereira, "A Precipitation-Based Regionalization for Western Iran and Regional

Drought Variability," *Hydrology and Earth System Sciences*, Vol. 12, No. 6, 2008, pp. 1309-1321.

[39] J. T. Shiau and R. Modarres, "Copula-Based Drought Severity-Duration-Frequency Analysis in Iran," *Meteorological Applications*, Vol. 16, No. 4, 2009, pp. 481-489.

[40] S. Morid, V. Smakhtin and M. Moghaddasi, "Comparison of Seven Meteorological Indices for Drought Monitoring in Iran," *International Journal of Climatology*, Vol. 26, No. 7, 2006, pp. 971-985.

[41] F. Rahimzadeh, A. Asgari and E. Fattahi, "Variability of Extreme Temperature and Precipitation in Iran during Recent Decades," *International Journal of Climatology*, Vol. 29, No. 3, 2008, pp. 329-343.

[42] H. Van den Dool, J. Huang and Y. Fan, "Performance and Analysis of the Constructed Analogue Method Applied to U.S. Soil Moisture over 1981-2001," *Journal of Geophysical Research*, Vol. 108, No. D16, 2003, p. 8617.

[43] E. J. Janowiak and P. P. Xie, "CAMS–OPI: A Global Satellite-Rain Gauge Merged Product for Real-Time Precipitation Monitoring Applications," *Journal of Climate*, Vol. 12, No. 11, 1999, pp. 3335-3342.

[44] R. W. Reynolds, "A Real-Time Global Sea Surface Temperature Analysis," *Journal of Climate*, Vol. 1, No. 1, 1988, pp.75-86.

[45] R. W. Preisendorfer, "Principal Component Analysis in Meteorology and Oceanography," Edited by C. D. Mobley, Elsevier, Amsterdam, 1988.

[46] D. S. Wilks, "Statistical Methods in the Atmospheric Sciences," Academic Press, California, 1995.

[47] T. Murat, K. Telat and S. Faize, "Spatiotemporal Variability of Precipitation Total Series over Turkey," *International Journal of Climatology*, Vol. 29, No. 8, 2008, pp. 1056-1074.

[48] A. Dai, E. K. Trenberth and T. T. Qian, "A Global Dataset of Palmer Drought Severity Index for 1870-2002: Relationship with Soil Moisture and Effects of Surface Warming," *Journal of Hydrometeorology*, Vol. 5, No. 6, 2004, pp. 1117-1130.

[49] A. Soltani and M. Gholipoor, "Teleconnections Between El Nino/Southern Oscillation and Rainfall and Temperature in Iran," *International Journal of Agricultural Research*, Vol. 1, No. 6, 2006, pp. 603-608.

[50] M. J. Nazemossadat, "Winter Drought in Iran: Associations with ENSO," *Drought Network News*, Vol. 13, No. 1, 2001, pp. 10-13.

[51] M. J. Nazemosadat and I. Cordery, "On the Relationships Between ENSO and Autumn Rainfall in Iran," *International Journal of Climatology*, Vol. 20, No. 1, 2000, pp. 47-62.

[52] S. A. Masoodian, "On Relationship Between Precipitation of Iran and North Atlantic Oscillation," *Geographical Research*, Vol. 23, No. 4, 2009, pp. 3-18.

Appendix 1

A station observation based global land monthly mean surface air temperature dataset at 0.5×0.5 latitude-longitude resolution for the period from 1948 to the present was developed recently at the Climate Prediction Center, National Centers for Environmental Prediction. This data set is different from some existing surface air temperature data sets in: 1) using a combination of two large individual data sets of station observations collected from the Global Historical Climatology Network version 2 and the Climate Anomaly Monitoring System, so it can be regularly updated in near real time with plenty of stations and 2) some unique interpolation methods, such as the anomaly interpolation approach with spatially-temporally varying temperature lapse-rates derived from the observation based Reanalysis for topographic adjustment. You can find the complete manuscript through: ftp://ftp.cpc.ncep.noaa.gov/wd51yf/GHCN_CAMS/Resource/cpc_globalT.pdf

Appendix 2

The yearly average of the spatial-temporal PCA proc-essed through the leading mode during the worst drought period (1999-2002).

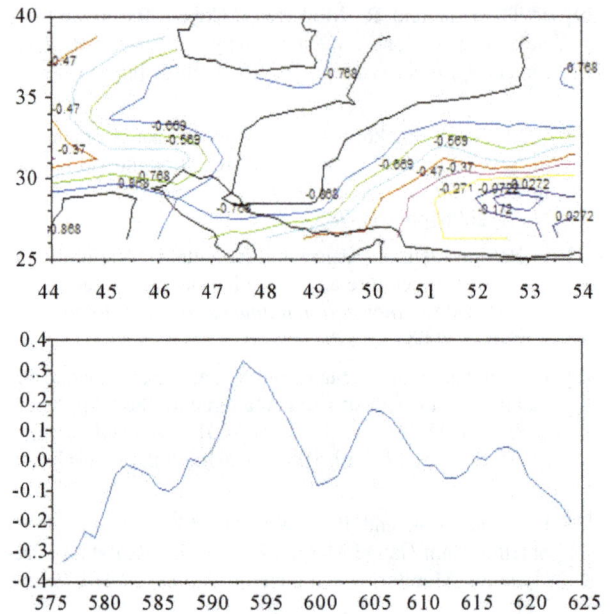

13

A Simulation Study of the Effect of Geomagnetic Activity on the Global Circulation in the Earth's Middle Atmosphere

Igor Mingalev, Galina Mingaleva, Victor Mingalev[*]
Polar Geophysical Institute, Kola Scientific Center of the Russian Academy of Sciences,
Apatity, Russia

ABSTRACT

To investigate how geomagnetic activity affects the formation of the large-scale global circulation of the middle atmosphere, the non-hydrostatic model of the global wind system of the Earth's atmosphere, developed earlier in the Polar Geophysical Institute, is utilized. The model produces three-dimensional global distributions of the zonal, meridional, and vertical components of the wind velocity and neutral gas density in the troposphere, stratosphere, mesosphere, and lower thermosphere. Simulations are performed for the winter period in the northern hemisphere (16 January) and for two distinct values of geomagnetic activity (Kp = 1 and Kp = 4). The simulation results indicate that geomagnetic activity ought to influence considerably on the formation of global wind system in the stratosphere, mesosphere, and lower thermosphere. The influence on the middle atmosphere is conditioned by the vertical transport of air from the lower thermosphere to the mesosphere and stratosphere and vice versa. This transport may be rather distinct under different geomagnetic activity conditions.

Keywords: Middle Atmosphere; Global Circulation; Numerical Simulation

1. Introduction

An investigation of the dynamical processes in the Earth's atmosphere is a very important problem. It is well known that the atmospheric processes influence considerably on the human activity and health. Unfortunately, modern scientific facility does not allow somebody to measure the momentary global three-dimensional wind system of the Earth's atmosphere. However, to investigate the planetary wind system of the Earth's atmosphere, mathematical models may be utilized.

During the last three decades, several general circulation models of the lower and middle atmosphere have been developed (e.g., see [1-11]). It can be noticed that the existing general circulation models of the lower and middle atmosphere may be successfully utilized for simulation of the slow climate changes.

Not long ago, in the Polar Geophysical Institute (PGI), the non-hydrostatic model of the global neutral wind system in the Earth's atmosphere has been developed [12,13]. This model enables to calculate three-dimen-

sional global distributions of the zonal, meridional, and vertical components of the neutral wind at levels of the troposphere, stratosphere, mesosphere, and lower thermosphere, with whatever restrictions on the vertical transport of the neutral gas being absent. This model has been utilized in order to simulate the global circulation of the middle atmosphere for conditions corresponding to different seasons [12-15] and to investigate numerically how solar activity affects the formation of the large-scale global circulation of the mesosphere and lower thermosphere [16]. The purpose of the present work is to continuer these studies and to investigate numerically, using the non-hydrostatic model of the global neutral wind system, developed earlier in the Polar Geophysical Institute, how geomagnetic activity affects the formation of the large-scale global circulation of the stratosphere, mesosphere, and lower thermosphere.

2. Mathematical Model

In the present study, the non-hydrostatic model of the global neutral wind system in the Earth's atmosphere,

---

[*]Corresponding author.

developed earlier in the PGI [12,13], is utilized. This model produces three-dimensional global distributions of the zonal, meridional, and vertical components of the neutral wind and neutral gas density in the troposphere, stratosphere, mesosphere, and lower thermosphere. The peculiarity of the utilized model consists in that the internal energy equation for the neutral gas is not solved in the model calculations. Instead, the global temperature field is assumed to be a given distribution, *i.e.* the input parameter of the model, and obtained from the NRL-MSISE-00 empirical model [17]. Moreover, in the model calculations, not only the horizontal components but also the vertical component of the neutral wind velocity is obtained by means of a numerical solution of a generalized Navier-Stokes equation for compressible gas, so the model is non-hydrostatic.

The mathematical model, utilized in the present study, is based on the numerical solution of the system of equations containing the dynamical equation and continuity equation for the neutral gas. For solving the system of equations, the finite-difference method is applied. The dynamical equation for the neutral gas in vectorial form can be written as

$$\rho\left(\frac{\partial V}{\partial t}+(V,\nabla)V\right)=\rho F+\nabla\cdot\widehat{P} \qquad (1)$$

where ρ is the neutral gas density, V is the neutral wind velocity, F is the acceleration comprising the gravity acceleration, Coriolis acceleration, acceleration of translation, and acceleration due to elastic collisions with the ion gas, and $\widehat{P}$ is the total stress tensor. The latter tensor can be decomposed as follows:

$$\widehat{P}=-p\widehat{I}+\widehat{\tau} \qquad (2)$$

where p is the pressure, $\widehat{I}$ is the unit tensor, and $\widehat{\tau}$ is the extra stress tensor whose components are given by the rheological equation of state or the law of viscous friction. A spherical coordinate system rotatable together with the Earth is utilized in model calculations. Therefore, from the dynamical equation, Equation (1), momentum equations for the zonal, meridional, and vertical components of the neutral gas velocity may be derived. These equations include not only the pressure gradients but also partial derivatives of components of the extra stress tensor, $\widehat{\tau}$. The latter tensor is composed of a Newtonian part, $\widehat{\tau}_0$, and a complementary part, $\widehat{\tau}_1$, namely,

$$\widehat{\tau}=\widehat{\tau}_0+\widehat{\tau}_1 \qquad (3)$$

the former tensor, $\widehat{\tau}_0$, is given by the well-known Newton's law of viscous friction,

$$\widehat{\tau}_0=2\mu\widehat{\varepsilon} \qquad (4)$$

where μ is the coefficient of molecular viscosity, whose dependence on the temperature is assumed to

obey the Sutherland's law, and ε is the tensor defined as

$$\widehat{\varepsilon}=\widehat{D}_0-\frac{1}{3}\widehat{I}\,Tr(\widehat{D}_0) \qquad (5)$$

where $\widehat{D}_0$ is the strain rate tensor and $Tr(\ )$ denotes the trace of a tensor. The complementary stress tensor, $\widehat{\tau}_1$, is supposed to be conditioned by a small-scale turbulence having the scales equal and less than the steps of the finite-difference approximations. It is assumed that this tensor represents the effect of the turbulence on the mean flow and is given by an expression, analogous to the Newton's law of viscous friction, Equation (4), with the scalar coefficient of viscosity, μ, being replaced by three distinct coefficients describing the eddy viscosities in the directions of the basis vectors of the utilized spherical coordinate system. For computing the eddy viscosities, the turbulence theory of Obukhov [18] is applied.

Thus, the momentum equations for the zonal, meridional, and vertical components of the neutral gas velocity acquire ultimately a form of a generalized Navier-Stokes equation for compressible gas on scales which are more than the steps of the finite-difference approximations, with the effect of the turbulence on the mean flow being taken into account by using an empirical subgrid-scale parameterization. The steps of the finite-difference approximations in the latitude and longitude directions are identical and equal to 1 degree. A height step is non-uniform and does not exceed the value of 1 km.

The simulation domain is the layer surrounding the Earth globally and stretching from the ground up to the altitude of 126 km at the equator. Upper boundary conditions provide the conservation law of mass in the simulation domain. The Earth's surface is supposed to coincide approximately with an oblate spheroid whose radius at the equator is more than that at the pole. More complete details of the utilized model may be found in the studies of I. Mingalev, and V. Mingalev, [12] and Mingalev *et al.* [13].

3. Presentation and Discussion of Results

The utilized mathematical model of the global neutral wind system can be used for different seasonal, solar cycle, and geomagnetic conditions. In the present study, simulations are performed for the winter period in the northern hemisphere (16 January) and for conditions corresponding to moderate 10.7 cm solar flux ($F_{10.7}$ = 101). To investigate the influence of geomagnetic activity on the global circulation of the atmosphere, we made calculations for conditions corresponding to two different values of geomagnetic activity: low and considerable, namely, Kp = 1 and Kp = 4. The variations of the atmospheric parameters with time were calculated until they become stationary. The steady-state distributions of the atmospheric parameters were obtained on condition that inputs to the model and boundary conditions correspond to 10.30 UT. The temperature distributions, corresponding to this moment, were calculated using the NRLMSISE-00 empirical model [17].

It turns out that atmospheric temperatures, calculated with the help of the NRLMSISE-00 empirical model for two distinct values of geomagnetic activity (Kp = 1 and Kp = 4), are very similar below approximately 80 km, while, above this altitude, they may be rather different. **Figure 1** shows the global distributions of the atmospheric temperature at 50 km altitude, obtained from the NRLMSISE-00 empirical model for 16 January, UT = 10.30 and calculated for two distinct values of geomagnetic activity: Kp = 1 and Kp = 4. It is seen no distinctions between the results obtained for two different values of geomagnetic activity.

On the contrary, from **Figure 2**, in which the global distributions of the atmospheric temperature at 110 km altitude are present, one can see that differences between temperatures, obtained for two considered values of geomagnetic activity, can achieve a few tens of degrees at identical points of the globe. Thus, the application of the NRLMSISE-00 empirical model shows that the influence of level of geomagnetic activity on the global

distribution of the atmospheric temperature ought to be absent at altitudes of the troposphere, stratosphere, and mesosphere, while this influence ought to be appreciable at altitudes of the lower thermosphere for the winter period in the northern hemisphere.

Distributions of the atmospheric parameters, calculated with the help of the mathematical model and obtained for 16 January for two different values of geomagnetic activity, are shown in **Figures 3 - 12**. The results of modeling illustrate both common characteristic features and distinctions caused by different values of geomagnetic activity.

The calculated global distributions of the atmospheric parameters display the following common features. At levels of the stratosphere, mesosphere, and lower thermosphere, the horizontal and vertical components of the wind velocity are changeable functions of latitude and longitude. Maximal absolute values of the horizontal and vertical components of the wind velocity are larger at higher altitudes. The horizontal domains exist where the steep gradients in the horizontal velocity field take place. The horizontal wind velocity can have various directions which may be opposite at the near points. Moreover, the horizontal domains exist in which the vertical neutral wind component has opposite directions. The horizontal

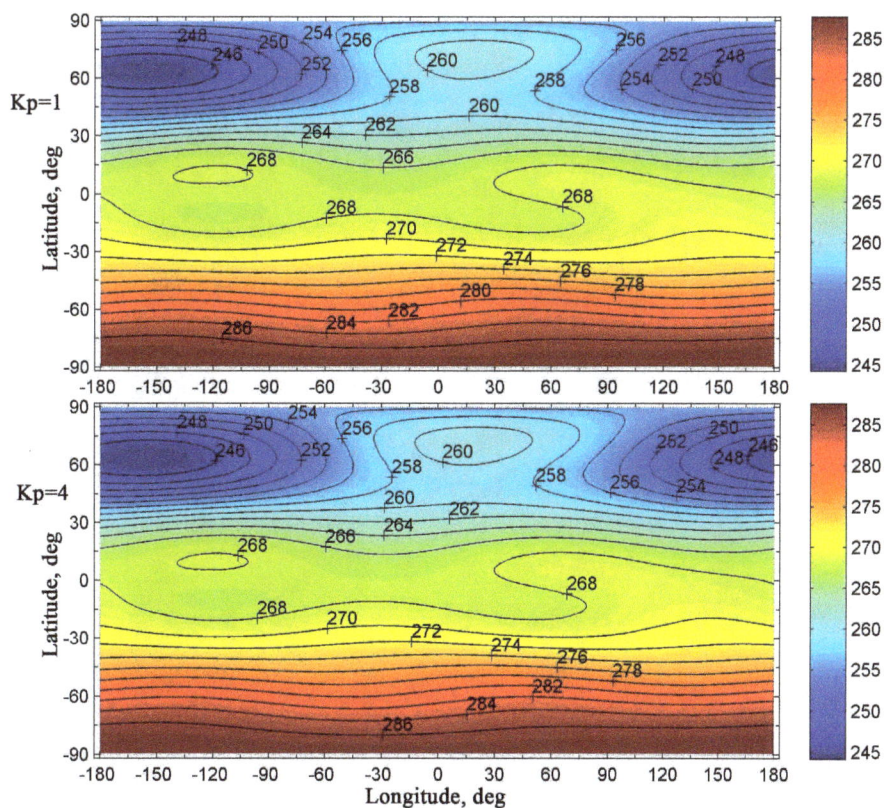

Figure 1. The global distributions of the atmospheric temperature (K) at 50 km altitude, obtained from the NRLMSISE-00 empirical model for 16 January, UT = 10.30 and calculated for two distinct values of geomagnetic activity: Kp = 1 (top panel) and Kp = 4 (bottom panel).

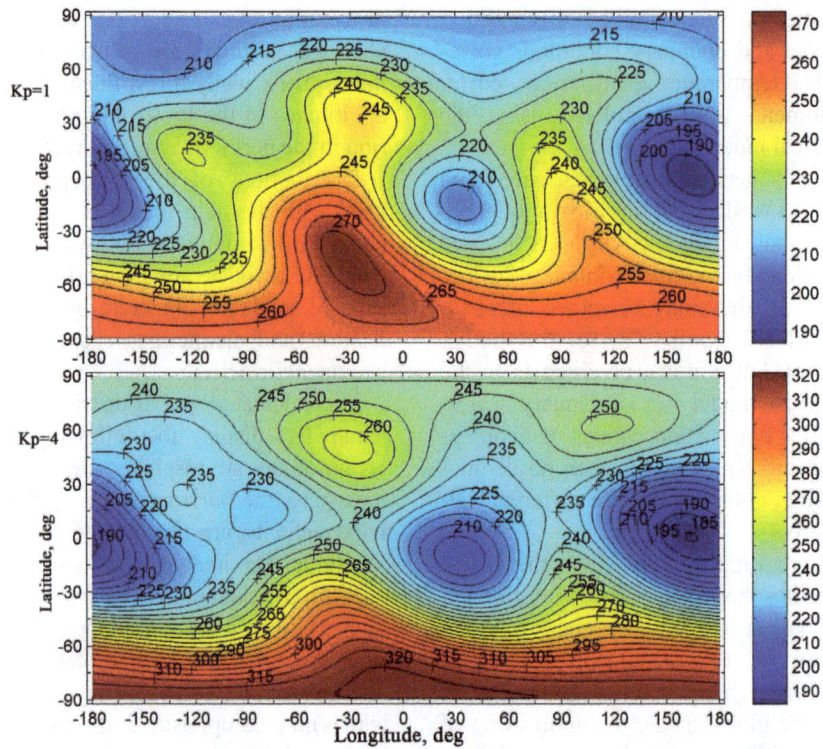

Figure 2. The global distributions of the atmospheric temperature (K) at 110 km altitude, obtained from the NRLMSISE-00 empirical model for 16 January, UT = 10.30 and calculated for two distinct values of geomagnetic activity: Kp = 1 (top panel) and Kp = 4 (bottom panel).

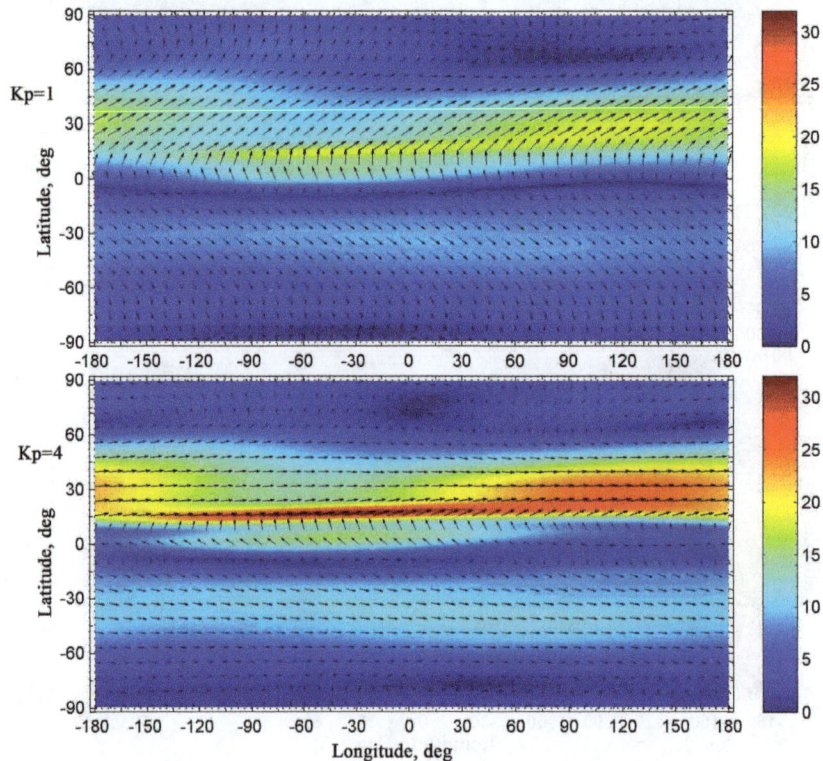

Figure 3. The global distributions of the vector of the simulated horizontal component of the neutral wind velocity at the altitude of 10 km, obtained for 16 January and calculated for two distinct values of geomagnetic activity: Kp = 1 (top panel) and Kp = 4 (bottom panel). The colouration of the figures indicates the module of the velocity in m/s.

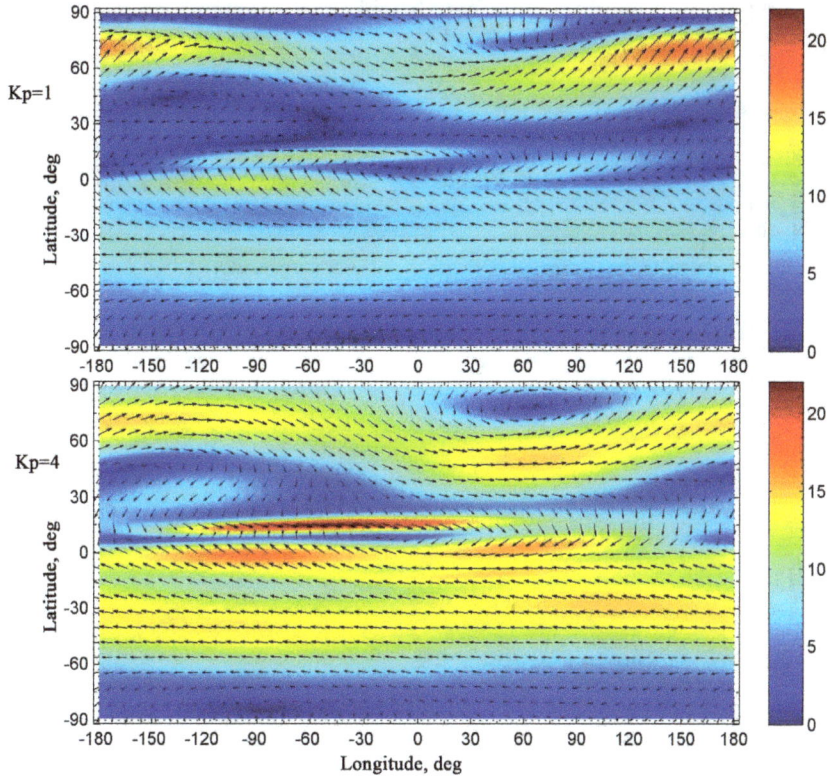

Figure 4. The same as in Figure 3 but at the altitude of 20 km. The colouration of the figures indicates the module of the velocity in m/s.

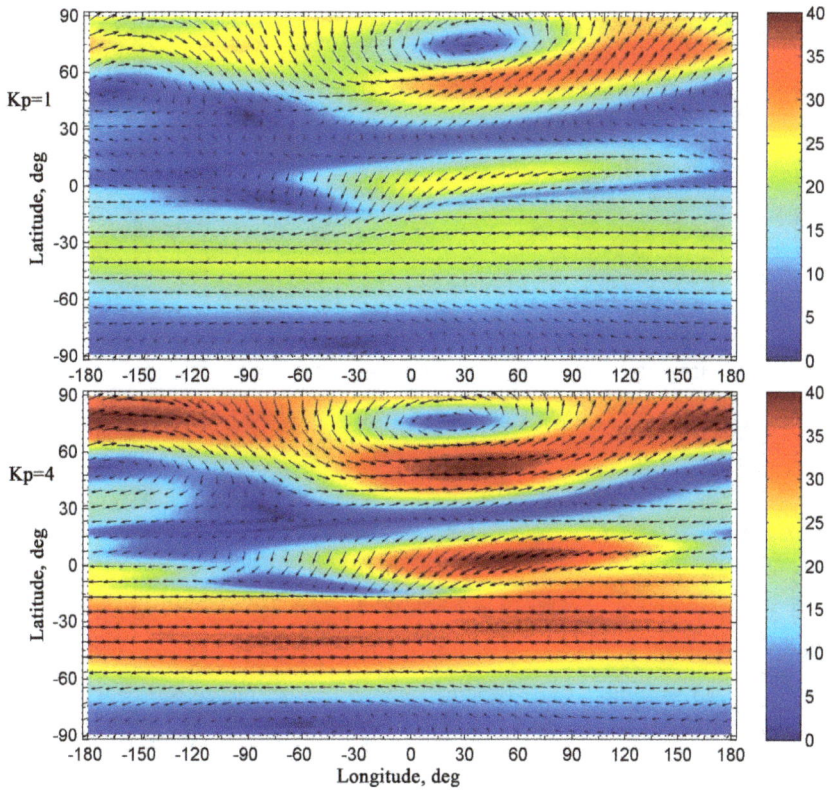

Figure 5. The same as in Figure 3 but at the altitude of 30 km. The colouration of the figures indicates the module of the velocity in m/s.

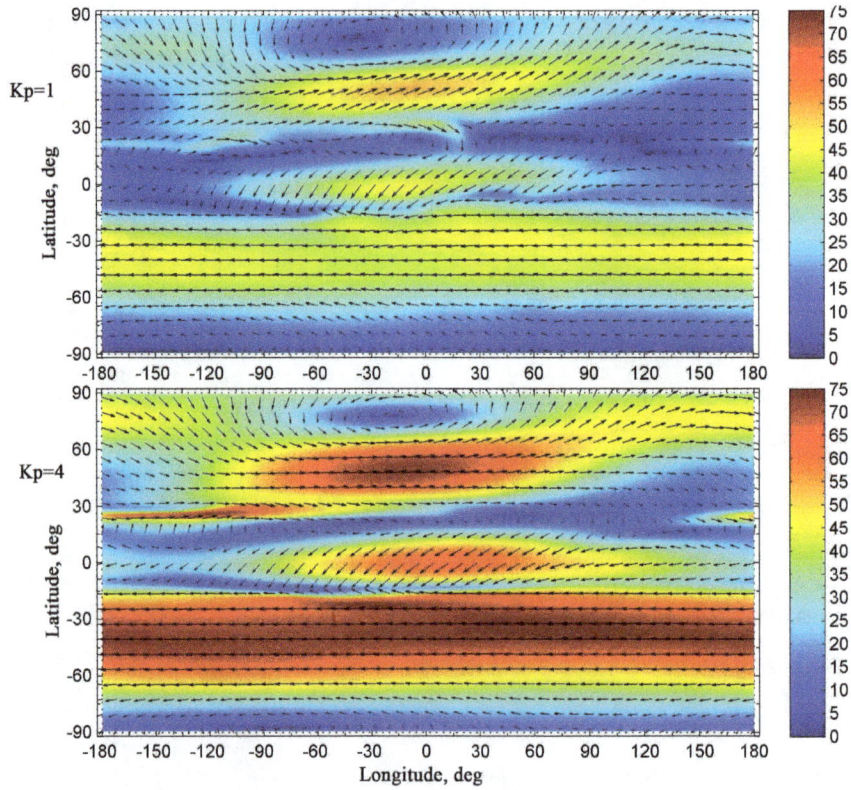

Figure 6. The same as in Figure 3 but at the altitude of 50 km. The colouration of the figures indicates the module of the velocity in m/s.

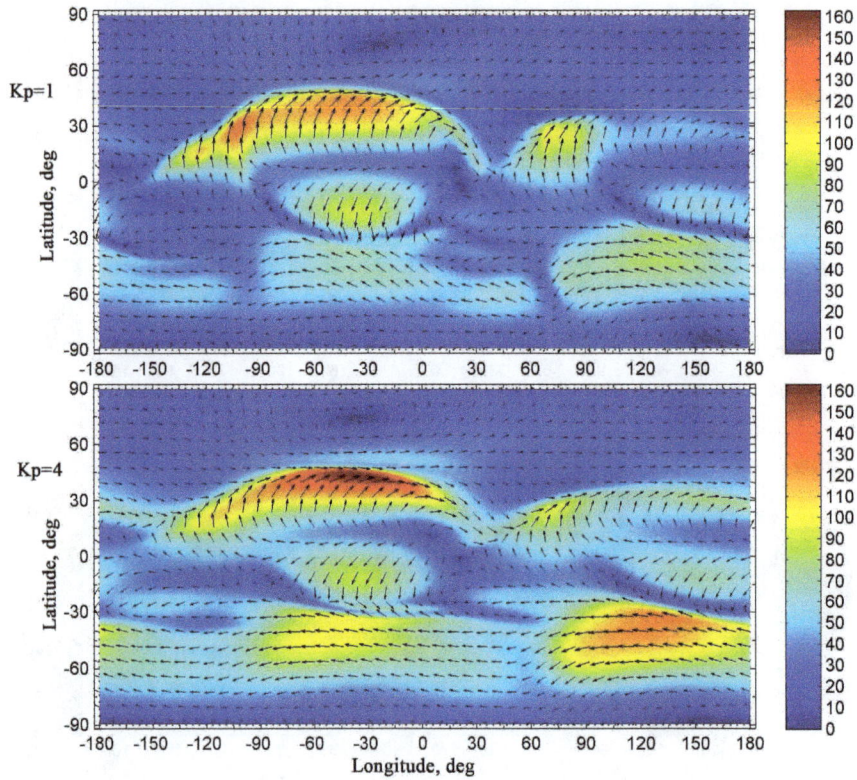

Figure 7. The same as in Figure 3 but at the altitude of 70 km. The colouration of the figures indicates the module of the velocity in m/s.

domains, where the vertical neutral wind component is upward, have a great length and large width, as a rule. Unlike, the horizontal domains, where the vertical neutral wind component is downward, have a large length and little width. Usually, the latter domains have a configuration like a long narrow band and coincide, as a rule, with the regions, where the steep gradients in the horizontal velocity field take place. Maximal absolute values of the upward vertical wind component are less than the maximal module of the downward vertical wind component. At levels of the mesosphere, the horizontal wind velocity can achieve values of more than 150 m/s.

It is well known from numerous observations that the global atmospheric circulation can contain sometimes so called circumpolar vortices that are the largest scale inhomogeneities in the global neutral wind system. Their extent can be very large, sometimes reaching the latitudes close to the equator. It is known that circumpolar vortices are often formed at heights of the stratosphere and mesosphere in the periods close to summer and winter solstices. The circumpolar cyclone arises in the northern hemisphere under winter conditions, while the circumpolar anticyclone arises in the southern hemisphere under summer conditions.

From the numerically obtained results for winter period in the northern hemisphere, we can see that, at levels of the stratosphere and mesosphere, the motion of the neutral gas in the northern hemisphere is primarily eastward, so a circumpolar cyclone is formed (**Figures 4 - 7**). It can be noticed that the center of the northern cyclone may be displaced from the pole. Simultaneously, the motion of the neutral gas is primarily westward in the southern hemisphere at levels of the stratosphere and mesosphere, so a circumpolar anticyclone is formed for summer period of the southern hemisphere (**Figures 4 - 7**). It can be seen that the circumpolar vortices of the northern and southern hemispheres, numerically simulated in the present study at levels of the stratosphere and mesosphere, correspond qualitatively to the global circulation, obtained from observations.

Let us consider simulation results, obtained for distinct values of magnetic activity, and their distinctions. It is easy to see that, at levels of the lower thermosphere, maximal absolute values of the horizontal components of the wind velocity, obtained for low geomagnetic activity, are less than those, obtained for considerable geomagnetic activity (**Figure 9**). It is obvious that these distinctions are conditioned by the differences of the temperature distributions, obtained for two distinct values of geomagnetic activity (**Figure 2**).

It can be seen that, at levels of the upper troposphere, stratosphere, and mesosphere, the global distributions of the vector of the simulated horizontal component of the neutral wind velocity, calculated for two distinct values of geomagnetic activity, are rather different (**Figures 3 - 7**). These differences can not be explained by the distinctions of the temperature distributions because of the absence of these distinctions below approximately 80 km, as was noted earlier. The simulation results indicate that the effect of geomagnetic activity on the global circulation of the atmosphere below 80 km is conditioned by the vertical transport of air from the thermosphere to the lower levels and vice versa. As can be seen from **Figures 10 - 12**, such vertical transport does exist. This transport may be rather distinct under different geomagnetic activity conditions. It can be noticed that the utilized mathematical model is able to simulate this vertical transport due to the fact that the model is non-hydrostatic.

Thus, the simulation results indicate that geomagnetic activity ought to influence considerably on the formation of global neutral wind system not only in lower thermosphere, but also in the mesosphere, stratosphere, and upper troposphere. In particular, from the results shown in **Figures 5 - 7**, one can see that the horizontal wind velocity in the circumpolar cyclone of the northern hemisphere, obtained for low geomagnetic activity, is less than that, obtained for considerable geomagnetic activity. Similarly, the horizontal wind velocity in the circumpolar anticyclone of the southern hemisphere, obtained for low geomagnetic activity, is less than that, obtained for considerable geomagnetic activity.

The level of geomagnetic activity can influence not only on the magnitude of the horizontal wind velocity in the circumpolar cyclone of the northern hemisphere, but also on the vertical dimension of this cyclone. From the results shown in **Figure 4**, one can see that the altitude of the lower edge of this circumpolar cyclone depends of the level of geomagnetic activity. Pronounced circumpolar cyclone is present in the northern hemisphere at 20 km altitude under considerable geomagnetic activity, with its center being displaced from the pole. On the contrary, under low geomagnetic activity, the pronounced circumpolar cyclone is absent in the northern hemisphere at 20 km altitude (**Figure 4**). From **Figure 5**, it can be seen that the pronounced circumpolar cyclone is present in the northern hemisphere at 30 km altitude under two distinct values of geomagnetic activity.

From the obtained results, we can see that, at levels near to the stratopause, the vertical wind velocity can have opposite directions in the horizontal domains having different configurations. Maximal absolute values of the downward vertical wind component are commensurable with the maximal module of the upward vertical wind component for conditions of low geomagnetic activity. On the contrary, for conditions of considerable geomagnetic activity, maximal absolute values of the downward and upward vertical wind components can be rather different (**Figure 10**).

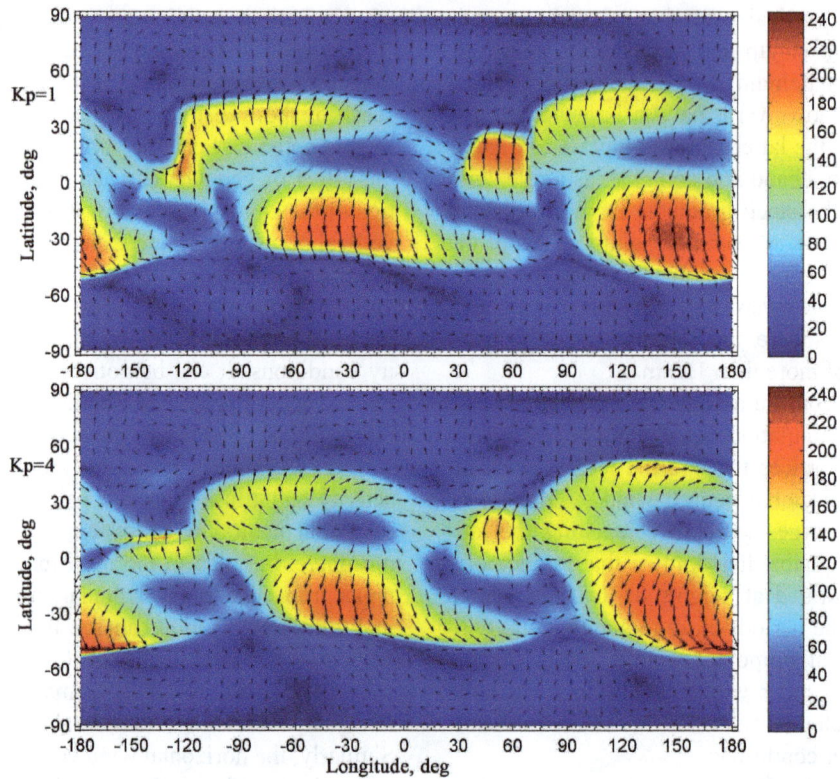

Figure 8. The same as in Figure 3 but at the altitude of 90 km. The colouration of the figures indicates the module of the velocity in m/s.

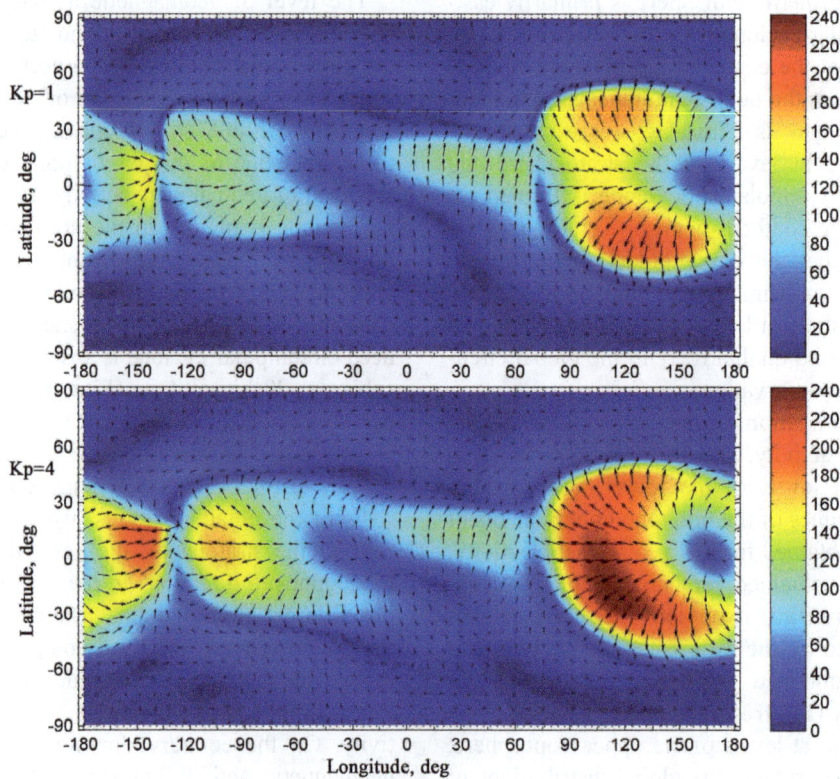

Figure 9. The same as in Figure 3 but at the altitude of 110 km. The colouration of the figures indicates the module of the velocity in m/s.

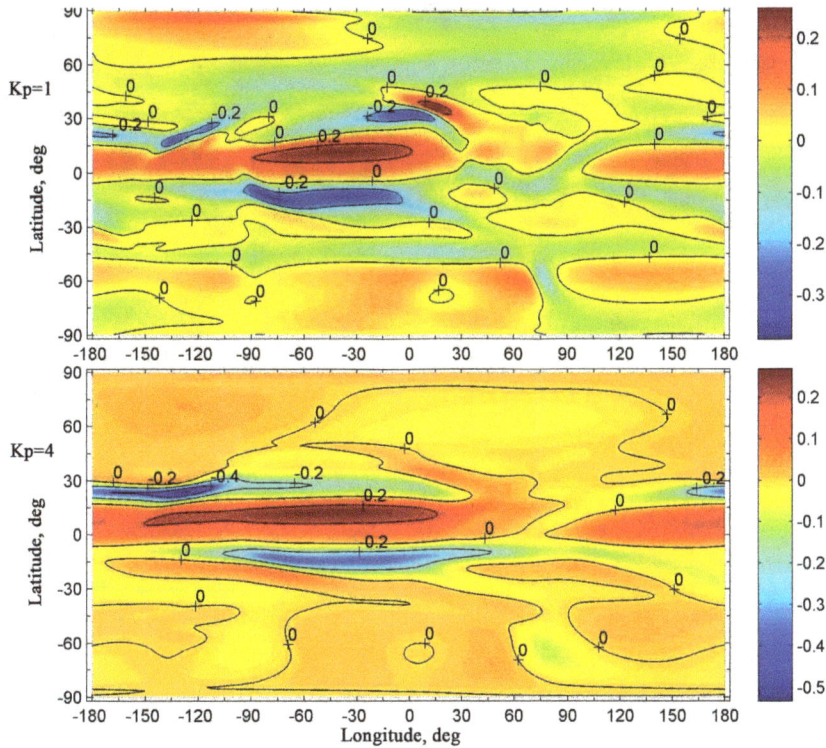

Figure 10. The global distributions of the simulated vertical component of the neutral wind velocity at the altitude of 50 km, obtained for 16 January and calculated for two distinct values of geomagnetic activity: Kp = 1 (top panel) and Kp = 4 (bottom panel). The colouration of the figures indicates the velocity in m/s, with the positive direction of the vertical velocity being upward.

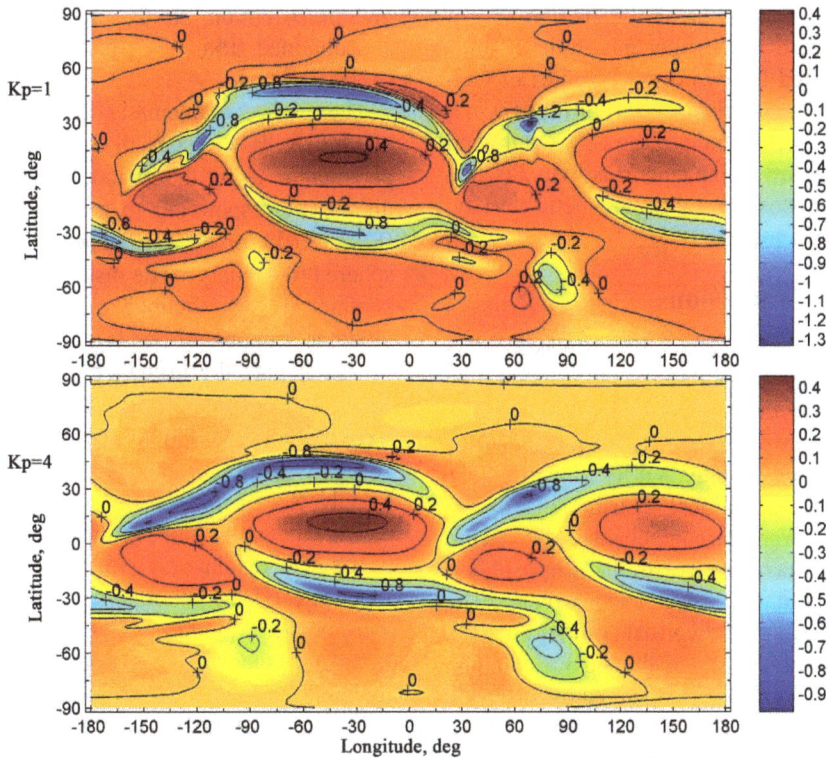

Figure 11. The same as in Figure 10 but at the altitude of 70 km. The colouration of the figures indicates the velocity in m/s, with the positive direction of the vertical velocity being upward.

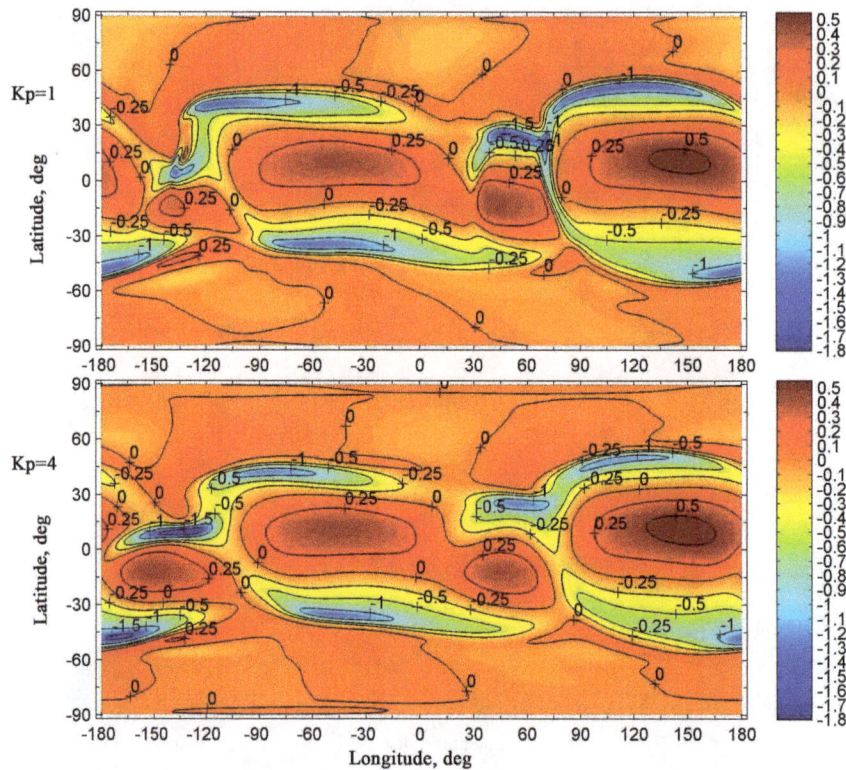

Figure 12. The same as in Figure 10 but at the altitude of 90 km. The colouration of the figures indicates the velocity in m/s, with the positive direction of the vertical velocity being upward.

The simulation results indicate that, at levels of mesosphere at latitudes close to the equator, the distributions of the vector of the horizontal component of the neutral wind velocity, calculated for two distinct values of geomagnetic activity, are qualitatively similar (**Figures 4 - 6**). However, the maximal module of the horizontal wind velocity, obtained for low geomagnetic activity, is less than that, obtained for considerable geomagnetic activity, at latitudes close to the equator.

4. Summary and Conclusions

To investigate how geomagnetic activity affects the formation of the large-scale global circulation of the middle atmosphere, the non-hydrostatic model of the global wind system of the Earth's atmosphere, developed earlier in the Polar Geophysical Institute, is utilized. The model produces three-dimensional global distributions of the zonal, meridional, and vertical components of the wind velocity and neutral gas density in the troposphere, stratosphere, mesosphere, and lower thermosphere. The peculiarity of the utilized model consists in that not only the horizontal components but also the vertical component of the neutral wind velocity is obtained by means of a numerical solution of a generalized Navier-Stokes equation for compressible gas, so the model is non-hydrostatic. Moreover, the internal energy equation for the neutral gas is not solved in the model calculations. In-

stead, the global temperature field is assumed to be a given distribution, *i.e.* the input parameter of the model, and obtained from the NRLMSISE-00 empirical model [17].

The applied mathematical model was utilized for obtaining the steady-state distributions of the atmospheric parameters, using the method of establishment, for conditions, corresponding to moderate 10.7 cm solar flux ($F_{10.7} = 101$) for the winter period in the northern hemisphere (16 January). The distributions of the atmospheric parameters were obtained on condition that inputs to the model and boundary conditions correspond to 10.30 UT. To investigate the influence of geomagnetic activity on the global circulation of the atmosphere, calculations were made for conditions corresponding to two different values of geomagnetic activity: low and considerable, namely, Kp = 1 and Kp = 4.

The calculated global distributions of the atmospheric parameters display the common characteristic features. At levels of the stratosphere, mesosphere, and lower thermosphere, the horizontal and vertical components of the wind velocity are changeable functions of latitude and longitude. It turned out that the calculated global distributions of the horizontal wind velocity, obtained for different values of geomagnetic activity, contain large-scale circumpolar vortices of the northern and southern hemispheres. The circumpolar vortices of the northern and southern hemispheres, numerically simulated in the present study at levels of the stratosphere and mesosphere, correspond qualitatively to the global circulation, obtained from observations.

The simulation results indicate that geomagnetic activity ought to influence considerably on the formation of global neutral wind system in the stratosphere, mesosphere, and lower thermosphere. However, this influence is not straightforward.

Undoubtedly, at levels of the lower thermosphere, this influence is conditioned by the differences of the temperature distributions, obtained for various values of geomagnetic activity at these levels. However, from the simulation results obtained, we can see that the atmospheric temperature, calculated with the help of the NRLMSISE-00 empirical model, does not depend on the geomagnetic activity below approximately 80 km. Nevertheless, the effect of geomagnetic activity on the global circulation of the atmosphere below 80 km exists. This effect is conditioned by the vertical transport of air from the lower thermosphere to the mesosphere, stratosphere, and upper troposphere, eventually, and vice versa. The simulation results indicate that this vertical transport may be rather distinct under different geomagnetic activity conditions.

Thus, the influence of geomagnetic activity on the global circulation in the Earth's middle and lower atmospheres for January conditions is a consequence of a relationship between large-scale global circulation of the lower thermosphere and large-scale planetary circulations of the middle and lower atmospheres, with the vertical transport of air playing a significant role.

It can be noticed that the utilized mathematical model was able to simulate the effect of geomagnetic activity on the global circulation in the Earth's middle and lower atmospheres due to the fact that the model is non-hydrostatic.

5. Acknowledgements

This work was partly supported by Grant No. 13-01-00063 from the Russian Foundation for Basic Research.

REFERENCES

[1]   S. Manabe and D. G. Hahn, "Simulation of Atmospheric Variability," *Monthly Weather Review*, Vol. 109, No. 11, 1981, pp. 2260-2286.

[2]   D. Cariolle, A. Lasserre-Bigorry, J.-F. Royer and J.-F. Geleyn, "A General Circulation Model Simulation of the Springtime Antarctic Ozone Decrease and Its Impact on Mid-Latitudes," *Journal of Geophysical Research*, Vol. 95, No. 2, 1990, pp. 1883-1898.

[3]   P. J. Rasch and D. L. Williamson, "The Sensitivity of a General Circulation Model Climate to the Moisture Transport Formulation," *Journal of Geophysical Research*, Vol. 96, No. D7, 1991, pp. 13123-13137.

[4]   H. F. Graf, I. Kirchner, R. Sausen and S. Schubert, "The Impact of Upper-Tropospheric Aerosol on Global Atmospheric Circulation," *Annales Geophysicae*, Vol. 10, No. 9, 1992, pp. 698-707.

[5]   P. A. Stott and R. S. Harwood, "An implicit time-stepping scheme for chemical species in a Global atmospheric circulation model", *Annales Geophysicae*, Vol. 11, 1993, pp. 377-388.

[6]   B. Christiansen, A. Guldberg, A. W. Hansen and L. P. Riishojgaard, "On the Response of a Three-Dimensional General Circulation Model to Imposed Changes in the Ozone Distribution," *Journal of Geophysical Research*, Vol. 102, No. D11, 1997, pp. 13051-13078.

[7]   V. Y. Galin, "Parametrization of Radiative Processes in the DNM Atmospheric Model," *Izvestiya Akademii Nauk, Physics of Atmosphere and Ocean*, Vol. 34, No. 3, 1997, pp. 380-389.

[8]   A.-L. Gibelin and M. Deque, "Anthropogenic Climate Change over the Mediterranean Region Simulated by a Global Variable Resolution Model," *Climate Dynamics*, Vol. 20, No. 4, 2002, pp. 327-339.

[9]   M. Mendillo, H. Rishbeth, R. G. Roble and J. Wroten, "Modelling F2-Layer Seasonal Trends and Day-To-Day Variability Driven by Coupling with the Lower Atmosphere," *Journal of Atmospheric and Solar-Terrestrial Physics*, Vol. 64, No. 18, 2002, pp. 1911-1931.

[10] M. J. Harris, N. F. Arnold and A. D. Aylward, "A Study into the Effect of the Diurnal Tide on the Structure of the Background Mesosphere and Thermosphere Using the New Coupled Middle Atmosphere and Thermosphere (CMAT) General Circulation Model," *Annales Geophysicae*, Vol. 20, No. 2, 2002, pp. 225-235.

[11] U. Langematz, A. Claussnitzer, K. Matthes and M. Kunze, "The Climate during Maunder Minimum: A Simulation with Freie Universitat Berlin Climate Middle Atmosphere Model (FUB-CMAT)," *Journal of Atmospheric and Solar-Terrestrial Physics*, Vol. 67, No. 1-2, 2005, pp. 55-69.

[12] I. V. Mingalev and V. S. Mingalev, "The Global Circulation Model of the Lower and Middle Atmosphere of the Earth with a Given Temperature Distribution," *Mathematical Modeling*, Vol. 17, No. 5, 2005, pp. 24-40.

[13] I. V. Mingalev, V. S. Mingalev and G. I. Mingaleva, "Numerical Simulation of Global Distributions of the Horizontal and Vertical Wind in the Middle Atmosphere Using a Given Neutral Gas Temperature Field," *Journal of Atmospheric and Solar-Terrestrial Physics*, Vol. 69, No. 4-5, 2007, pp. 552-568.

[14] I. V. Mingalev, O. V. Mingalev and V. S. Mingalev, "Model Simulation of Global Circulation in the Middle Atmosphere for January Conditions", *Advances in Geo-*

sciences, Vol. 15, No. 4, 2008, pp. 11-16.

[15] I. V. Mingalev, V. S. Mingalev and G. I. Mingaleva, "Numerical Simulation of the Global Neutral Wind System of the Earth's Middle Atmosphere for Different Seasons," *Atmosphere*, Vol. 3, No. 1, 2012, pp. 213-228.

[16] I. V. Mingalev and V. S. Mingalev, "Numerical Modeling of the Influence of Solar Activity on the Global Circulation in the Earth's Mesosphere and Lower Thermo-

sphere," *International Journal of Geophysics*, Vol. 2012, 2012, Article ID: 106035.

[17] J. M. Picone, A. E. Hedin, D. P. Drob and A. C. Aikin, "NRLMSISE-00 Empirical Model of the Atmosphere: Statistical Comparisons and Scientific Issues," *Journal of Geophysical Research*, Vol. 107, No. A12, 2002, pp. 1-16.

[18] A. M. Obukhov, "Turbulence and Dynamics of Atmosphere," Hydrometeoizdat, Leningrad, 1988.

14

Decadal Changes in the Near-Surface Air Temperature in the Western Side of the Antarctic Peninsula

Jorge F. Carrasco[1,2,3]

[1]Dirección Meteorológica de Chile, Estación Central, Chile
[2]Centro de Estudios Científicos, Valdivia, Chile
[3]Universidad de Magallanes, Punta Arenas, Chile

ABSTRACT

An analysis of the minimum air temperature behavior was carried out for the southern tip of South America and the western side of the Antarctica Peninsula. Punta Arenas shows an overall annual warming of 0.15°C per decade during the 1960-2010 period, although this occurred mainly in the summer and winter seasons. The trend of the air temperature in the western side of the Antarctic Peninsula shows an increase until around 2000, but the warming rate during the last 2001-2010 decade has been less than previous decades; in particular, meteorological stations in King George Island show slight cooling. The lineal annual warming per decade as shown by Bellingshausen, Verndsky/Faraday and Rothera stations are 0.26°C ± 0.75°C, 0.55°C ± 1.26°C and 0.69°C ± 1.31°C; for the respectively, 1969-2010, 1951-2010 and 1978-2010 periods. These rates of warming are slightly lower than those found for the same stations but for the 1969-2000, 1951-2000 and 1978-2000 periods.

Keywords: Air Temperature; Exponential Filter; Global Warming; Antarctic Peninsula

1. Introduction

During the last decades global warming has becoming an increasing issue around the world. The physical science basis behind the observed and predicted changes, as well as, the implication of the climate change has been a permanent preoccupation since the eighties. On one hand, the science community investigates and produces the assessment reports around every 6 years about the state-of-the-art of what is known about climate change [1-3]; and on the other, govern representatives meet every year (the United Nations Framework Convention on Climate Change) with the aim of stabilizing the greenhouse gas concentrations in the atmosphere to prevent dangerous and irreversible anthropogenic changes in the climate system. Overall, the continental air temperature has increased by 0.84°C - 1.12°C over the period 1901-2010, with the rate of warming increasing during the last 2 - 3 decades [4], or 0.13°C ± 0.13°C per decade according with Hansen *et al.* [5]. The West Antarctica region which includes the Antarctic Peninsula (AP, hereafter) and the Marie Byrd Land is the region where the increased air temperature is the highest registered on the planet during the last few decades [6-10], as a consequence of the an-

thropogenic greenhouse increased. In contrast, the rest of the Antarctic continent does not show the same rate of warming but rather a slight cooling [11,12]. However, an analysis of the mid-tropospheric air (above the inversion layer) the whole Antarctica shows an increase in temperature [13].

Most of the Antarctic continent is located southern of 70°S surrounded by the southern oceans with average near-surface air temperatures below 0°C all year-round. The AP is a relative narrow mountain barrier that extends northward from Ellsworth Land for about 1300 km until near 63°S, leaving to the north the Drake Passage, to the west the southern Pacific Ocean and to the east the southern Atlantic Ocean. The AP is a complex orographic feature with several ice shelves adjacent to its coast and several glaciers draining into the ice shelves or directly to the ocean.

The mean sea-level pressure [14] in the southern polar region indicates that the AP is within the surface circumpolar trough affected by the atmospheric cyclonic circulations yielding a prevailing north/northwesterly airflow in its western side, and a prevailing southerly wind in its eastern side. The steep mountains and the nearly permanent surface circulations result into two distinct climatic environments at each side of the Peninsula. The north/northwesterly winds in the western side bring relatively warmer and moister air, defining a *sub-polar climate*. While, the southerly winds in the eastern coast bring cold air from the continent and define a *polar continental climate*. This climatic difference is also manifested by other parameters like cloudiness, relative humidity, and precipitation/accumulation which are higher over the western side [15].

As a consequence of the regional warming in the AP, ice shelves adjacent to the peninsula have collapsed and retreated during recent decades [16]. The ice shelves do not contribute directly to sea level rise; however, their disintegration can change the dynamic of glaciers that before the collapse drained onto solid ice shelves, and after they can directly drain to the ocean contributing to sea level rise [17]. Here is, therefore, the importance of monitoring and understanding the atmospheric changes in the AP.

The worldwide increase in the air temperature, with higher rate in the last 2 - 3 decades [5] has not been observed in the northern tip of the AP during the last 2001-2010 decade [12,18,19]; therefore, it remains to be investigated if this behavior is a signal of a natural mechanism overcoming the anthropogenic forcing, or a response to local environmental changes. Here, an analysis based on selected stations located in the western side of the AP is presented to evaluate the warming of the region (**Figure 1**) on decadal and long-term time scales. The air temperature series from Punta Arenas is included in the analysis to compare with the changes in the southern tip of South America.

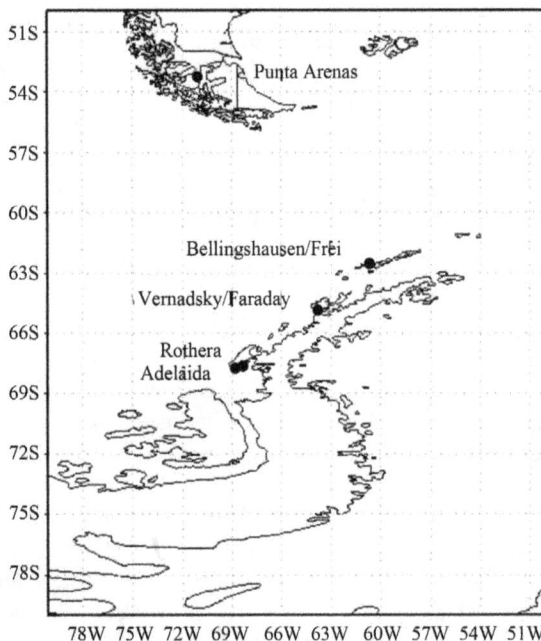

Figure 1. Location map with the geographical distribution of the meteorological stations.

2. Methods

Daily and monthly data of air temperature for Punta Arenas (53.0S, 70.9W, 37 m asl), Bellingshausen/Frei (62.2S, 58.9W, 10 m asl), Vernadsky/Faraday (65.4S, 64.4W, 11 m asl), Rothera (67.5S, 68.1W, 32 m asl) and Adelaida (67.8S, 68.9W, 26 m asl); were obtained from the Dirección Meteorológica de Chile and from the British Antarctic Survey. This last institute allows downloadding meteorological data from the scar_egoma google interface and the Met READER Project (www.antarctica.ac.uk/met/READER/) [8,20]. The daily minimum air temperature recorded in Punta Arenas and Frei, and the daily air temperatures recorded at 12:00 UTC (Universal Time Coordinate) were used to calculate daily-decadal averages. Then the data were smoothed by applying an exponential filter [21] given by:

$$y_t = cx_t + (1-c)y_{t-1} \quad t = 2,3, \qquad n \qquad (1)$$

$$z_t = cy_t + (1-c)z_{t+1} \quad t = (n-1),(n-2), \qquad 1 \qquad (2)$$

where Equations (1) and (2) are respectively the first forward and second backward smoothing. The first value y_t is the average of the first 10 values of the series and z_t corresponds to the final value of the first smoothing. The degree of smoothing (c) can range from 1 (maximum, reproduced the original data) to 0 (it gives a straight line). Here, it was used $c = 0.05$ for smoothing the decadal average annual cycle, this allows to filter out the high inter-daily variability. For the time series analysis, $c = 0.2$ was preferred in order to filter out the high inter-annual variability but to keeps the longer term behavior. Graphs in **Figure 2** were constructed using the minimum or the 12 UTC air temperatures, while the trends in **Table 1** were calculated using the monthly means.

3. Results

The air temperature analysis shows an overall warming of 0.73°C during 1960-2010 period, in the southern tip of South America as revealed by Punta Arenas station. However, analyzing the time series for the periods before and after the 1976/1977 climate shift [22], it can be seen that for the 1961-1976 period the air temperature behavior does not reveal a trend while the 1978-1995 period shows a cooling. Only after the mid-90s the time series reveals a positive trend. The analysis indicates that warming observed in Punta Arenas in the last 5 decades is mostly associated with the shift of the atmospheric-ocean system over the North Pacific Ocean, in other words, by the Pacific Decadal Oscillation (PDO) [22], also called the Pacific Decadal Variability (PDV) [21].

Decadal average of the daily air temperatures and the annual long-term are shown in **Figure 2**. All stations show an overall increase in air temperature, however, the

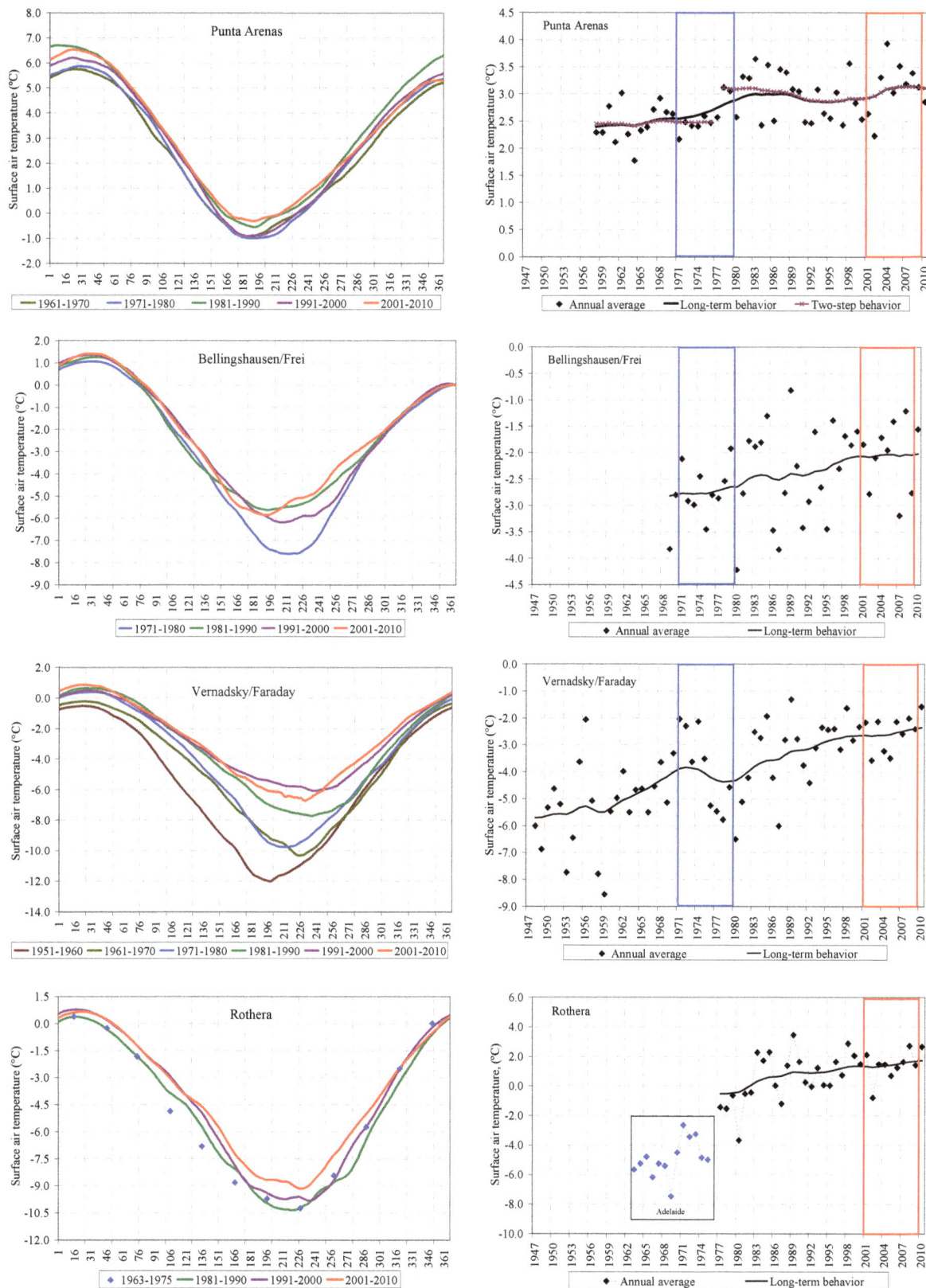

Figure 2. Decadal average of the annual cycle (left side) and annual average long-term (right side) behavior of the near-surface air temperature, after applying the exponential filter. Blue rhombuses in Rothera graph are monthly and annually average from the nearby Adelaide station.

Table 1. Annual and seasonal linear trends (per decade) for stations located in the Antarctic Peninsula updated from Turner *et al.* **[8]. Italics (bold) numbers are significant at 10% (5%), while those italic/bold are significant at 1%. Added is Frei station [20]. [#]Values are from Bromwich** *et al.* **[10].**

Station	Annual	Spring	Summer	Autumn	Winter	Period
Frei	0.21 ± 0.35	0.22 ± 0.46	0.05 ± 0.09	$\mathbf{0.39 \pm 0.33}$	0.62 ± 2.40	1970-2000
	0.11 ± 0.71	-0.06 ± 0.87	-0.06 ± 0.43	0.17 ± 0.96	0.39 ± 1.86	1970-2010
Rothera	$\mathbf{1.01 \pm 1.42}$	1.06 ± 1.53	0.36 ± 0.57	$\mathbf{1.37 \pm 1.46}$	1.73 ± 2.79	1978-2000
	$\mathbf{0.69 \pm 1.31}$	$\mathbf{0.76 \pm 1.51}$	0.18 ± 0.56	$\mathbf{0.67 \pm 1.32}$	1.09 ± 2.87	1978-2010
Vernadsky/Faraday	$\mathbf{0.56 \pm 0.43}$	0.25 ± 0.44	$\mathbf{0.24 \pm 0.17}$	$\mathbf{0.63 \pm 0.60}$	$\mathbf{1.09 \pm 0.88}$	1951-2000
	$\mathbf{0.55 \pm 1.26}$	0.39 ± 1.43	0.23 ± 0.63	0.52 ± 1.62	$\mathbf{1.06 \pm 2.79}$	1951-2010
Bellingshausen	$\mathbf{0.35 \pm 0.46}$	-0.10 ± 0.47	$\mathbf{0.30 \pm 0.20}$	0.51 ± 1.05	0.58 ± 0.97	1969-2000
	0.26 ± 0.75	0.10 ± 0.91	0.13 ± 0.45	0.33 ± 1.04	0.48 ± 1.89	1969-2010
Esperanza	0.41 ± 0.42	-0.07 ± 0.57	$\mathbf{-0.07 \pm 0.57}$	$\mathbf{0.43 \pm 0.34}$	0.51 ± 0.82	1961-2000
	$\mathbf{0.33 \pm 1.08}$	0.14 ± 1.52	0.14 ± 1.52	0.37 ± 0.80	0.32 ± 2.20	1961-2010
Marambio	<90%	-0.80 ± 10.5	<90%	<90%	0.81 ± 1.53	1971-2000
	$\mathbf{0.45 \pm 1.17}$	0.34 ± 1.87	$\mathbf{0.51 \pm 0.87}$	0.62 ± 2.43	0.40 ± 2.72	1971-2010
Orcadas	$\mathbf{0.20 \pm 0.10}$	$\mathbf{0.15 \pm 0.14}$	$\mathbf{0.15 \pm 0.06}$	$\mathbf{0.21 \pm 0.16}$	$\mathbf{0.27 \pm 0.24}$	1904-2000
Central West Antarctica[#]	$\mathbf{0.47 \pm 0.23}$	$\mathbf{0.82 \pm 0.40}$	$\mathbf{0.30 \pm 0.27}$	-	$\mathbf{0.54 \pm 0.51}$	1958-2010

Antarctic ones reveal that the significant warming took place in winter months and before the 1991-2000 decade. In fact, the annual long-term behavior reveals no significant trend during the 2001-2010 period as suggested by Setzer *et al.* [18,19]. This also concurs with the results of Bromwich *et al.* [10] who found a substantial linear increase in annual temperature of 2.4°C ± 1.2°C for the Central West Antarctica during the 1958-2010 period, but no positive trend for the last decade can be inferred from their **Figure 2(a)**) [10]. No relation between the behavior of the surface air temperature for the Antarctic stations and the PDO was found.

Table 1 shows a partial update of the linear regression of **Table 1** from Turner *et al.* [8] for the stations located in the AP. Included are the trends for Frei station and the decadal trends found by Bromwich *et al.* [10]. Winter is the season with larger positive trends, except for Central West Antarctica region which experience the larger warming in spring season [10]. Trends obtained for periods ending in 2000 [8] are lower than those extended for the periods ending in 2010 [20]. This is because the air temperature does not show the previous rate of warming for the 2001-2010 decade (**Figure 2**).

4. Discussion

The southern tip of South America shows an overall warming but the significant increase of the air temperature concurs with the 1976/1977 climate shift. This relation is not found for Antarctic stations. The AP is one of the regions on Earth where annual air surface temperatures have experienced substantial warming at higher rate than global average [6-10]. However, decadal behavior

of the air surface temperature reveals that this warming occurred before the last decade (2001-2010), at least in the northern tip of the AP. The warming has been related with the strengthening of the Antarctic Annular Oscillation (AAO) [23-25], *i.e.*, a positive AAO trend, resulting in a southward displacement of the westerlies, which allows more relatively warm air passing over the AP. However, the summer warming in the Marie Byrd Land is not supported by the AAO as indicated by Bromwich *et al.* [10].

Results from this analysis show the same overall upward trend of the surface air temperature in the western side of the AP from previous works, but one intriguing result is that the air temperature reveals insignificant changes during the last decade in the northern tip of the AP as previously suggested by Setzer *et al.* [18,19]. In fact, Vernadsky/Faraday and Bellingshausen stations show a cooling in some spring-summer and winter months, respectively. The lineal annual warming per decade as shown by Bellingshausen, Verndsky/Faraday and Rothera stations are 0.26°C ± 0.75°C, 0.55°C ± 1.26°C and 0.69°C ± 1.31°C; for the respectively, 1969-2010, 1951-2010 and 1978-2010 periods. These rates of warming are slightly lower than those found for the same stations but for the 1969-2000, 1951-2000 and 1978-2000 periods.

Figure 3 shows the long-term behavior of the global air temperature anomaly obtained from Climatic Research Unit (CRU) time series datasets of the University of East Anglia [26], along with those obtained for the Antarctic stations (upper part). Vernadsky/Faraday and Rothera stations show larger air temperature increases than the global average, but the annual trend is nearly flat during the last decade (2001-2010) at Bellingshausen/

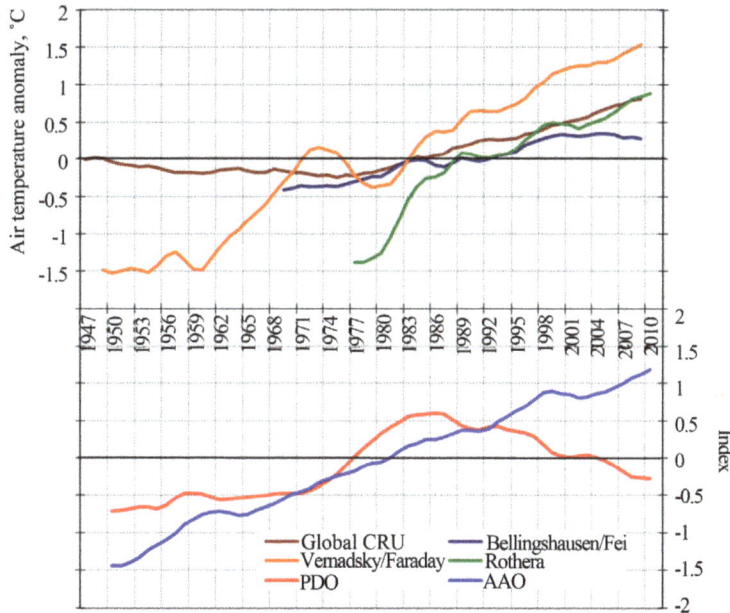

Figure 3. Long-term behavior using annual anomalies of the global air temperature and those obtained for Bellingshausen/ Frei, Vernadsky/Faraday and Rothera stations, after applying the exponential filter.

Frei station. On the other hand, the southernmost stations, Verndsky/Faraday and Rothera, show a continuing positive trend for the last decade. **Figure 3** also includes the long-term behavior of the PDO and AAO as revealed by their respective indexes, after applying the exponential smoothing to filter out the inter-annual variability. A positive PDO is associated with a high pressure anomaly region centered to the west of the AP [27], and therefore, favors the relative cold air advection toward the western side of the AP. The fact that the PDO shows a positive (negative) trend until (after) mid 80s might suggest that cold air advection over the western side of the AP was becoming weaker (stronger) or less (more) frequent. However, the correlations between the PDO and the annual air temperature at each station are low and not statistically significant (**Table 2**), implying that the PDO does not play a significant role as mechanism supporting the observed warming (or eventual cooling). While on the other hand, positive AAO values indicate a predominant negative pressure anomalies in the Antarctic continent, surrounded by positives anomalies around 40°S - 60°S [24,25]. This results in a southward displacement and strengthening of the westerly airflow favoring the warm air advection on the AP. The correlations between AAO index and the annual surface air temperature at each station are above 0.65 (**Table 2**) and statistically significant at $\alpha = 0.05$. Therefore, the overall behavior of the AAO supports the warming of the AP [12].

In summary, the lineal trend that includes the surface air temperature data until 2010 in the western side of the AP, mainly in its northern tip; shows warming rate less

Table 2. Matrix of correlation (Pearson) among the stations and the AAO and PDO indexes, obtained from the original data without filtering. Bold numbers are statistically significant at $\alpha = 0.05$ (B = Bellinghausen, V/F = Vernadsky/ Faraday, R = Rothera).

	B	V/F	R
AAO	**0.68**	**0.66**	**0.78**
PDO	−0.31	−0.33	−0.30
Vernadsky/Faraday	**0.57**	-	**0.79**
Bellingshausen	-	-	**0.76**

than those obtained when the series ended in 2000 [8]. This is a result of less warming rate during 2001-2010 period, or even a slight cooling in the northern tip of the AP. However, this different behavior, observed in the northern tip of the AP during the last decade, dose not necessary support the assertion that the trend or the rate of global warming is less after mid 90s [28], so that the explanation can be found in the local environmental changes and interaction among atmosphere-ocean-sea ice, rather than large scale forcing like stratospheric water vapor concentrations even though this can be an important driver for a global climate change [28].

5. Acknowledgements

This work was supported by the Project Bicentennial Science and Technology Programme of CONICYT (Scientific and Technological National Commission of Chile), Proyecto Anillo Bicentenario ARTG02-2006.

REFERENCES

[1] IPCC, "Climate Change 2007: The Physical Science Basis. Contribution of Working Group I to the Fourth Assessment Report of the Intergovernmental Panel on Climate Change," In: S. Solomon, D. Qin, M. Manning, Z. Chen, M. Marquis, K. B. Averyt, M. Tignor and H. L. Miller, Eds., Cambridge University Press, Cambridge and New York, 2007.

[2] IPCC, "Climate Change 2007: Impacts, Adaptation and Vulnerability. Contribution of Working Group II to the Fourth Assessment Report of the Intergovernmental Panel on Climate Change," In: M. L. Parry, O. F. Canziani, J. P. Palutikof, P. J. van der Linden and C. E. Hanson, Eds., Cambridge University Press, Cambridge, 2007.

[3] IPCC, "Climate Change 2007: Mitigation, Contribution of Working Group III to the Fourth Assessment Report of the Intergovernmental Panel on Climate Change," In: B. Metz, O. R. Davidson, P. R. Bosch, R. Dave, L. A. Meyer, Eds., Cambridge University Press, Cambridge and New York, 2007.

[4] P. D. Jones, D. H. Lister, T. J. Osbom, C. Harphan, M. Salmon and C. P. Morice, "Hemispheric and Large-Scale Land-Surface Air Temperature Variations: An Extensive Revision and Update to 2010," *Journal of Geophysical Research*, Vol. 117, 2012, Article ID: D05127.

[5] J. Hansen, R. Ruedy, M. Sato and K. Lo, "Global Surface Temperature Change," *Review of Geophysics*, Vol. 48, No. 4, 2010, Article ID: RG4004.

[6] D. G. Vaughan, G. J. Marshall, W. M. Connolley, C. Parkinson, N. R. Mulvaney, D. A. Hodgson, J. C. King, C. J. Pudsey and J. Turner, "Recent Rapid Regional Climate Warming on the Antarctic Peninsula," *Climatic Change*, Vol. 60, No. 3, 2003, pp. 243-274.

[7] Q. Din, D. S. Batisti and M. Kuttel, "Winter Warming in West Antarctica Caused by Central Tropical Pacific Warming," *Nature Geoscience*, Vol. 4, No. 6, 2011, pp. 398-403.

[8] J. Turner, J. C. Comiso, G. J. Marshall, T. A. Lachan-Cope, M. A. Carleton, P. D. Jones, V. Lagun, P. A. Reid and S. Iagovkina, "Antarctic Climate Change during the Last 50 Years," *International Journal of Climatology*, Vol. 25, No. 8, 2005, pp. 279-294.

[9] Q. Din, D. S. Batisti and M. Kuttel, "Winter Warming in West Antarctica Caused by Central Tropical Pacific Warming," *Nature Geoscience*, Vol. 4, No. 6, 2011, pp. 398-403.

[10] D. H. Bromwich, J. P. Nicolas, A. Monaghan, M. A. Lazzara. L. M. Keller, G. A. Weidner and A. B. Wilson, "Central West Antarctica among the Most Rapidly Warming Regions on Earth," *Nature Geoscience*, Vol. 6, No. 2, 2012, pp. 139-145.

[11] P. T. Doran, J. C. Priscu, W. B. Lyons, J. E. Walsh, A. G. Fountain, D. M. McKnight, D. L. Moorhead, R. A. Virginia, D. H. Wall, G. D. Clow, C. H. Fritsen, C. P. McKay and A. N. Parsons, "Antarctic Climate Cooling and Terrestrial Ecosystem Response," *Nature*, Vol. 415, No.

6871, 2002, pp. 517-520.

[12] A. J. Monaghan, D. H. Bromwich, W. Chapman and J. C. Comiso, "Recent Variability and Trends of Antarctic Near-Surface Temperature," *Journal of Geophysical Research*, Vol. 113, No. 457, 2008, Article ID: D04105.

[13] J. Turner, T. A. Lachan-Cope, S. Colwell, G. J. Marshall, and W. M. Connolley, "Significant Warming of the Antarctic Winter Troposphere," *Science*, Vol. 311, No. 5769, 2006, pp. 1914-1917.

[14] W. Schwerdtfeger, "Weather and Climate of the Antarctic," Elsevier, New York, 1984.

[15] D. G. Vaughan, J. L. Bamber, M. Giovinetto, J. Russell and A. P. R. Cooper, "Reassessment of Net Surface Mass Balance in Antarctica," *Journal of Climate*, Vol. 12, No. 4, 1999, pp. 933-946.

[16] A. J. Cook and D. G. Vaughan, "Overview of Areal Changes of the Ice Shelves on the Antarctic Peninsula over the Past 50 Years," *The Cryosphere*, Vol. 4, No. 1, 2010, pp. 77-98.

[17] E. Rignot, G. Casassa, P. Gogineni, W. Krabill, A. Rivera and R. Thomas, "Accelerated Ice Discharge from the Antarctic Peninsula Following the Collapse of Larsen B Ice Shelf," *Geophysical Research Letters*, Vol. 31, No. 18, 2004, Article ID: L18401.

[18] A. W. Setzer, F. E. Aquino and R. M. O. Romao, "Climate Tendencies in the South Shetlands: Was 1998 a Climate Divider?" *Antarctic Peninsula Climate Variability: Observed, Models and Plans for IPY Research*, University of Colorado, Boulder, 14-16 May 2006.

[19] A. W. Setzer and R. M. O. Romao, "Recent Cooling in the North of the Antarctic Peninsula," *4th SCAR Open Science Conference*, Buenos Aires, 2010.

[20] J. F. Carrasco, "Red de Estaciones de Observación Atmosférica en la Antártica, una Colaboración Internacional Para la Investigación del Cambio Climático," *Anales Instituto Patagonia*, Vol. 40, No. 1, 2012, pp. 57-63.

[21] B. Rosenbluth, H. A. Fuenzalida and P. Aceituno, "Recent Temperatures Variations in Southern South America," *International Journal of Climatology*, Vol. 17, 1997, pp. 67-85.

[22] B. S. Giese, S. C. Urizar and N. S. Fuckar, "Southern Hemisphere Origins of the 1976 Climate Shift," *Geophysical Research Letter*, Vol. 29, No. 2, 2002, p. 1014.

[23] C. Wang, C. Deser, J.-Y. Yu, P. DiNezio and A. Clement, "El Niño Southern Oscillation (ENSO): A Review," In: P. Glymn, D. Manzello and I. Enochs, Eds, *Coral Reefs of the Eastern Pacific*, Spring Science Publisher, 2012.

[24] D. W. J. Thompson and S. Solomon, "Interpretation of recent Southern Hemisphere Climate Change," *Science*, Vol. 296, No. 5569, 2002, pp. 895-899.

[25] D. W. J. Thompson and J. M. Wallace, "Annular Mode in the Extratropical Circulation. Part I: Monthly-to-Month Variability," *Journal of Climate*, Vol. 13, 2000, pp. 1000-1006.

[26] I. Harris, P. D. Jones, J. T. Osborn and D. H. Lister, "Update High-Resolution Grids of Monthly Climatic Observations," *International Journal of Climatology*, 2012, in Press.

[27] R. D. Garreaud and D. S. Battisti, "Interannual (ENSO) and Interdecadal (ENSO-Like) Variability in the Southern Hemisphere Tropospheric Circulation," *Journal of Climate*, Vol. 12, No. 7, 1999, pp. 2113-2123.

[28] S. Salomon, K. H. Rosenlof, R. W. Portmann, J. S. Daniel, S. M. Davis, T. J. Sanford and G.-K. Plattner, "Contributions of Stratospheric Water Vapor to Decadal Changes in the Rate of Global Warming," *Science*, Vol. 327, No. 5970, 2010, pp. 1219-1223.

15

Calculated and Experimental Regularities of Cloud Microstructure Formation and Evolution

Nikolai Romanov, Vasiliy Erankov

Institute of Experimental Meteorology, FBSI RPA "Typhoon", Obninsk, Russia

ABSTRACT

Based on the model of regular condensation it was found that at low concentrations of CN (LC mode) at a height of about 10 m from the condensation level narrow spectra of cloud drop are formed. Their dispersion quickly decreases with increasing height. For high concentrations (HC mode) broad spectra are formed immediately due to the absence of separation into growing drops and CN covered with water. The process of spectra evolution at a constant height results, in all the cases, in the appearance of asymptotic spectra with a relative width $rb \geq 0.215$. To approximate these calculated asymptotic spectra, the modified gamma-distribution with the fixed parameter $\alpha = 3$ and a variable parameter γ are most suitable. For the intermediate spectra applicable are the simpler mirror-transformed known distributions. The comparison of the above distributions with the experimental spectra of a fog artificially formed in the Big Aerosol Chamber (BAC) of RPA "Typhoon" and the spectra of the morning fog and super cooled stratiform clouds demonstrated their good agreement. The phenomenon of multimodal spectra formation at a sharp rise of stratiform clouds with the velocity more than 0.1 - 0.3 m/s was theoretically shown and experimentally confirmed. The effect of CN high concentrations, evolution processes and sharp fluctuations of vertical velocities on the formation of cloud spectra observed in nature is discussed.

Keywords: Clouds; Fogs; Cloud Microstructure; Spectra Approximation

1. Introduction

This Studying of the processes of cloud and fog microstructure formation during their initiation plays an important role in the understanding of the mechanisms of the microstructure formation and interaction of the drops formed with the environmental pollutants. The connection of a cloud microstructure with atmospheric aerosol composition influences the climate forming conditions. Of importance is also the problem of precipitation enhancement at cloud modification with hygroscopic agents. But up to the present, there is no generally accepted mechanism of experimentally observed in natural conditions rather broad cloud spectra [1], because the mechanism of regular condensation does not explain this aspect. In the monograph [2] devoted to the initial stage of condensation it is shown that within the frameworks of the theory of regular condensation at a constant rate of updraft the drop spectrum becomes narrower with altitude. In the same way, by analyzing dominant estimations for the Junge condensation nuclei (CN) the regularities of cloud drop spectra were obtained. These are the exponential function of maximum oversaturation and the concentration of the drops formed unambiguously connected with it on the velocity of updraft and CN concentration. Corresponding expressions for the exponents are given. In [3] broad drop spectra, formed due to evaporation of fine drops, were obtained for high concentrations of CN. But this result has not been analyzed from the physical viewpoint and was not considered later on. In the recently published work [4] with the use of a numerical model of regular condensation confirmed is the dependence, obtained in [2] of drop concentration on the velocity of air mass updraft, and its corrected dependence on CN concentration is given. The dependence of drop concentration on air mass temperature and pressure is given as well. But the author limited himself by the case of small concentrations at which the cloud drop concentration formed at the initial moment remained constant at a further updraft

Analytical solutions were used [5] to study the dependences of integral values of cloud spectra on environmental parameters and the velocity of air mass updraft for a lognormal distribution of CN. The proposal to use some new parameters of drop spectra was made in [6]. But a comprehensive understanding of cloud spectra can be obtained only with the development of an adequate model of processes involved in the formation of cloud spectra.

The attempts to find an answer in solving this problem, including the problem of broad spectra, are described in detail in the review [7]. It is emphasized in this review that the model of regular condensation with a constant velocity of air mass updraft results in narrowing with the increasing updraft height cloud spectra in contrast to the broad and even miltimodal spectra observed in nature. In the works considered by them earlier, where an answer to this question was sought, most frequently proposed were the mechanisms of shifting of several cloud portions with different drop spectra, including those with pure air. Special attention is paid at the situation when the spectra of separate cells are mixed—these spectra are formed at different velocities of updrafts at the cloud base. In more detail this situation is analyzed in [8] where the situation with increasing updraft velocities from zero to 1.5 m/s is considered.

A possibility of bimodal spectra formation in the cells of a low velocity at the cloud base and its strong increase in the middle section are considered in [7]. But for this case there exist doubts in the sufficiency of the amounts of condensation nuclei left after the formation of an initial cloud drop spectrum for the formation of additional drops. These doubts are especially enhanced after the analysis of the aerosol-cloud system [9], where it is concluded that at the formation of cloud drops almost 80% of aerosol go into the drops.

When searching analytic approximations of cloud spectra [7] was thinking that at a modern level of computer technologies more adequate for studies of cloud processes is the direct computation of cloud drop spectra. Numerous researchers are likely to support the same position. Therefore, at present a three-parametric modified gamma-distribution, introduced in the monograph [10], is used (in particular) [11] to analyze optical properties of clouds. Because of a great number of parameters in this distribution, an unambiguous choice of the latter for concrete cloud structures presents difficulties.

To remove the above-mentioned gaps in the present paper a series of studies on the cloud microstructure formation at the stage of air mass updraft and its further evolution at a constant height based on the numerical model of regular condensation constructed by the authors is presented. As a result, a qualitative difference of cloud spectra formation at low and high concentrations of condensation nuclei (the LC and HC) has been found. The formation of spectra with a negative asymmetry was also

found both at the updraft and evolution, that required the search of new approximation relationships for them. Qualitative and quantitative regularities of cloud drop concentration formation, relative spectrum widths and other characteristics depending on CN concentrations and also the updraft velocity and environmental temperature were obtained.

The calculated regularities were compared with the observed ones during the process modeling in the cloud chamber and also with the stratified clouds and morning fogs spectra measured by the authors. As a result of such studies, an identification of processes occurring in real clouds leading to broad and even multi-modal spectra is given. A recommendation is given for an analytic description of cloud spectra with the use of the known mirror symmetric distributions and a modified gamma-distribution with the parameter $\alpha = 3$.

2. Calculated Regularities of Cloud Spectra Formation

2.1. Basic Relationships and General Pattern of Calculation Spectra Formation

First, For studying calculated regularities of cloud spectra formation we shall limit ourselves by a two-parametric inverse power distribution of the condensation nuclei over their equivalent radii r_c.

$$F(r_c) = C_c \frac{(v-1)}{r_{c0}} \left(\frac{r_c}{r_{c0}} \right)^{-v} \tag{1}$$

where C_c determines the number of particles with $r_c > r_{c,0}$ per 1 mg of air. This distribution is transformed at a 100% humidity into an equilibrium distribution over the radii r_w of CN covered with water being droplets of a salt particles dissolved in water [2]

$$F(r_c) = C_c \frac{(v_0-1)}{r_{w0}} \left(\frac{r_w}{r_{w0}} \right)^{-v_0} \quad v_0 = \frac{2v+1}{3} \tag{2}$$

Under an approximation of the equality of the surface tension σ_w and the solution density ρ with the corresponding characteristics of pure water, the value of r_w is determined by the number of salt molecules n_c multiplied by the Van't-Hoff factor i_c with the relationship [12].

$$r_w^2 = \frac{3m_w R_w T}{8\pi\sigma_w} i_c n_c \tag{3}$$

Here m_w is the mass of a water molecule, T is temperature, R_n is the gas constant of water vapor.

The use of distribution (3) makes it possible to avoid the numerical procedure of studying the process of dry CN transformation at increasing relative humidity of air. For calculating the processes of drop formation the CN interval chosen is separated into n channels with an equal number of CCN in them, that is not necessarily to be an integer. At a growth of a drop the number of salt molecules in it remains constant, and its growth depends on the precipitation of the number of water molecules at increasing relative humidity f. The Van't-Hoff factor is considered constant, the distribution of salt is homogeneous. The formulae for drop growth are taken from [2]. The value of vapor molecules accommodation is taken equal to 0.98.

At the first stage during the time interval Δt is determined by the variation of thermodynamic parameters due to the air mass updraft. At the second stage an increment of mass that occurred during this time is calculated for every nucleus. Then, the value of f is corrected for the total growth of mass on drops. It has been stated experimentally that stable results of calculations were obtained at $\Delta h \leq 1$ cm and $n \geq 100$. For the drop spectrum formation the condensation nuclei from Equation (2) were included, the radii of which were greater than r_{wb}. Its value was chosen from the condition that there would stay about 5% of nuclei cowered with water after the drop spectrum formation. At varying CN concentration it appeared that during the air mass updraft two different modes of cloud spectra formation (that will be named the modes of low (LC) and high (HC) concentrations) may be realized

The difference of these regimes is illustrated by the curves in **Figure 1**, where for the process of air mass updraft with the velocity of 1 m/s to the height of 30 m and at its further evolution at a ceased updraft are the trajectories of drop growth at the separation of the whole spectrum into 100 channels (the upper figures). In the lower figures present the trend of the integral values: asymmetry coefficients k_{as}, relative breadth rb, drop concentration C_d, oversaturation $\varepsilon = \ln f$ and $u_m = \max (u = r/\langle r \rangle)$.

The vertical lines in the upper figures denote the moments of updraft stopping at $t = 30$ s, and the numbers at the trajectories correspond to the serial number of the channels. From **Figure 1(a)** the process of growing drops separation from the CN covered with water at the moment when maximum oversaturation occurred at the 10-th second is distinctly seen. For the case presented in **Figure 1(b)** the concept of a drop is already smeared because in the spectrum different from the stationary spectra of CN covered with water there exist both the growing and evaporating drops. Therefore, all the drops covered with water with non-stationary sizes or those exceeding any size (in the given case—0.4 micrometer) can be considered drops.

From the figures presented it is seen that under the regime of a constant temperature fine drops are evaporating and large drops are growing. Such a process takes place at an air mass updraft at small velocities and high concentrations of CN. The distribution function formed in this case has a negative asymmetry coefficient and a maximum shifted to the right relatively to an average value.

2.2. Approximation of Calculated Spectra

The well-studied and frequently used for the representation of cloud spectra two-parameter gamma-distribution, as is seen from the supplement, has positive values of k_{as}

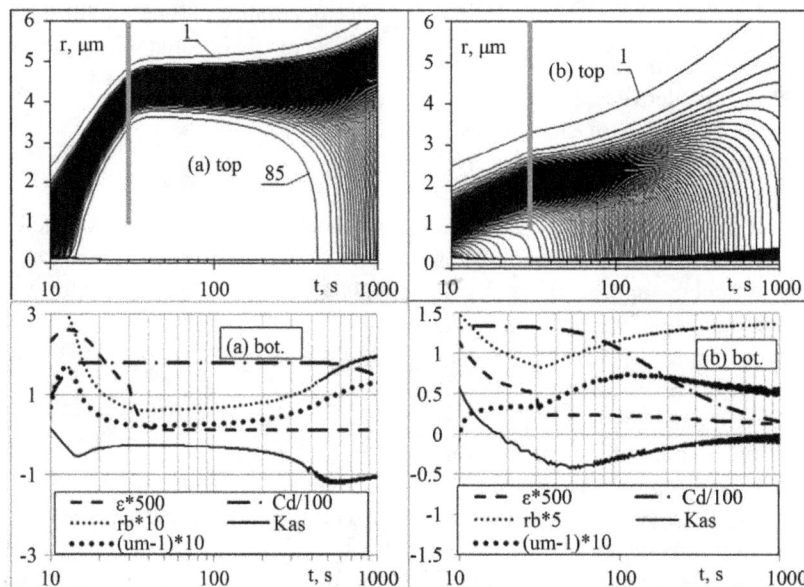

Figure 1. Trajectories of drop growth (the top figures) and the course of ε, C_d, rb, k_{as} and u_m values with multipliers indicated in the legend (the bottom figures). (a) In the mode of LC, $C_c = 10^2$ mg^{-1}; (b) In the mode of HC, $C_c = 10^4$ mg^{-1} ($v = 5$, $T = 20°C$, $n = 100$).

and a maximum shifted to the left. The same peculiarities belong to the log-normal distribution and to that proposed in [13]. Therefore, they in principle are inapplicable for the approximation of the calculated evolution spectra. The selection of expressions in the form of the power law with the exponent from 2.6 to 3 in the left and the inverse power law with the exponent from 8 to 10 in the right wings and parabolic in the spectrum center has demonstrated that such an approach can satisfactorily describe calculation spectra but is practically inconvenient. The for-parameter modified gamma-distribution at a certain combination of the parameters α and γ used in it may lead to negative values of k_{as}. But it should be mentioned that the above-mentioned dimensions are not physically substantiated.

The only one known to us asymptotic distribution [14] f_{lif}, obtained at solving the differential equations describing the growth and evolution of crystals in an oversaturated solution without external crystallization centers. In work [15] was physically substantiated the similarity of these processes with the evolution processes of cloud drop spectra not containing salt. This normalized by unity solution with the use of a relative drop radius $u = r/\langle r \rangle$ is written in the form of

$$f_{jif}(u) = \begin{cases} \dfrac{3^4 e}{2^{\frac{5}{3}}} \dfrac{u^2 \exp\left(-\dfrac{1}{1-2u/3}\right)}{(u+3)^{\frac{7}{3}} (3/2-u)^{\frac{11}{3}}}, & u < u_0 = \dfrac{3}{2}, \\ 0, & u > u_0 = \dfrac{3}{2} \end{cases} \quad (4)$$

The main parameters of this distribution have the following values: $rb = 0.215$, $k_{as} = -0.92$, $u_m = 1.135$, $f(u_m) = 2.135$. It coincides qualitatively with the above properties of numerical calculations, i.e. it gives broad asymptotic spectra with negative asymmetry. The attempts to introduce into this distribution additional parameters, that could lead to a possibility to parameterize the calculated spectra formed on CN, failed.

From the expression it is seen that the value of u operates in the asymptotic solution as an algebraic combination with several constants. The introduction of such combinations in the above-mentioned distributions can be made by the method of their mirror transformations with the use of

$$\delta = u_b - u \quad (5)$$

where u_b is similar to u_0 in Equation (4) and determines the boundary of the non-zero value of the function.

For the value of γ and S-distribution characteristics of which are given in the Supplement the mirror functions ($f_{\gamma m}(u)$ and $f_{Sm}(u)$) normalized by unity are written in the form of

$$f_{\gamma m}(u)\,du = \begin{cases} -\dfrac{\mu^{\mu}}{\Gamma(\mu)} \delta^{\mu-1} \exp(-\mu\delta)\,d\delta, & 0 \le \delta < u_{b\gamma} \\ 0, & \delta \ge u_{b\gamma} \end{cases}$$

$$(6a)$$

$$f_{Sm}(u)\,du = \begin{cases} -\dfrac{q^{q+1}}{\Gamma(q+1)} \dfrac{1}{\delta^{q+2}} \exp\left(\dfrac{-q}{\delta}\right) d\delta, & 0 \le \delta < u_{bS} \\ 0, & \delta \ge u_{bS} \end{cases}$$

$$(6b)$$

In spectra (6a and b) the locations of maxima for the value of δ are correspondingly equal to

$$\delta_{\gamma\,max} = \dfrac{\mu-1}{\mu}; \quad \delta_{S\,max} = \dfrac{q}{q+2} \quad (6c)$$

For the locations of maximum values of functions (6a and b) for the approximated spectra to have the values of u_m the condition of

$$u_b = \delta_{max} + u_m \quad (6d)$$

should be fulfilled.

The substitution of expressions (6c) into (6a and b) give the expressions for maximal values $maxf$ of the function:

$$\max f_{\gamma m} = \dfrac{\mu(\mu-1)^{\mu-1}}{\Gamma(\mu)\exp(\mu-1)}$$

$$(6e)$$

$$\max f_{Sm} = \dfrac{(q+2)^{q+2}}{q\Gamma(q+1)\exp(q+2)}.$$

The problem of choosing the parameters of these distributions is set in the following way. First the location of maximum u_m and the value of $maxf$ for a calculated (experimental) spectrum are determined. After this the values of μ and q can be determined by solving the equation of the direct dependence (6e) with one variable. As far as the inverse dependence in these relationships does not have exact analytic expressions, the approximation relationships giving an error of no more than unity of the second sign after the point at the values of $q \ge 10$ were obtained. They have the form of

$$\mu = 6.29(\max f)^2 - 0.9; \quad q = \mu - 3.2 \quad (6f)$$

After the substitution of the obtained from Equation (6f) values μ or q into Equations (6c) and (6d) the corresponding value of u_b is determined. Thus the problem of finding both parameters of the approximation functions (6a) and (6b) is solved.

Both distributions are the two-parametric ones that turn to zero at the boundary value of $u = u_b$ and do not turn to zero at $u = 0$. The latter obstacle should not embarrass one, because physics of cloud spectra formation points to the presence of the lower boundary of cloud spectra determined by the maximal equilibrium radius of the remaining CN covered with water. The moments of these functions may (with a certain error) be expresses by the moments of the main functions at the substitution of the u_b value by ∞ in the relationship

$$M_m(n,u) = \int_{u_b}^{0} f_m(\delta)(u_b - \delta)^n \, d\delta \qquad (7)$$

For narrow distributions this substitution will not result in a great error. Therefore, below we give the formulae for expressing the first two moments and the relationships for rb obtained with them.

$$M_{\gamma m}(1) = M_{Sm}(1) = u_b - 1; \qquad (8a)$$

$$M_{\gamma m}(2) = (u_b - 1)^2 + \mu^{-1}$$
$$M_{Sm}(2) = (u_b - 1)^2 + (q-1)^{-1}; \qquad (8b)$$

$$rb_{\gamma m} \equiv \frac{1}{(u_b - 1)\sqrt{\mu}}, \quad rb_{Sm} \equiv \frac{1}{(u_b - 1)\sqrt{q-1}}. \qquad (8c)$$

With the use of (relationships) correlations 8(c) and (6c) and (6d) an iteration scheme for determining the parameters of distributions (6a) and (6b) can be constructed on the basis of the known parameters u_m and rb. Such a procedure may be useful for a description of experimental spectra for which due to their cut-up character the definition of max$f$ presents difficulties.

It has been mentioned earlier that a more complex tree-parametric modified gamma-distribution $f_{m\gamma}(u)$ with the parameters α and γ, the general form of which is given in the Supplement, has also the regions (at $\gamma > 3$) with negative k_{as} values. When studying the problem of choosing the appropriate parameters of this distribution to approximate the calculation spectra it appeared that for all the spectra with max$f(u)$ within the limits from 2.3 to 1.5 the calculation spectra at the evolution stage at a constant height are well approximated by the function $f_{m\gamma}(u)$ with the parameter $\alpha = 3$ at the known value of u_m. As a result, the case with the distribution $f_{m\gamma}(u)$ written as

$$f_{m\gamma}(u)\,du = \frac{\gamma}{u_m\Gamma(4/\gamma)}\left(\frac{3}{\gamma}\right)^{4/\gamma}\left(\frac{u}{u_m}\right)^3 \exp\left[-\frac{3}{\gamma}\left(\frac{u}{u_m}\right)^{\gamma}\right] du \qquad (9)$$

So this distribution is reduced to the two-parametric one as well. In particular, its form coincides with the spectrum (4) at the substitution of $u_m = 1.1345$ and $\gamma = 16.3$. The maximum of function (9) occurs at $u = u_m$. It is expressed by a rather complex dependence on the parameter γ not having an exact analytic expression for the inverse function. That is why it is not given here. The approximation expression for the inverse function was

obtained in the form having an error no more than unity of the first sign after the point at $2 < \gamma < 20$

$$\gamma = 2 + 4.3(u_m \max f - 1) + 2.85(u_m \max f - 1)^3 \qquad (9a)$$

The approximation expression between rb and γ with account for the asymptotic value from the Supplement for rb for high values of γ was obtained in the form of

$$\gamma(rb, \alpha = 3) = \frac{2}{3\left(rb - 1/\sqrt{24}\right)^{0.707}} \qquad (9b)$$

Relationship (9b) can be applied directly for determining the parameter γ at approximating experimental spectra by distribution (9).

2.3. Results of Calculated and Approximated Spectra Comparison

The results of calculated and approximated spectra comparison for updraft beginning and ceasing of evolution process at a constant temperature are given in **Figures 2(a)** and **(b)**. For convenience of formation of the figure here and later on the Greek γ is substituted by the Latin g.

From this figure it is seen that the set of distributions f_{mg}, f_{gm} and f_{Sm} makes it possible to adequately reproduce the shape of the calculation spectra during the whole process of formation and evolution of cloud drop spectra. From comparison at other updraft heights, CN concentrations and other parameters it appeared that at all the regimes studied the set of three functions f_{mg}, f_{gm} and f_{Sm} describes rather well the calculated spectra in the range of $0.3 < u < 1.4$. In a general case it is possible to recommend after the definition of parameters for these distributions to choose out of three the best coincidence k_{as} of the approximation spectrum. The existing difference at $u > 1.4$ can be removed by applying the inverse power law to this part.

2.4. Approximation Relationships for Integral Parameters of the Initial Stage of Condensation for Regime of LC

The studies of maximum oversaturation ε_m and other parameters effect on the updraft velocity V, CN concentration C_c, the index v, air mass temperature T and pressure p were performed at the subdivision into 1000 intervals. Studied were the ranges of $V = 0.1 \div 10$ m/s, $C_c = 10^2 \div 10^4$ (1/mg) at $r_{w,0} = 0.1$ μm, v from 4 to 8, $T_c = -10 \div +30°C$ and $p = 0.8 \div 1.25$ atm. In the relationships presented below the measurement units are valid.

As a result, it appeared that at LC ε_m can be approximated via the following relationships:

$$\varepsilon_m(C_c, V, T) = \varepsilon_{m0}(T)\left(\frac{C_c}{C_{c0}}\right)^{-x}\left(\frac{V}{V_0(T)}\right)^{y}$$

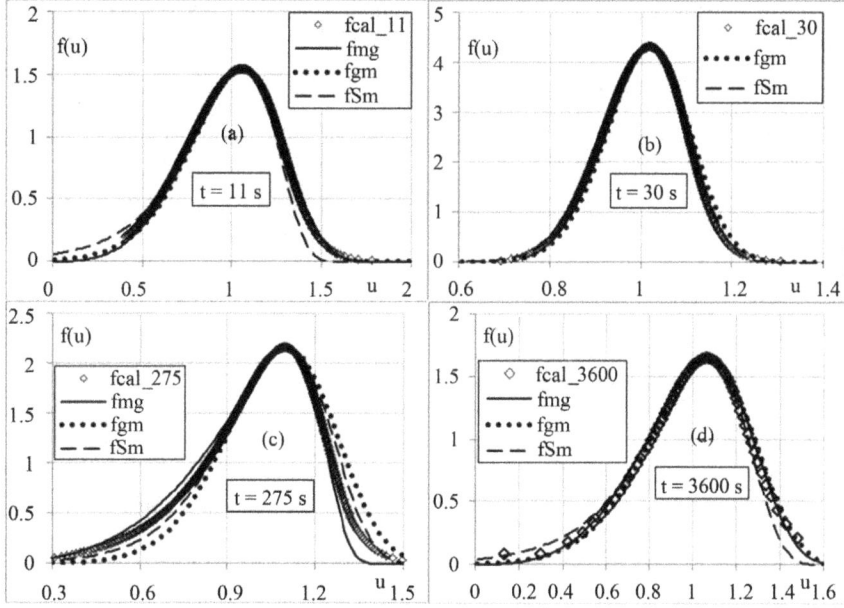

Figure 2. Comparison of calculated f_{cal} and the approximation spectra for the time moments (a) 11 s, (b) 30 s (termination of updraft), (c) 275 s (the moment of reaching $rb = 0.215$) and (d) 3600 s after the beginning of the process.

$$y = \frac{1.05}{(v+1)^{2/3}}; \quad x = \frac{1}{v+1}$$

$$V_0\left(T, C_{c0} = 100\right) = 0.82 + 0.009 T_c$$

$$\varepsilon_{m0}\left(T, C_{c0} = 100\right) = 0.62 - 0.005 T_c. \quad (10)$$

The height of the updraft h_m at which ε_m is attained is practically proportional to ε_m and is determined by the relationship similar to (10). Let us write this relationship in the form of

$$h_m\left(C_c, V, T\right) = h_{m0}\left(T\right)\left(\frac{C_c}{C_{c0}}\right)^{-x}\left(\frac{V}{V_{h0}(T)}\right)^{y} \quad (11)$$

where the exponents x and y are determined by formulae (10) and the values of h_{m0} (20°C) = 10 m and V_{h0} (20°C) = 0.42 m/s at C_{c0} = 100 mg^{-1}.

At the heights $h = 2h_m$ and $3h_m$ the drop liquid content makes 80% and 90% as compared with the thermodynamic one, correspondingly. Note that for the mode of LC an approximation error (11) does not exceed the units of percent.

The analytic regularities for the drop concentration C_d (1/mg) for the mode of LC can be obtained by integration over ε from 0 to ε_m of the nucleus distribution function over oversaturation (10). But for the errors from there not to be included here, an approximation for the drop concentration was obtained on the basis of its own calculations. It has the form of:

$$C_d = C_{k0}\left(\frac{C_c}{C_{ck}}\right)^{s}\left(\frac{V}{V_{k0}}\right)^{z}$$

$$s = \frac{v+5}{3(v-1)}$$

$$z = 0.707 \frac{v-1}{(v+1)^{2/3}} - 0.01(v-3.5)$$

$$C_{k0}\left(T\right) = 114 + 0.4 T_c$$

$$V_{k0}\left(T\right) = 0.26 + 0.07 T_c$$

$$C_{ck} = 200\left(\text{mg}^{-1}\right). \quad (12)$$

We shall note that in our calculations the dependence relation (10)-(12) on p has not been found.

The comparison of the above-mentioned relationships for the indices x, y, s and z with the corresponding values [2] has shown an extremely good numerical agreement (despite the difference in expressions) for the values of q and y obtained without corrections for hygroscopicity, the jumps of vapor concentration and temperature. The values of x and especially of s are noticeably different from our results. The relationship presented in [4] for the parameter z coincides with [2] and for the parameter s it numerically corresponds to our relationships with a difference in the analytic expression.

The condition of LC, at which relations (10)-(12) are valid, can be expressed by the following rather approximate relationship of drop concentration dependence on air mass updraft velocity (m/s) and temperature (°C).

$$C_d^{LC}\left(1/\text{mg}\right) \le \left(950 - 5 T_C\right) V \quad (13)$$

2.5. Evolution of Cloud Parameters at a Constant Height

A calculated character of integral parameters behavior at a constant temperature depending on the conditions of their formation at an ascending air mass is illustrated in **Figures 3(a)-(c)**.

From the data presented in these figures and calculation for other conditions it follows that under evolution at a constant temperature a relative dispersion increase with time and tends to the theoretical limit $rb = 0.215$ at low concentrations of CN ($C_c = 10^2$ in **Figure 4(b)**) and high liquid water content ($h = 100$ m in **Figure 4(a)**). At increasing CCN the asymptotic limit rb grows.

From the figures it is also seen that in some time after the air mass ascent ceased depending on the value of h, the growth of the drop average mass starts to acquire a

linear regularity determined in [14] with time counted from the moment t_0. The coefficient a of this dependence weakly depends on the updraft height (drop liquid content) and increases insignificantly with a growing CN concentration and becomes especially noticeable at a temperature increase. Considering only the dependence on temperature and taking into consideration the range of the value of a changes at varying h and C_c, this dependence can be written in the form of

$$\bar{m}_d(t) = \bar{m}_d(t_0) + a(T, h, W_T)(t - t_0),$$
$$a \cong (1/2) \cdot 10^{-10} (1 + 0.05 T_c)(g/s). \tag{14}$$

The moment when the regularity comes into action t_0 strongly depends on the initial breadth of spectra and may last several hours at $rb \approx 0.01$. We shall not analyze these dependences in detail in this work.

2.6. Formation of Multi-Modal Drop Spectra

A fast growth of updraft velocity can result in the appearance in the spectrum the drops of the second kind. An illustration of such a condition is given in **Figure 4**, where for two CN concentrations given are the trajectories of particle sizes at an increase of updraft velocity by 5 times.

From the results of calculations for the parameters, given in p. 2.3, it appeared that the condition for the appearance of the second mode $V2 \geq 5V1$ does not practically depend on the C_c and the velocity $V1$. For a cloud medium remaining for some time at a constant height, the appearance of the second mode is possible at the updraft velocities of $V \approx 0.2$ m/s.

3. Experimental Studies of Cloud Microstructure Formation Processes in the Chamber and Natural Conditions

3.1. Description and Results of Experiments in Chamber

Experimental studies of regularities of cloud spectra formation and transformation were performed in the BAC (Big Aerosol Chamber) of RPA "Typhoon" [16] (cylinder of the height of 18 and diameter of 15 m), where an air mass updraft was simulated by dropping excessive pressure created earlier. The Chamber is equipped with the temperature and humidity sensors with the use of standard instrumentation and a photo-electric counter of water drops FIROK operating in the range of cloud drop radii from 0.5 to 50 micrometer [17].

A change of updraft equivalent velocity is ensured by the change of a corresponding pressure dropping rate. The conditions of isothermality were ensured by a termination of the process of pressure dropping at equal temperature in the Chamber and the temperature of its walls.

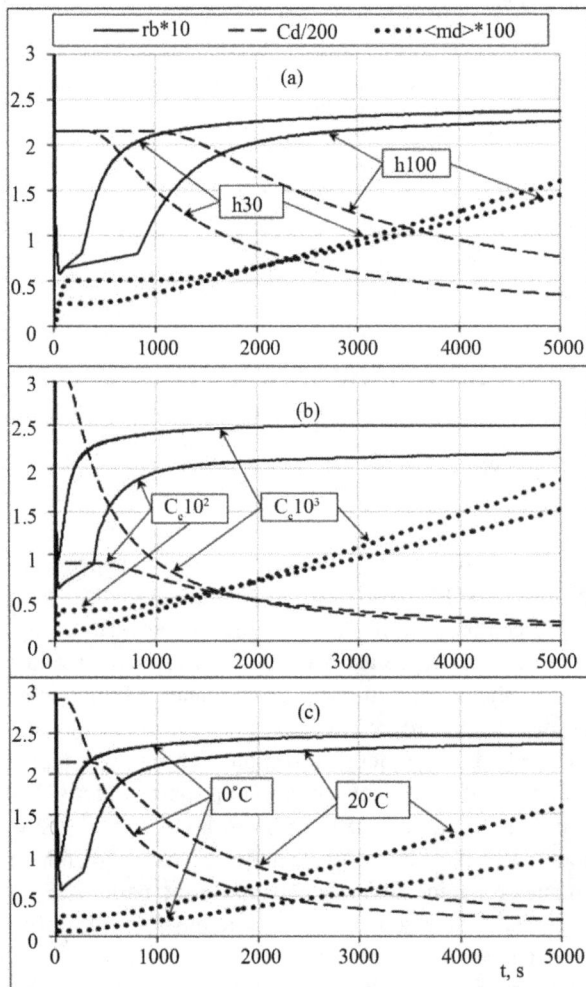

Figure 3. Dependence ($V = 1$ m/s) of an average drop mass $<m_d>$ (10^{-10} g), C_d (mg^{-1}) and relative breadth rb on time at different: (a) Updraft heights ($C_c = 500$, $T = 20°$C); (b) CN concentrations ($h = 30$ m, $T = 20°$C) and (c) Temperatures ($h = 30$ m, $C_c = 500$). The data are multiplied by the coefficients from the legend.

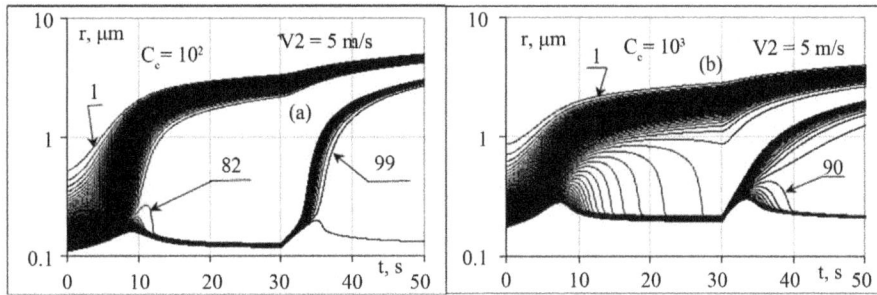

Figure 4. Trajectories of particles at a strong growth of updraft velocity at moment $t = 30$ s ($V1 = 1$ m/s, $T = 20°C$, $v = 5$).

The composition of CN was determined by their content in the environmental air. CN spectra and composition was not controlled, but the stability of their characteristics was checked by the reproducibility of characteristics of cloud drop spectra formed under similar conditions of pressure dropping. Some experimental results obtained with the use of the BAC were reported at the 13-th Conference on Cloud Physics [18] without a comparison with the calculation data. The present paper gives new results with a corresponding comparison.

At the first stage studied was a dependence of cloud drop concentration on the equivalent velocity of air mass updraft. It appeared that the experimental points coincide best of all with the calculated ones at $v = 5$, $C_c = 10^4$ (1/mg). Such high concentrations of condensation nuclei are explained by the fact that the air is pumped into the chamber from the height of only 20 m. The formation of spectra at such CN concentration occurs in the HC mode. Therefore, in the experiments in the BAC no narrow spectra were observed. Their form is stabilized in some time. The dependence of cloud parameters on time connected with drop spectra is shown in **Figure 5**, where for an updraft cycle to 200 m with the velocity of 1 m/s (after which the air temperature in the chamber becomes equal to the wall temperature) shown is the curve of an average drop mass, and their concentration and a relative width of spectra depending on time.

From the figure it is seen that the relative spectrum width changes little during the whole process, the average drop mass grows linearly just after the updraft termination with the growth coefficient, that is almost twice as that in the calculated relationship (14). The observation of spectra form made during the process showed its sufficient stability.

A comparison of spectra obtained in two different experiments (without and with cleaning of the air from aerosols) is shown in **Figure 6**.

From the comparison it is seen that for the approximation of the experimental spectra most reasonable is the function f_{sm}. Without taking into consideration the differences in the range of small values of u, we shall note that at $u > 1.4$ the experimental spectra are a little greater

than the approximation ones. They are higher for the spectrum a) than for the spectrum b). Such an excess was noted for the calculated spectra as well. The experiments with the formation of the second mode demonstrated that for the updraft velocities $V1 = 0.3 - 1$ m/s the appearance of the second mode was noted at $V2 = (4.5 - 5)V1$, that corresponds to the calculated values of this relationship. For cloud spectra after 10 min of evolution at a constant temperature the appearance of the second mode was noted at $V = 0.2$ m/s, that qualitatively agrees with the calculated results obtained. **Figure 7** is an illustration of spectra formation and transformation.

From this figure it is seen that already at the 20-th second of the updraft beginning a fine-drop spectrum fraction is formed. Before the 260-th second the spectrum becomes a bimodal one. Later both modes merge.

3.2. Comparison with Field Measurements of Cloud and Fog Drop Spectra

Headings Measurements of natural fog drop spectra were made by one of the authors [19,20]. In [19] appearance of two-modal spectra over the hill-like risings of the fog upper boundary level was noted. In view of the conditions of the second mode formation created above at small updraft velocities in stationary clouds we have now an explanation of the appearance of this mode. It is the presence of a convective updraft causing a local increase of the fog upper boundary. Due to a low resolution of the drop spectrum counter, the spectra shown are not reliably comparable with the approximation relationships. The registration of two- and even of three-modal spectra in China was mentioned in the review [21].

In [20] the results of measurements of drop spectra of a warm fog formed as a result of morning dew evaporation into the cold morning humid air are presented. During these measurements an improved instrumentation described in this paper was used. But because of the absence of an adequate model no approximation was made. The next figure gives the comparison of one of this fog spectrum with the approximations proposed above.

The measurements of stratiform super cooled cloud drop spectra were carried out from a helicopter in 2000

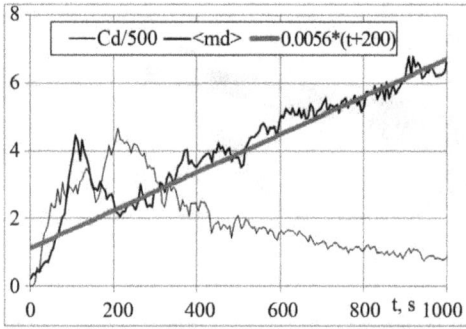

Figure 5. Dependence of C_d (cm^{-3}), $<m_d>$ (microgram) and its approximation by the linear law for a cycle of updraft to the altitude of 200 m and a further evolution of the cloud medium, $t = 200$ s is the time of the discharge termination. The data are multiplied by the coefficients from the legend. An average temperature of the process is equal to 20°C.

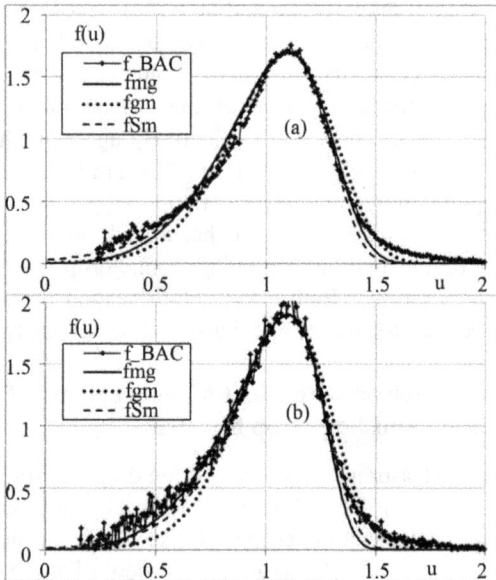

Figure 6. Comparison of an experimental in BAC and approximation spectra with $u_m = 1.1$ for the conditions: (a) without cleaning ($<r>$ 5.1 μm, $C_d = 800$ cm^{-3}, $\gamma = 7.6$, $\mu = 17$, $q = 14$) and (b) with the cleaning of the air medium ($<r> = 8$ μm, $C_d = 300$ cm^{-3}, $\gamma = 10.4$, $\mu = 22$, $q = 19$).

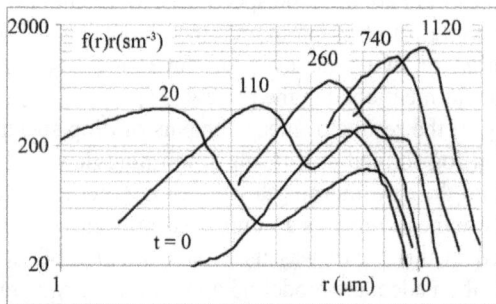

Figure 7. Formation of the second mode at the updraft velocity of 0.2 m/s after fog evolution at a constant height (temperature). The numbers at the curves indicate time in seconds from the updraft beginning.

in the Northern Caucasus Mountains. An average cloud thickness was 250 - 300 with the upper level boundary of 1300 m. The cloud medium temperature was about minus 10°C at the lower level; it decreased with height by about 0.5 degree with the inversion at the upper boundary (UB) with about 2 degrees. An average drop radius increased with height and reached the value of 3.0 - 3.5 μm near the UB. The results of these measurements have not been published because the differences connected with their analytic description. A comparison of two spectra measured near the UB with the approximation ones is shown in **Figure 8**.

The spectra a and b differ only in the time of their measurements at the same heights. The jagged form of spectrum lines is explained by a short time of spectrum sampling (5 s at the helicopter horizontal flight velocity of about 50 m/s). One can note a plateau in the left-hand portion of experimental spectra. For its description it is possible to use the spectrum of CN covered with water (2), but for a reliable determination the number of measurements of small sizes is insufficient. Therefore, we shall only make a conclusion on an adequate description of the drop fraction of the investigated cloud spectra with the distributions like f_{mg} and f_{Sm}.

4. Main Results and Conclusions

A On the basis of the calculations made with the model of cloud spectra formation and evolution it has been found that the formation processes differ qualitatively depending on CN concentrations. At small CN concentrations (the mode of LC) narrow spectra of cloud drops

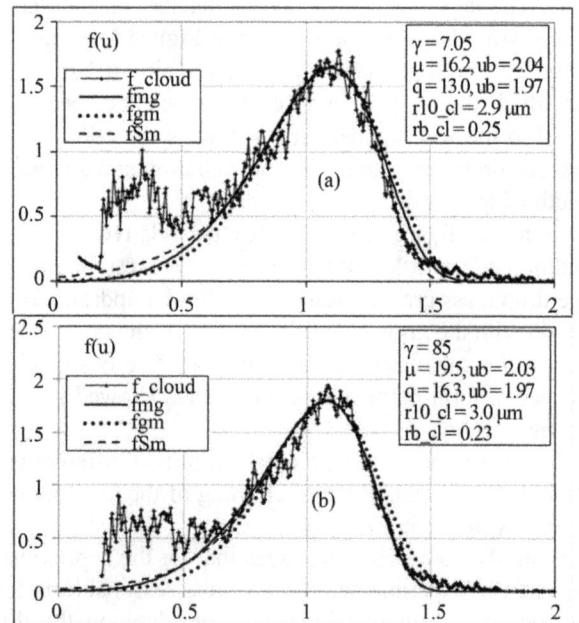

Figure 8. A drop spectrum of a super cooled stratiform cloud and its approximation.

are formed already at the height of about 10 meters from the condensation level, a relative drop dispersion of which quickly decreases with increasing height. For high concentrations (the mode of HC) no distinct separation of growing drops and CN covered with water is observed due to the immediately beginning evaporation of the drops. The boundary of separation of these modes in view of cloud drop concentrations is almost proportional to the air mass updraft velocity with the coefficient, decreasing with increasing temperature. But when the updraft is terminated, the evolution of spectra both in the first and the second case leads to the appearance of asymptotic spectra, a relative dispersion of which depends generally only on the condensation nuclei concentration and changes from the theoretical value of $rb = 0.215$ for the spectra at very low to $rb = 0.25 \div 0.28$ at high CN concentrations

At all the stages of formation and evolution, the calculation spectra have the negative asymmetry coefficient and a shifted from the mean value (towards high positive values) location of maxima. At the results of analysis of analytical relations it is assumed that for the mirror functions f_{mg} the well-known gamma distributions and less-known Smirnov's formulae f_{Sm} and also a modified gamma-spectrum f_{gm} with the fixed parameter $\alpha = 3$ are applicable. The reconstruction of the parameters of the above-mentioned distributions is made with the use of location and amplitude of the spectrum maximum. The choice of a most-applicable from those mentioned above (and also of other possible mirror distributions) can be made with the use of best coincidence of asymmetry coefficients. Here, most reasonable for us is the use in the numerator of expression (3s) of the value u_m instead of u_{10}.

The dependences of integral parameters (drop concentration, maximum supersaturation, etc.) on the updraft velocity, CN concentration and temperature given in this paper make the results obtained previously more accurate and broad. It is shown that the conditions of appearance of additional modes in drop spectra during updraft at an excess of additional increase of updraft velocity by 4.5 - 5

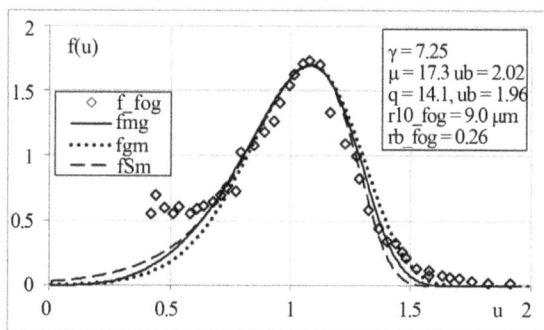

Figure 9. Approximation of the morning fog drop spectrum mentioned in [20].

times are studied. During spectra evolution at a constant height the process of additional mode formation can occur at updraft velocities of 0.1 - 0.2 m/s. This conclusion was confirmed by the experiment in the BAC.

Experimental spectra obtained in the Big Aerosol Chamber of FSBI RPA "Typhoon" and the natural fog and supercooled stratiform cloud spectra obtained by the authors are well approximated by Smirnov's distribution f_{Sm}. The modified gamma-distribution f_{mg} in the frameworks of uncertainties in the experimental spectra can be considered applicable as well.

One can draw the following conclusions on the peculiarities of natural cloud spectra formation with the use of the results obtained:

• Broad spectra in natural cloud spectra can be explained both by a great CN concentration and evolution of cloud microstructure under steady meteorological conditions;

• The first argument can be supported by the following consideration: drop concentrations of 300 - 800 cm^{-3} existing in continental stratiform clouds at equivalent velocities of cloud clusters and fog formation of 0.1 m/s and less, according to the condition of (13), can be formed only in the mode of HC;

• An effect of cloud drop spectrum broadening under evolution, with the exception of the evident case of stratiform cloud formation, can be possible also in cumulus clouds (especially at their upper boundary) in the presence in them of regions with zero and even negative air mass updraft velocities [7,8];

• The multimodality of natural spectra is likely to be caused by the fluctuation of air mass vertical motions. This conclusion is supported by the bimodal spectra observed in the hilly formations at the fog upper boundary [19]. The formation of an additional mode at vertical fluctuations of a cloud layer even at the velocities of several tenths of m/s may lead in reality to a considerable drop spectrum broadening and to stimulation of coagulation processes in clouds. They may serve as a trigger mechanism for precipitation formation from warm clouds;

• The spectra of natural fogs and stratified clouds (see **Figures 8** and **9**) have the form described by Smirnov's mirror distribution and the modified gamma-distribution with the parameter $\alpha = 3$. The spectra of various forms of natural objects are likely to be presented by these analytic expressions.

REFERENCES

[1] P. Mazin and V. M. Merkulovich, "Stochastic Condensation and Its Possible Role in the Formation of Cloud Drop Microstructure (Review). Some Problems of Cloud Physics," In: *Collected Papers*: "*Memorial Issue Dedicated to Prof. Solomon M. Shmeter*", Moscow, 2008, pp. 217-267

(in Russian).

[2]   Yu. S. Sedunov, "Physics of Liquid-Drop Phase Formation in the Atmosphere," Leningrad, Gidrometeoizdat, 1972 (in Russian).

[3]   E. L. Alexandrov, N. V. Klepikova and Yu. S. Sedunov, "Some Results of Cloud Drop Spectra Formation," Parts I and II. Trudy IEM, Gidrometeoizdat, Moscow, 1976, p. 19.

[4]   A. S. Drofa, "Formation of Cloud Microstructure during Hygroscopic Seeding," *Izvestiya, Atmospheric and Oceanic Physics*, Vol. 42, No. 3, 2006, pp. 326-336.

[5]   V. I. Khvorostyanov and J. A. Curry, "Kinetics of Cloud Drop Formation and Its Parameterizationfor Cloud and Models," *Journal of the Atmospheric Sciences*, Vol. 65, No. 9, 2008, pp. 2784-2802.

[6]   N. Kivekas, V.-M. Kerminen, T. Anttila, H. Korhonen, H. Lihavainen, M. Komppula and M. Kulmala, "Parameterization of Cloud Droplet Activation Using a Simplified Treatment of the Aerosol Number Size Distribution," *Journal of Geophysical Research*, Vol. 113, No. D15, 2008.

[7]   A. Khain, M. Ovchinnikov, M. Pinsky, A. Pokrovsky and H. Kruglak, "Notes o State-of-the-Art Numerical Modeling of Cloud Microphysics," *Atmospheric Research*, Vol. 55, No. 3-4, 2000, pp. 159-224.

[8]   C. Erlick, A. Khain, M. Pinsky and Y. Segal, "The Effect of Wind Velocity Fluctuations of Drop Spectrum Broadening in Stratocumulus Clouds," *Atmospheric Research*, Vol. 75, No. 1-2, 2005, pp. 15-45.

[9]   A. I. Flossmann and W. Wobrock, "A Review of Our Understanding of the Aerosol-Cloud Interaction from the Perspective of a Bin Resolved Cloud Scale Modeling," *Atmospheric Research*, Vol. 97, No. 4, 2010, pp. 478-497.

[10] D. Deirmenjian, "Electromagnetic Scattering on Spherical Polydispersion," Amerikan Elsevier Publishing Company, New York, 1969.

[11] S. M. Prigarin, K. B. Bazarov and U. G. Oppel, "Looking for a Glory in A-Water Clouds," *Atmospheric and Oceanic Optics*, Vol. 25, No. 4, 2012, pp. 256-262.

[12] N. P. Romanov, "Analytic Representation of Van'-Hoff's Factor for Aqueous Solutions of Strong Electrolytes," *Izvestiya, Atmospheric and Oceanic Physics*, Vol. 41, No. 5, 2005, pp. 641-651.

[13] V. I. Smirnov, "About of the Approximation of Empirical Distributions on a Size Clouds Drops and Other Aerosol Particles," *Izvestiya, Atmospheric and Oceanic Physics*, Vol. 9, No. 1, 1973, pp. 54-65.

[14] I. M. Lifshitz and V. V. Slyosov, "On the Kinetics of Oversaturated Solid Solution Diffusion Decay," *ZETF*, Vol. 35, No. 2, 1958, pp. 479-492 (in Russian).

[15] A. S. Kabanov, "Kinetics of Condensation Growth of a Drops at Late Stage of Adiabatic Process," *Izvestiya, Atmospheric and Oceanic Physics*, Vol. 8, No. 12, 1972, pp. 1279-1288.

[16] N. P. Romanov and G. P. Zhukov, "Thermodynamic Relationships for the Fog Chamber," *Russian Meteorology and Hydrology*, No. 10, 2000, pp. 37-52.

[17] A. S. Drofa, V. N. Ivanov, D. Rosenfeld and A. G. Shilin, "Studying an Effect of Salt Powder Seeding Used for Precipitation Enhancement from Convective Clouds," *Atmospheric Chemistry and Physics*, Vol. 10, No. 16, 2010, pp. 8011-8023.

[18] N. P. Romanov, "Experimental Investigation of the Processes of Formation and Evolution of Natural and Artificial Fogs Microstructure," 13*th International Conference on Clouds and Precipitation*, Reno Area, 14-18 August 2000.

[19] N. P. Romanov, "Some Characteristics of Fog in the Center of European Territory of Russia in October 1987," *Soviet Meteorology and hydrology*, No. 4. 1990, pp. 49-55.

[20] N. P. Romanov, "Evolution of Morning Fogs Microstructure," *Meteorologiya i gidrologiya*, No. 2, 2001, pp. 36-45 (in Russian).

[21] S. J. Niu, C. S. Lu, H. Y. Yu, L. J. Zhao and J. J. Lu, "Fog Research in China: A Review," *Advances in Atmospheric Sciences*, Vol. 27, No. 3, 2010, pp. 639-661.

[22] I. P. Mazin and S. M. Smeter, "Clouds, Their Structure and Formation," Gidrometeoizdat, Leningrad, 1983 (in Russian).

Supplement

Moments of the n-th order

$$M(n,u) = \int_0^\infty f(u)u^n dr \; ; \; u_{n,n-1} = \frac{M(n)}{M(n-1)}. \qquad (1s)$$

Relative Breadth

$$rb = \left(\frac{M(2)M(0)}{M^2(1)} - 1 \right)^{1/2} \qquad (2s)$$

Asymmetry coefficients

$$k_{as} = \frac{M(3) - 3M(2)u_{10} + 3M(1)u_{10}^2 - M(0)u_{10}^3}{rb^3}. \qquad (3s)$$

Gamma-distribution described, in particular, in [22].

$$f_\gamma(u)\,du = \frac{\mu^\mu}{\Gamma(\mu)}u^{\mu-1}\exp(-\mu \cdot u)\,du \; ; \qquad (4s)$$

$$M_\gamma(n,u) = \frac{\Gamma(\mu+n)}{\mu^n \Gamma(\mu)} = \frac{\mu(\mu+1)\cdots(\mu+n-1)}{\mu^n} \; ; \qquad (4sa)$$

$$M_\gamma(0) = M_\gamma(1) = 1, \; M_\gamma(2) = \frac{\mu+1}{\mu} \; ; \qquad (4sb)$$

$$u_{10} = 1, \; rb = \frac{1}{\sqrt{\mu}}, \; k_{as} = \frac{2}{\sqrt{\mu}} \; ; \qquad (4sc)$$

$$u_m = \frac{\mu-1}{\mu}, \; f_\gamma(u_m) = \frac{\mu(\mu-1)^{\mu-1}}{\Gamma(\mu)}\exp(1-\mu). \qquad (4sd)$$

Distribution proposed in [13] at $s = 1$:

$$f_S(u)\,du = \frac{q^{q+1}}{\Gamma(q+1)}\frac{1}{u^{q+2}}\exp\left(-\frac{q}{u}\right)du \; ; \qquad (5s)$$

$$M_S(u,n) = \frac{q^n \Gamma(q-n+1)}{\Gamma(q+1)} = \frac{q^n}{q(q-1)\cdots(q-n+1)} \; ; \qquad (5sa)$$

$$u_m = \frac{q}{q+2}, \; u_{10} = 1, \; rb = \frac{1}{\sqrt{q-1}}, \; k_{as} = \frac{4\sqrt{(q-1)}}{q-2}. \qquad (5sb)$$

Modified gamma-distribution $f_{m\gamma}$:

$$f_{m\gamma}(r) = \frac{\gamma}{r_m}\frac{(\alpha/\gamma)^{\frac{\alpha+1}{\gamma}}}{\Gamma\left(\frac{\alpha+1}{\gamma}\right)}\left(\frac{r}{r_m}\right)^\alpha \exp\left(-\frac{\alpha}{\gamma}\left(\frac{r}{r_m}\right)^\gamma\right) \; ; \qquad (6s)$$

$$M_{m\gamma}(n) = \frac{r_m^n}{\left(\frac{\alpha}{\gamma}\right)^{\frac{n}{\gamma}}}\frac{\Gamma\left(\frac{\alpha+n+1}{\gamma}\right)}{\Gamma\left(\frac{\alpha+1}{\gamma}\right)}. \qquad (6sa)$$

Direct expressions for rb and k_{as} for this distribution are rather complex and therefore it would be better simply to calculate them with the use of moments (6sa) and expressions (2s) and (3s), correspondingly.

Let us note only two limited cases. $i.e.$ at $\gamma = 1$ distribution (6s) reduce to similar (4s) with $\mu = \alpha + 1$. When $\gamma \gg \alpha$ with the account for the tendency of the functions $\Gamma(z) \approx 1/z$ at $z \to 0$ for rb we obtain the following asymptotic relationship

$$rb(\alpha, \gamma \approx \infty) = \frac{1}{\sqrt{\alpha^2 + 4\alpha + 3}}. \qquad (6sb)$$

16

Granger Causality Analyses for Climatic Attribution

Alessandro Attanasio[1], Antonello Pasini[1*], Umberto Triacca[2]

[1]Institute of Atmospheric Pollution Research, National Research Council (CNR), Monterotondo Stazione, Rome, Italy
[2]Department of Computer Engineering, Computer Science and Mathematics, University of L'Aquila, L'Aquila, Italy

ABSTRACT

This review paper focuses on the application of the Granger causality technique to the study of the causes of recent global warming (a case of climatic attribution). A concise but comprehensive review is performed and particular attention is paid to the direct role of anthropogenic and natural forcings, and to the influence of patterns of natural variability. By analyzing both in-sample and out-of-sample results, clear evidences are obtained (e.g., the major role of greenhouse-gases radiative forcing in driving temperature, a recent causal decoupling between solar irradiance and temperature itself) together with interesting prospects of further research.

Keywords: Granger Causality; Climatic Attribution; Global Warming; Forcings; Greenhouse Gases; Solar Radiation; Natural Variability

1. Introduction

The climate is a complex system characterized by several subsystems and many bidirectional relations between them. At present, the standard strategy to catch the complex behavior of climate is the application of dynamical modeling, using Global Climate Models (GCMs) and Regional Climate Models (RCMs): see [1] for the description of this dynamical approach and the conceptual and practical relevance of these simulations.

The problem of understanding and weighting the main causes of recent climate change is generally faced by numerical experiments within this modeling framework. The final aim of these studies is to evaluate if one is able to attribute this change to some specific causes out of a number of possibilities. The situation is quite complex but, at least as far as the attribution of global temperature changes is concerned (a case of climatic attribution), the results coming from dynamical models are quite clear and indicate that the fundamental causes of recent global warming are anthropogenic forcings (especially the increase of greenhouse gases in the atmosphere): a comprehensive review is provided in [2].

However, these dynamical models are very complex. In particular, just a limited number of processes, interactions and feedbacks can be considered and there are unavoidable uncertainties in attempting to simulate all of them in these standard climate models. The study of other complex systems has however shown that one often benefits from a change in viewpoint when analyzing them. There are complementary approaches in a number of other fields: e.g., in biology, the molecular biology approach vs. a more systemic point of view; in economy, the application of "traditional structural" models vs. the use of vector autoregressive (VAR) models.

Thus, a more data-driven approach can be fruitful in studies of climatic attribution, e.g. in assessing cause-effect relationships between external forcings and temperature behavior. In the past, for instance, neural network modeling has been applied for the attribution of global temperature (T) [3] and its results confirm the major role of anthropogenic forcings in driving T. Further researches have shown the usefulness of neural investigations for the attribution of temperature and precipitation at a regional scale, too [4,5].

In this framework, during the last years, analyses using the concept of Granger causality [6] have been performed to investigate the possible causal relations between external forcings and temperature behavior. In this paper we review the studies of climatic atttribution *via* this inferential method.

2. The Concept of Granger Causality

The concept of Granger causality is quite simple. Suppose that we have two variables, *x* and *y*. First, we at-

*Corresponding author.

tempt to forecast y_{t+1} using past terms of y. We then try to forecast y_{t+1} using past terms of x and y. We say that x Granger causes y, if the second forecast is found to be more successful, according to standard cost functions. If the second prediction is better, then the past of x contains a useful information for forecasting y_{t+1} that is not in the past of y. Clearly, Granger causality is based on precedence and predictability.

In a more formal way, we consider the vector time series $\left(y_t, x_t\right)'$ and the following information sets: $I_{yx}(t) = \{y_t, x_t, y_{t-1}, x_{t-1}, \cdots\}$ and $I_y(t) = \{y_t, y_{t-1}, \cdots\}$. We denote with $P\left(y_{t+1} \middle| I(t)\right)$ the optimal (minimum mean square error) linear forecast of the variable y_{t+1} based on the information set $I(t)$. We say that x does not Granger cause y, in a bivariate system, if $P\left(y_{t+1} \middle| I_y(t)\right) = P\left(y_{t+1} \middle| I_{yx}(t)\right)$ for any t.

In literature, the causal relationship between the variables x and y has often been investigated in a bivariate system. However, it is well known that in a bivariate framework problems of spurious causality and of noncausality due to omission of a relevant variable can arise. These problems can be solved if an auxiliary variable z is considered in the analysis, specifying a trivariate system.

We have that x does not Granger cause y, in a trivariate system, if $P\left(y_{t+1} \middle| I_{yz}(t)\right) = P\left(y_{t+1} \middle| I_{yxz}(t)\right)$ for any t, where $I_{yz}(t) = \{y_t, z_t, y_{t-1}, z_{t-1}, \cdots\}$ and $I_{yxz}(t) = \{y_t, x_t, z_t, y_{t-1}, x_{t-1}, z_{t-1}, \cdots\}$.

Suppose that the trivariate time series $\left(y_t, x_t, z_t\right)'$ follows a vector autoregressive (VAR) model of finite order k:

$$\begin{bmatrix} y_t \\ x_t \\ z_t \end{bmatrix} = \begin{bmatrix} c_1 \\ c_2 \\ c_3 \end{bmatrix} + \sum_{j=1}^{k} \begin{bmatrix} \phi_{11,j} & \phi_{12,j} & \phi_{13,j} \\ \phi_{21,j} & \phi_{22,j} & \phi_{23,j} \\ \phi_{31,j} & \phi_{32,j} & \phi_{33,j} \end{bmatrix} \cdot \begin{bmatrix} y_{t-j} \\ x_{t-j} \\ z_{t-j} \end{bmatrix} + \begin{bmatrix} u_{yt} \\ u_{xt} \\ u_{zt} \end{bmatrix} \quad (1)$$

where $c = \left(c_1, c_2, c_3\right)'$ is a vector of constants, $\phi_{il,j}$ are fixed coefficients and $u_t = \left(u_{yt}, u_{xt}, u_{zt}\right)'$ is a trivariate white noise process with nonsingular covariance matrix. In this framework, we have that x does not Granger cause y, with respect to the information set $I_{yxz}(t)$, if and only if $\phi_{12,j} = 0$ for $j = 1, 2, \cdots, k$. This characterization of the condition of noncausality is often used in literature to conduct the Granger causality tests.

In what follows we mainly review the studies of climatic attribution performed by Granger causality analyses.

3. Granger Analyses in Specific Climatic Problems

During the last decade the notion of Granger causality has been used quite frequently in addressing specific causality problems in the climate system.

For instance, Diks and Mudelsee [7] analyzed the results of an ocean drilling program in order to estimate the causal relationships and directions among data about

insolation, $\delta^{18}O$ (a proxy for global ice volume) and $\delta^{13}C$ (which reflects mainly the strength of formation of the so-called North Atlantic deep water).

Kaufmann et al. [8] used satellite data and a Granger causality analysis for estimating causal influences of snow cover and vegetation on temperatures in different seasons. In a further study, considered that the strength of Atlantic hurricanes is related to the sea surface temperatures (SST) of this ocean, Elsner [9,10] applied a Granger causality analysis to time series of global temperatures (GT) and SST and found a causal link from GT to SST, thus corroborating the hypothesis of changes induced by global warming.

Mosedale et al. [11] investigated SST effects on North Atlantic Oscillation (NAO)—an index which substantially drives the European winter climate—using data from simulations made with a coupled Global Climate Model (GCM). They showed that the so-called SST tripole index provides additional predictive information for the NAO than that available by using only past values of NAO, i.e. the SST tripole is Granger causal for the NAO.

Kaufmann et al. [12] studied the effect of urbanization and enlargement of towns on precipitation in a Chinese case study. They applied Granger causality and clearly found that, generally, urbanization causes a deficit in precipitation, even if differences for distinct seasons are detectable. Finally, Mohkov et al. [13] analyzed the relationship between El Niño Southern Oscillation (ENSO) and the strength of Indian monsoons. They found a bidirectional coupling which varies with time and this result shall be certainly useful for better understanding the dynamical mechanism behind this interaction.

The examples of application of Granger causality analyses just sketched show the potentiality of this technique in addressing causality problems in the climate system. Actually, however, in the realm of climate research there is a causality problem which overwhelms all other ones. It can be summarized in the question: what did cause the recent climate change or, at least, the recent global warming? Even considered the complexity of the climate system, which is the main external forcing that primarily induced the increase of temperature observed in the last century? Obviously, this is the main problem of attribution studies.

Given the potentialities of Granger analyses, it should not be a surprise that several studies have been performed by this technique in the framework of climatic attribution. In the next section we will describe those analyses conducted by a standard in-sample approach.

4. In-Sample Granger Analyses for Climatic Attribution and Their Problems

As a matter of fact, as the problem of global warming began to be recognized outside the field of climatologists, several analyses have been performed on the link between greenhouse gases and temperatures by several experts of statistical methods: see, for instance, [14,15] for two pioneering works.

More recently, even Granger causality has been specifically used by several researchers in order to analyze the causes of the recent rise in global temperature.

At our knowledge, the first paper dealing with this problem was written by Sun and Wang [16]. They analyzed time series of global CO_2 emissions and global temperature anomalies, finding a strong numerical evidence that the increase in CO_2 emissions causes global temperature change. Their approach is based on both direct Granger causality and cross-spectral analysis and the results of these two methods corroborate each other.

Kaufmann and Stern [17] assessed, in both directions, the linear causality between Southern Hemisphere temperature and Northern Hemisphere temperature, finding a Granger causation from South to North. After having included natural and anthropogenic forcings in the regressive models, they arrived at the conclusion that human activity played a major role in driving the historical record of temperature. In a strictly logical sense, however, other conclusions are possible in their study (see [18]). In fact, when the bivariate analysis is combined with a multivariate analysis, as in Kaufmann and Stern's study, the results must be analyzed with great care.

Another study in which CO_2 radiative forcing is considered in its causal relationship with temperature is that of Triacca [19], where, using the methodology of Toda and Yamamoto [20], he did not find any detectable linear Granger causality from CO_2 to global temperature.

Even the influence of Sun on temperatures has been studied by the Granger technique. For instance, Reichel et al. [21] used a smoothed solar cycle length (SCL) as an index of long-term variability of Sun, estimated by spectral analysis of sunspot counts at different data frequencies. Another index of solar activity used by these researchers was total solar irradiance (TSI). In both cases they found a significant Granger causality from indices of solar activity to temperatures in their in-sample tests.

Even Mohkov and Smirnov [22] considered the problem of weighting Sun influences on temperatures, here in terms of solar radiation flux. They applied Granger causality for several in-sample tests with different periods, also adopting a moving window approach. The final results showed that the influence of solar activity on the Earth's climate varies widely over time, but a sensible influence is detectable in the second half of the 20th century, even if it seems to decrease at the end of this period, since the 80s. Another period of a significant, but weaker, influence of solar flux variations on global surface temperature has been recognized to be 1896-1939.

In a recent study, Kodra et al. [23] introduced an alternative test of causality which evidenced that the strength of linear causality from CO_2 to global temperature is stronger than that in the opposite direction. They also performed a forecasting test, considering ten out-of-sample observations, AR and VAR models, that confirmed the previous in-sample results.

Attanasio [24] faced the problem of testing Granger causality from CO_2 radiative forcing (RF) to temperatures, using the same Toda-Yamamoto technique applied by Triacca [19]. Here, however, the deterministic component of the model was characterized only by a constant term (vs. the linear trend used by Triacca). Furthermore, several time windows were explored, by expanding them from present to past: this approach allowed the author to estimate the parameters of the model by considering always the most recent observations. In this paper a clear Granger causality from CO_2 RF to temperatures has been recognized since 1850. Replacing this anthropogenic forcing with natural forcings led to discover no Granger-causal link with temperatures.

More recently, Triacca et al. [25] extended the work by Attanasio [24] by considering several trivariate systems with the presence of a context variable–a natural forcing or an index of natural variability. Their results show that the Granger causal link between the radiative forcing of greenhouse gases and global temperature persists even in these cases, so reflecting its robustness.

This review of in-sample studies shows that several different approaches have been performed for application of Granger causality tests and that the corresponding results are sometimes contrasting.

Obviously, not all these studies used the same regressive models. Furthermore, some pioneering research was based on the use of variables that, probably, are not directly influencing temperatures, as one should require for the application of a linear method. At present, for instance, the direct influence of greenhouse gases is generally described by their radiative forcings, rather than by their concentrations or, even, their emissions, as done in [16]. Finally, in other studies [17] Granger causality has been applied in a multivariate framework where a problem of dimensionality clearly rises: they had too many free parameters in the models if compared with the time series length, so that the efficiency of the estimate parameters is not assured and overfitting becomes more probable.

However, it seems to us that even other problems affect the in-sample approach and this fact can weaken the robustness of the results obtained in this framework. In what follows we briefly discuss this situation.

First of all, before performing in-sample tests for Granger causality, it is important to establish the stochastic properties of the time series involved, by analyzing whether these series are stationary, non-stationary or co-integrated, because, for instance, the use of non-stationary time series can lead to spurious causality results [26-28]. Of course, the weakness of this approach is that incorrect conclusions drawn by this preliminary analysis may affect the results of causality tests and their reliability.

A way to overcome this situation can be the use of the Toda-Yamamoto technique [20], that is robust to the integration and possible co-integration properties of the variables. In fact, one can apply it whether the variables are stationary, integrated or co-integrated of an arbitrary order: this procedure requires only the knowledge of the maximum order of integration of the series. On the other hand, due to the further delays introduced, this technique emphasizes the problem of overfitting.

As a matter of fact, significant in-sample Granger causality does not guarantee significant out-of-sample predictability. Out-of-sample tests are often recommended because they are able to catch the true forecasting ability of one variable for another, and the results are more robust in terms of overfitting [29-31].

In order to overcome these problems, according to the analysis of Ashley et al. [32], in recent papers [33,34] we used a technique that relies on the out-of-sample comparison of the forecasting performance of two linear models. This may be more robust in terms of model selection biases and overfitting [30,31]. Furthermore, according to Granger's definition, Granger causality builds upon the notion of incremental predictability, so that our out-of-sample approach is more keeping the spirit of the original definition by Granger [6]. In the next section a review of this approach will be presented.

5. Out-of-Sample Granger Analyses for Climatic Attribution

In this section, we will briefly sketch the method used and the results obtained in two studies recently published [33,34]. For further details, the reader may refer directly to these papers.

The final aim of the first paper [33] was to establish which external forcings can be considered Granger causal for global temperature. We analyzed the influence of many natural and anthropogenic forcings in a bivariate manner.

Total solar irradiance (TSI) describes quite well the direct effect of Sun on Earth's climate in terms of radiative forcing; cosmic ray intensity (CRI) can be considered as an indirect effect of our star (by means of solar wind) on some processes of climatic interest, such as the formation of clouds; stratospheric aerosol optical thickness (SAOT) summarizes the impact of strong volcanic eruptions and their interference with climate due to the emission and persistence in the low stratosphere of volcanic ash composed by sulfates. All these forcings can be

considered as natural.

As far as anthropogenic forcings are concerned, CO_2, CH_4 and N_2O concentrations data were taken into account for these major greenhouse gases (GHG), their single radiative forcings (RF) were calculated and considered as effective forcings, and also a GHG-total RF has been estimated.

By taking all these data into account, we were able to test the influence of a wide range of forcings on global temperature, even of forcings never considered before in causality analyses but at present very discussed in the arena of the climate debate, such as CRI, whose role is very controversial.

If we consider $y = T$ and $x_i (i = 1, \cdots, 7)$ = one of the external forcings, in our application we compared the predictive ability one step ahead (in terms of mean square errors—MSE) of the two following nested regression models:

$$\text{VAR} : y_t = \delta_1^{(i)} + \sum_{j=1}^{k} \alpha_j^{(i)} y_{t-j} + \sum_{j=1}^{k} \beta_j^{(i)} x_{i,t-j} + \varepsilon_t^{(i)} \qquad (2)$$

$$\text{AR} : y_t = \delta_2 + \sum_{j=1}^{k} \gamma_j y_{t-j} + \eta_t \qquad (3)$$

Here, $\delta_1^{(i)}$ and δ_2 are constants included as deterministic terms, x_i is the i-th forcing, $\alpha_j^{(i)}$, $\beta_j^{(i)}$ and γ_j are coefficients of our regressions, $\varepsilon_t^{(i)}$ and η_t are univariate white noises. The order k of the models was kept low $(k = 1, \cdots, 4)$, so that the models are parsimonious and the residuals are uncorrelated, and the models finally selected were those endowed with the best predictive performance on each test set.

The Granger out-of-sample tests were performed on five test sets which span the following periods: 1941-2007, 1951-2007, 1961-2007, 1971-2007, 1981-2007. For each test set, the correspondent training set is composed by data patterns since 1850 till the year before the beginning of the test set itself.

Fixed and recursive schemes were adopted for predicttions. Under the recursive scheme we used the training set for the first estimate and forecast out of sample one-step ahead; then we added an annual pattern to our training set, obtained a second estimate and forecast for the next year; and so on, iteratively. Under the fixed scheme the parameters were estimated only once on the original training set and every one-step ahead forecast has been obtained using just these fixed parameters.

The statistical significance of results has been evaluated by MSE-t and MSE-REG tests, as described in [35]. However, we were not able to use critical values of these test statistics, as reported in [35], because our series are not stationary. So, we performed a bootstrap procedure to calculate our critical values: see [33] for further details.

The results obtained by this out-of-sample Granger

analysis are very clear. If we take TSI, CRI or SAOT as x variable, in every case (any natural forcing, scheme and test set considered) the null hypothesis of non-Granger causality on $y = T$ is never rejected (with only two exceptions), even just at 10% significance level. *Vice versa*, there is a clear general evidence of Granger causality from anthropogenic forcings to global temperature (see [33] for the complete results and other detailed considerations).

In short, this paper shows that a genuine Granger out-of-sample predictive approach permits to overcome problems and contrasting results shown by previous in-sample analyses and gives a clear contribution to the assessment of temperature attribution.

In the paper just discussed we limited our analysis to a bivariate framework. However, it is well known that Granger causal links are sensitive to the information set which is employed in the analysis. Changing the information set, by extending or reducing the number of time series in the study, may lead to different Granger causal links. In particular, it is possible to find Granger causality from x to y in a bivariate system although x does not Granger cause y when also the information contained in a third variable z is taken into account [36,37].

Furthermore, together with this technical note about the possible role of omitted variables on results coming from Granger causality analyses, also a more climatic argument must be taken into account, which leads to possibly extend the information set considered here. In [33] we considered the influence of external forcings in VAR forecast improvements with respect to the predictions of the AR model built on data about T only. But, as a matter of fact, the climate system shows its own internal variability which can contribute to changes in global temperature, at least at decadal scale. Thus, it seems a good idea to insert some index of this climate variability as a context variable z in the information set: this has been done in [34].

The scope of this new paper was to investigate the causal influence of natural and anthropogenic forcings in a trivariate framework, where z is represented by one of the following indices: Southern Oscillation Index (SOI), related to El Niño Southern Oscillation (ENSO); Pacific Decadal Oscillation (PDO); Atlantic Multidecadal Oscillation (AMO).

We considered the VAR unrestricted model described in Equation (1) and the following restricted model:

$$\begin{bmatrix} y_t \\ x_t \\ z_t \end{bmatrix} = \begin{bmatrix} a_1 \\ a_2 \\ a_3 \end{bmatrix} + \sum_{j=1}^{k} \begin{bmatrix} \theta_{11,j} & 0 & \theta_{13,j} \\ \theta_{21,j} & \theta_{22,j} & \theta_{23,j} \\ \theta_{31,j} & \theta_{32,j} & \theta_{33,j} \end{bmatrix} \cdot \begin{bmatrix} y_{t-j} \\ x_{t-j} \\ z_{t-j} \end{bmatrix} + \begin{bmatrix} v_{yt} \\ v_{xt} \\ v_{zt} \end{bmatrix}. \quad (4)$$

By adopting the same test sets as in the previous paper, the one-step-ahead forecast errors were calculated as:

$$\hat{u}_{yt} = y_t - \hat{c}_1 - \sum_{j=1}^{k} \hat{\phi}_{11,j} y_{t-j} - \sum_{j=1}^{k} \hat{\phi}_{12,j} x_{t-j} - \sum_{j=1}^{k} \hat{\phi}_{13,j} z_{t-j} , \quad (5)$$

$$\hat{v}_{yt} = y_t - \hat{a}_1 - \sum_{j=1}^{k} \hat{\theta}_{11,j} y_{t-j} - \sum_{j=1}^{k} \hat{\theta}_{13,j} z_{t-j} . \quad (6)$$

Then we evaluated the MSE of these predictions and used the MSE-t and MSE-REG tests in order to test the null hypothesis. As in the previous paper, a bootstrap procedure has been performed to calculate the critical values of these tests (see [34] for further details).

The results of this paper can be summarized as follows. If we take GHG-total RF as the x variable, in every case (all circulation patterns and test sets considered)—except one—the null hypothesis of Granger non-causality on T is rejected at the 5% significance level, and very often also at 1% significance. This is clear evidence that there is a causal link (in the Granger sense) between GHG-total RF and global temperature since 1941 up to the present day.

On the other hand, if TSI is considered as the x variable, a Granger causal link is significant only in the first test set when AMO is included in the information set, and in the first two sets when PDO and ENSO are considered. In more recent periods this causal link disappears.

The situation becomes even clearer if the p-values of tests are plotted for every test period, as in **Figure 1**: see [34] for other figures and detailed tables. Here, it is evident that, while the influence of GHG-total RF on global temperature remains important throughout all the periods, the Granger causal link between TSI and T becomes progressively less marked with time and completely disappears for the last two periods. In particular, the influences of GHG-total RF and TSI on T appear comparable till the 50s, but, after that decade, a clear causal decoupling be-

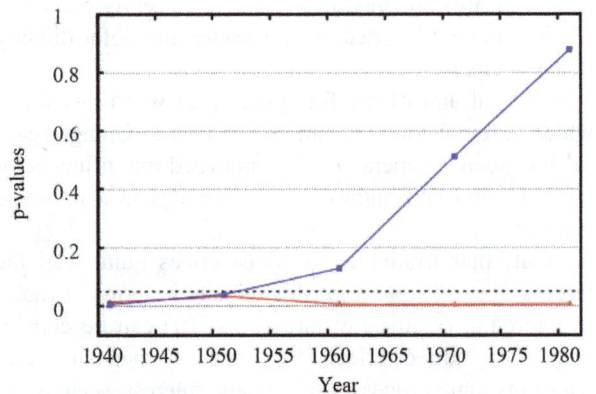

Figure 1. Plot of the p-values from the MSE-REG test when x = TSI (blue line) and x = GHG-total RF (red line) for z = ENSO. The significance threshold of 0.05 is shown (dashed line). The increase in p-values over the recent decades is evident for the performance of the model with TSI.

tween TSI and T is evident and very marked in the data of our Granger analysis. At the same time, the Granger causality from GHG-total RF to T remains robust and, possibly, becomes even more evident: the p-values, which are already very small, decrease further.

In particular, in this way we evidenced a causal decoupling between Sun and global temperatures which has been pictured previously just in terms of simple correlations and graphical methods [38,39].

6. Conclusions, Discussion and Prospects

As shown in previous sections, a number of attempts have been performed at applying the concept of Granger causality to climatic problems and, more specifically, to climatic attribution. After some pioneering works, where the choice of influencing variables is quite dubious or the dimensionality of the multivariate models probably exceeds the maximum number of parameters for obtaining reliable results, at present the application of Granger causality to the climate framework is quite well posed.

Nevertheless, our review and discussion at Section 4 show that in-sample approaches may crucially depend on preliminary analyses of the stochastic properties of time series involved: this could explain also the somewhat contrasting results obtained by these attribution studies.

Therefore, in order to overcome this critical situation, we performed out-of-sample Granger analyses for the attribution of recent global warming. This approach is less dependent on the preliminary assumptions, and more properly predictive, and more in the spirit of the original concept of Granger causality. The results obtained in this way are very clear: the radiative forcings of greenhouse gases appear as the main temperature drivers, while natural forcings do not Granger cause T in the last decades, in the case of Sun even if the principal patterns of climate variability are considered in an extended trivariate model. Furthermore, the direct Sun influence on T (via total solar irradiance) shows a recent causal decoupling since the 60s.

Obviously, even if these results represent a clear contribution to the problem of attribution of recent global warming, a discussion of methods and outcomes can show directions of future work. We briefly do this in what follows.

The first open problem is surely to test the robustness of these results when extended information sets are considered. Probably, due to the problem of dimensionality, this can be done effectively just in a trivariate framework. Here, anyway, it is possible to study several combinations of variables considered as x (the variable to be tested for Granger causality) and z (the context variable). Furthermore, in this framework, analyses about the rising of spurious or indirect causalities can be performed.

As a specific study inside a trivariate context, an in-teresting analysis can be also performed about the joint roles of direct and indirect Sun influences on T, where the direct forcing could be represented by solar irradiance and the indirect one by cosmic rays (modulated by solar wind).

In our opinion, however, a more basic problem concerns the application of a linear technique to studies of causation in a nonlinear system such as climate. In the majority of studies reviewed here, the variables used are averaged in space (the entire world) and time (one year). Thus, it is quite reasonable that, as a consequence of the central limit theorem [40], averaging can produce near-linear climate relations among variables of the climate system, even if we have to do with highly nonlinear relations at shorter space-time scales. In this context Granger causality may be applied with a good confidence; but what happens if the averaging is performed on reduced space-time scales?

With the final aim at approaching this general problem, Attanasio and Triacca [41] developed a nonlinear extension of a Granger causality model based on neural networks and applied it to the classical problem of CO_2 influences on T. Outcomes from this nonlinear Granger causality analysis are consistent with other results assessing that CO_2 radiative forcing causes recent global temperatures.

Even if the analysis of this research exceeds the scope of this paper, in our opinion this approach could show its usefulness and should be considered in analyses of attribution at reduced space-time scales and when the behavior of other variables of climatic importance, such as precipitation, are considered. It is well known, in fact, that many nonlinear processes are involved in the hydrological cycle and they cannot be easily "averaged away".

We hope to have shown that the research in this field is quite active and that future exciting studies can be surely envisaged.

7. Acknowledgements

We thank I. M. Hedgecock for useful hints.

REFERENCES

[1] A. Pasini, "From Observations to Simulations. A Conceptual Introduction to Weather and Climate Modeling," World Scientific, Singapore City, 2005.

[2] G. C. Hegerl, *et al.*, "Understanding and Attributing Climate Change," In: S. Solomon, *et al.*, Eds., *Climate Change 2007: The Physical Science Basis*, Cambridge University Press, Cambridge, 2007, pp. 663-745.

[3] A. Pasini, M. Lorè and F. Ameli, "Neural Network Modelling for the Analysis of Forcings/Temperatures Relationships at Different Scales in the Climate System," *Ecological Modelling*, Vol. 191, No. 1, 2006, pp. 58-67.

[4]   A. Pasini and R. Langone, "Attribution of Precipitation Changes on a Regional Scale by Neural Network Modeling: A Case Study," *Water*, Vol. 2, No. 3, 2010, pp. 321-332.

[5]   A. Pasini and R. Langone, "Influence of Circulation Patterns on Temperature Behavior at the Regional Scale: A Case Study Investigated via Neural Network Modeling," *Journal of Climate*, Vol. 25, No. 6, 2012, pp. 2123-2128.

[6]   C. W. J. Granger, "Investigating Causal Relations by Econometric Models and Cross-Spectral Methods," *Econometrica*, Vol. 37, No. 3, 1969, pp. 424-438.

[7]   C. Diks and M. Mudelsee, "Redundancies in the Earth's Climatological Time Series," *Physics Letters A*, Vol. 275, No. 5-6, 2000, pp. 407-414.

[8]   R. K. Kaufmann, L. Zhou, R. B. Myneni, C. J. Tucker, D. Slayback, N. V. Shabanov and J. Pinzon, "The Effect of Vegetation on Surface Temperature: A Statistical Analysis of NDVI and Climate Data," *Geophysical Research Letters*, Vol. 30, No. 22, 2003, p. 2147.

[9]   J. B. Elsner, "Evidence in Support of the Climate Change —Atlantic Hurricane Hypothesis," *Geophysical Research Letters*, Vol. 33, No. 16, 2006, Article ID: L16705.

[10]   J. B. Elsner, "Granger Causality and Atlantic Hurricanes," *Tellus A*, Vol. 59, No. 4, 2007, pp. 476-485.

[11]   T. J. Mosedale, D. B. Stephenson, M. Collins and T. C. Mills, "Granger Causality of Coupled Climate Processes: Ocean Feedback on the North Atlantic Oscillation," *Journal of Climate*, Vol. 19, No. 7, 2006, pp. 1182-1194.

[12]   R. K. Kaufmann, K. C. Seto, A. Schneider, Z. Liu, L. Zhou and W. Wang, "Climate Response to Rapid Urban Growth: Evidence of a Human-Induced Precipitation Deficit," *Journal of Climate*, Vol. 20, No. 10, 2007, pp. 2299-2306.

[13]   I. I. Mohkov, D. A. Smirnov, P. I. Nakonechny, S. S. Kozlenko, E. P. Seleznev and J. Kurths, "Alternating Mutual Influence of El Niño/Southern Oscillation and Indian Monsoon," *Geophysical Research Letters*, Vol. 38, 2011, Article ID: L00F04.

[14]   R. S. J. Tol and A. F. de Vos, "Greenhouse Statistics— Time Series Analysis," *Theoretical and Applied Climatology*, Vol. 48, No. 2-3, 1993, pp. 63-74.

[15]   R. S. J. Tol and A. F. de Vos, "A Bayesian Statistical Analysis of the Enhanced Greenhouse Effect," *Climatic Change*, Vol. 38, No. 1, 1998, pp. 87-112.

[16]   L. Sun and M. Wang, "Global Warming and Global Dioxide Emission: An Empirical Study," *Journal of Environmental Management*, Vol. 46, No. 4, 1996, pp. 327-343.

[17]   R. K. Kaufmann and D. I. Stern, "Evidence for Human Influence on Climate from Hemispheric Temperature Relations," *Nature*, Vol. 388, No. 6637, 1997, pp. 39-44.

[18]   U. Triacca, "On the Use of Granger Causality to Investigate the Human Influence on Climate," *Theoretical and Applied Climatology*, Vol. 69, No. 3-4, 2001, pp. 137-138.

[19]   U. Triacca, "Is Granger Causality Analysis Appropriate to Investigate the Relationship between Atmospheric Concentration of Carbon Dioxide and Global Surface Air Temperature?" *Theoretical and Applied Climatology*, Vol. 81, No. 3-4, 2005, pp. 133-135.

[20]   H. Y. Toda and T. Yamamoto, "Statistical Inference in Vector Autoregression with Possibly Integrated Processes," *Journal of Econometrics*, Vol. 66, No. 1-2, 1995, pp. 225-250.

[21]   R. Reichel, P. Thejll and K. Lassen, "The Cause-and-Effect Relationship of Solar Cycle Length and the Northern Hemisphere Air Surface Temperature," *Journal of Geophysical Research*, Vol. 106, No. A8, 2001, pp. 15635-15641.

[22]   I. I. Mohkov and D. A. Smirnov, "Diagnostics of Cause-Effect Relation between Solar Activity and the Earth's Global Surface Temperature," *Izvestiya, Atmospheric and Oceanic Physics*, Vol. 44, No. 3, 2008, pp. 263-272.

[23]   E. Kodra, S. Chatterjee and A. R. Ganguly, "Exploring Granger Causality between Global Average Observed Time Series of Carbon Dioxide and Temperature," *Theoretical and Applied Climatology*, Vol. 104, No. 3-4, 2011, pp. 325-335.

[24]   A. Attanasio, "Testing for Linear Granger Causality from Natural/Anthropogenic Forcings to Global Temperature Anomalies," *Theoretical and Applied Climatology*, Vol. 110, No. 1-2, 2012, pp. 281-289.

[25]   U. Triacca, A. Attanasio and A. Pasini, "Anthropogenic Global Warming Hypothesis: Testing Its Robustness by Granger Causality Analysis," *Environmetrics*, Vol. 24, No. 4, 2013, pp. 260-268.

[26]   J. Y. Park and P. C. B. Phillips, "Statistical Inference in Regressions with Integrated Processes: Part 2," *Economic Theory*, Vol. 5, No. 1, 1989, pp. 95-131.

[27]   J. H. Stock and M. W. Watson, "Interpreting the Evidence on Money-Income Causality," *Journal of Econometrics*, Vol. 40, No. 1, 1989, pp. 161-181.

[28]   C. A. Sims, J. H. Stock and M. W. Watson, "Inference in Linear Time Series Models with Some Unit Roots," *Econometrica*, Vol. 58, No. 1, 1990, pp. 113-144.

[29]   J. Chao, V. Corradi and N. Swanson, "Out-of-Sample Test for Granger Causality," *Macroeconomic Dynamics*, Vol. 5, 2001, pp. 598-620.

[30] T. Clark, "Can Out-of-Sample Forecast Comparisons Help Prevent Overfitting?" *Journal of Forecasting*, Vol. 23, No. 2, 2004, pp. 115-139.

[31] S. Gelper and C. Croux, "Multivariate Out-of-Sample Tests for Granger Causality," *Computational and Statistical Data Analysis*, Vol. 51, No. 7, 2007, pp. 3319-3329.

[32] R. Ashley, C. W. J. Granger and R. Schmalansee, "Advertising and Aggregate Consumption: An Analysis of Causality," *Econometrica*, Vol. 48, No. 5, 1980, pp. 1149-1167.

[33] A. Attanasio, A. Pasini and U. Triacca, "A Contribution to Attribution of Recent Global Warming by Out-of-Sample Granger Causality Analysis," *Atmospheric Science Letters*, Vol. 13, No. 1, 2012, pp. 67-72.

[34] A. Pasini, U. Triacca and A. Attanasio, "Evidence of Recent Causal Decoupling between Solar Radiation and Global Temperature," *Environmental Research Letters*, Vol. 7, No. 3, 2012, Article ID: 034020.

[35] M. W. McCracken, "Asymptotics for Out-of-Sample Tests of Granger Causality," *Journal of Econometrics*, Vol. 140, No. 2, 2007, pp. 719-752.

[36] H. Lütkepohl, "Non-Causality Due to Omitted Variables," *Journal of Econometrics*, Vol. 19, No. 2-3, 1982, pp. 367-378.

[37] J. D. Hamilton, "Time Series Analysis," Princeton University Press, Princeton, 1994.

[38] M. Lockwood and C. Fröhlich, "Recent Oppositely Directed Trends in Solar Climate Forcings and the Global Mean Surface Air Temperature," *Proceedings of the Royal Society A*, Vol. 463, No. 2086, 2007, pp. 2447-2460.

[39] P. Stauning, "Solar Activity-Climate Relations: A Different Approach," *Journal of Atmospheric and Solar-Terrestrial Physics*, Vol. 73, No. 13, 2011, pp. 1999-2012.

[40] A. Yuval and W. W. Hsieh, "The Impact of Time-Averaging on the Detectability of Nonlinear Empirical Relations," *Quarterly Journal of the Royal Meteorological Society*, Vol. 128, 2002, pp. 1609-1622.

[41] A. Attanasio and U. Triacca, "Detecting Human Influence on Climate Using Neural Networks Based Granger Causality," *Theoretical and Applied Climatology*, Vol. 103, No. 1-2, 2011, pp. 103-107.

17

Multi-Decadal Trends of Global Surface Temperature: A Broken Line with Alternating ~30 yr Linear Segments?

Vincent Courtillot[1], Jean-Louis Le Mouël[1], Vladimir Kossobokov[1,2],
Dominique Gibert[1], Fernando Lopes[1]

[1]Geomagnetism and Paleomagnetism, Institut de Physique du Globe de Paris, Université Paris Diderot,
Sorbonne Paris Cité, 1 rue Jussieu, Paris, France
[2]Institute of Earthquake Prediction Theory and Mathematical Geophysics, Russian Academy of Sciences,
Moscow, Russia

ABSTRACT

We investigate global temperature data produced by the Climate Research Unit at the University of East Anglia (CRU) and the Berkeley Earth Surface Temperature consortium (BEST). We first fit the 1850-2010 data with polynomials of degrees 1 to 9. A significant ~60-yr oscillation is accounted for as soon as degree 4 is reached. This oscillation is even better modeled as a broken line, more precisely a series of ~30-yr long linear segments, with slope breaks (singularities) in ~1904, ~1940, and ~1974 (±3 yr), and a possible recent occurrence at the turn of the 20th century. Oceanic indices PDO (Pacific Decadal Oscillation) and AMO (Atlantic Multidecadal Oscillation) have undergone major changes (respectively of sign and slope) roughly at the same times as the temperature slope breaks. This can be interpreted with a system of oceanic non-linear coupled oscillators with abrupt mode shifts. Thus, the Earth's climate may have entered a new mode (a new ~30-yr episode) near the turn of the 20th century: no further temperature increase, a dominantly negative PDO index and a decreasing AMO index might be expected for the next decade or two.

Keywords: Global Surface Temperature; Multi-Decadal Evolution; Linear Segments; ~60-Year Oscillation

1. Introduction

Global surface temperatures are one of the parameters most commonly used to discuss the evolution of climate. Databases of instrumental temperatures that cover the past century and a half have been compiled by four main groups (Climate Research Unit at the University of East Anglia—CRU, NASA Goddard Institute for Space Studies—GISS, National Oceanic and Atmospheric Administration—NOAA, Berkeley Earth Surface Temperature consortium—BEST). This is a huge, difficult task and these databases are necessarily faced with a number of limitations: the geographical distribution of stations is far from uniform and varies with time; also, there may be fundamental difficulties in establishing meaningful global temperatures for the Earth [1]. Climate is generally defined as the ~30 year average of weather and since only 100 to 150 years of instrumental data are available, this places severe limitations on the significance of global temperature trends and multi-decadal oscillations. A monotonic (low degree) trend can be fit to all global tem-

perature data sets over the period from 1850 to the present. This trend amounts to a secular rise of ~1 K over the period. Once this trend is removed, a ~60 yr oscillation stands out in most records. Such an oscillation has been discussed for instance by [2] and [3]. Lean and Rind [4,5] have built a model curve for 1889-2006 monthly mean global temperatures, in which they distinguish contributions from oceans (using the El Nino Southern Oscillation ENSO), volcanic aerosols, solar irradiance and anthropogenic forcing. Both observations and their empirical curve display the secular rise of approximately 1 K. But data are lower than model by some 0.2 K around 1910 and higher by 0.2 K around 1940: this is because of the significant ~60 year oscillation identified by previous authors, which is not accounted for by [4].

In this paper, we first check whether some of the main global temperature data sets can be reasonably fit by smooth polynomials of increasing degree from 1 (linear trend) up to 9, with particular focus on the "~60-yr oscillation". We next suggest that this oscillation may be fit by a series of ~30-yr long linear segments with rather abrupt changes in trend. We find correlations between the breaks in the multi-decadal trends in global surface temperature and changes in sign of the Pacific Decadal Oscillation and slope of the Atlantic Multidecadal Oscillation indices. We discuss our results in the frame of the dynamical mechanism for major climate shifts proposed by [6].

2. Polynomial Fits to the "~60-yr Oscillation" in Global Temperatures

We have used monthly temperature anomaly data from the CRU and BEST databases. We chose to use the "raw data" rather than other data-derived products that have been "homogenized" and have thus undergone rather extensive and not always transparent/reproducible data polishing. The CRUTem4 database is found at http://www.cru.uea.ac.uk/cru/data/temperature/. We have analyzed the most recent variance-adjusted (version 4) CRU data from 1850 through 2010 for the whole globe (CRUTem4vGL). We have also analyzed hemispheric and global series CRUTem4-SH, CRUTem4-NH, CRU-Tem4vSH, HadSST2GL, HadCRU3GL, and CRUTEM-3vGL (these series are described in http://www.metoffice.gov.uk/hadobs/crutem4/ and [7], and http://www.cru.uea.ac.uk/cru/data/temperature/crutem4vgl.txt), and the Berkeley Earth Surface Temperature (BEST) monthly global averages (found at http://berkeleyearth.org/dataset/, description in [8]). All yield similar results and mainly results based on CRU-Tem4vGL are shown in this paper.

We note in passing differences between the long-term linear or low-degree trends obtained in the various temperature anomaly databases (notwithstanding a possible overall difference in baseline due to the definition of the period with respect to which the anomalies are calculated); the CRU and GISS curves are about 0.3 K below the NOAA and BEST curves in the 2000-2010 decade, whereas all agree to better than 0.1 K between 1920 and 1970 [8]. We return to this in Section 5.

In order to check the significance of the "60-yr oscillation", we have very simply fit the data sets of monthly mean temperature anomalies (using least squares) with polynomials of increasing degree (**Figure 1**). The linear (n = 1) trend fit to the CRUTem4vGL data shows the secular rise of approximately 1 K. For n = 2 and 3 the curves fail to account for the "60-yr oscillation" (**Figure 1**, purple and blue curves). As soon as the degree of the polynomial reaches 4, the fit accommodates the "60-yr oscillation", after which the situation remains stable (shown up to n = 9 in **Figure 1**). Extrapolations of these trends outside of the data domain show quick divergence and are of course meaningless. A ~60 year oscillation would occur only 2.5 times over a 150 yr interval and could not be accounted for by a polynomial of degree

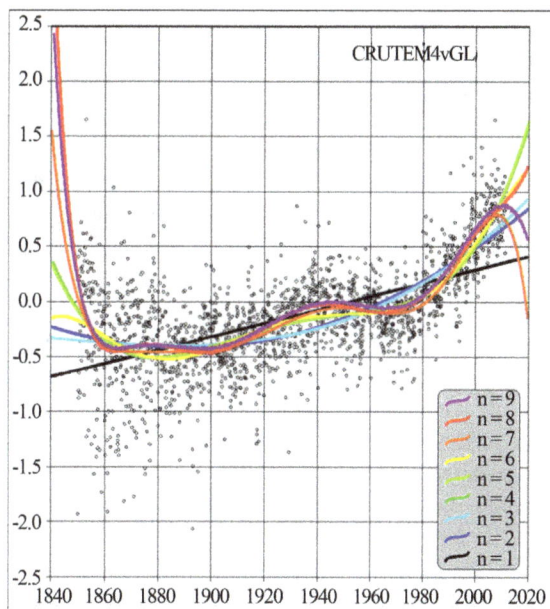

Figure 1. Monthly global temperature anomaly averages from 1850 to 2010 together with least squares polynomial fits with degree from 1 to 9 shown by the color code at lower right. Whole Earth, variance adjusted, CRUTem4vGL database (http://www.cru.uea.ac.uk/cru/data/temperature/crutem4vgl.txt).

less than ~4.

In order to quantify the goodness of fit, we have calculated the adjusted R-squared values not only for polynomial fits to the data with degrees 1 up to 9 but also for an exponential fit. The adjusted R-squared [9] is a statistically unbiased version (*i.e.* corrected for the number of degrees of freedom) of the R^2 "coefficient of determination", whose main purpose is the prediction of future outcomes on the basis of prior information. R^2 itself is 1 minus the ratio of variance of errors to variance of observations. The adjusted R-squared values take into consideration the number of free parameters of the model (the adjusted R-squared values may be negative when the parameters of the model do not improve the fit compared to the simple average value of the data, which happened in the case of one data set, CRUTem4vSH, that could not be adequately fit by an exponential). In the case of CRUTem4vGL, the adjusted R-squared increases significantly from 0.36 (n = 1) to 0.50 (n = 4) and remains stable thereafter (0.52 for n = 9). As soon as the degree exceeds 1 or 2, the adjusted R-squared values for polynomials of degree larger than 3 are larger than for any exponential fit.

3. Broken Line Fits to the "~60-yr Oscillation" in Global Temperatures

The possibility of fitting ~30-yr long linear segments as an alternative to polynomials or a 60-yr sinusoid to the

"~60-yr fluctuation" is the topic of this section. **Figure 2(a)** shows another display of the CRUTem4vGL monthly data set, limited to the period from 1900 to 2010 (the previous 50 years have significantly larger variance, which is likely due to insufficient, changing station distribution and also possibly to lesser data quality). We note that a series of continuous linear segments with abrupt changes in slope may provide a good fit: such fits have appeared in the literature (e.g. Figure 8 of [10], where linear regression lines are shown for intervals 1901-1934, 1934-1979 and 1979-2010). The most recent has been proposed and discussed by [11], using methods from the dynamics of synchronized chaotic systems.

The models we are seeking belong to the class of "change-point models" [12]. Vasko and Toivonen [13] for instance have published quantitative algorithms to estimate the number of linear segments to be fit to a time series using permutation tests. We have constructed an algorithm that produces such models, paying particular attention to the management of secondary minima using non-linear simulated annealing. Simulated annealing uses the Metropolis algorithm (see [14-16], for description of the algorithms and definition of terms; see similar use of simulated annealing in [17]). The method is outlined briefly in the Appendix.

We have first tested our method on synthetic data consisting of N = 2 to 10 linear segments with variable durations and slopes and varying noise levels from 5% to 25% of the total amplitude range of the data. For each value of N, and for each step of the control parameter, we ran 100 attempts at locating the nodes and displayed the histograms of node locations. For 4 segments and 5% noise, the 3 nodes are identified to within a year in a 2000-month (166 yr) long test series; for 15% noise, detection of the 3 nodes is unambiguous but the histograms widen to ~100 months (±4 years uncertainties). For 25% noise, the nodes are still detected but less prominently and with larger uncertainties (~±8 years).

We have then applied the method to the 1900-2010 CRUTem4vGL monthly data, using from 3 to 10 segments. Results are shown in **Figure 2(a)** for the best case, *i.e.* with 5 segments, and in **Figure 2(b)** with 4 segments for comparison. Two nodes are clearly identified at ~1940 and ~1974. Two other nodes at ~1904 and ~2006 are near the edges of the data distribution and could be artefacts (end-effects), though this has not been observed with the tests on synthetic data. The ~1904 node appears to be supported by visual inspection of some data predating 1900 (but data dispersion is significantly larger before 1900 than afterwards; see Section 5). The more recent one may require at least a decade of additional data to ascertain its validity. In the case of 4 segments (**Figure 2(b)**), the histogram does confirm the same 4 nodes with the same dates, although of course each individual fit out of the 500 realizations must miss one of the 4 clusters seen in the histogram, since only 3 nodes are allowed. The histogram with 6 segments (not shown) still

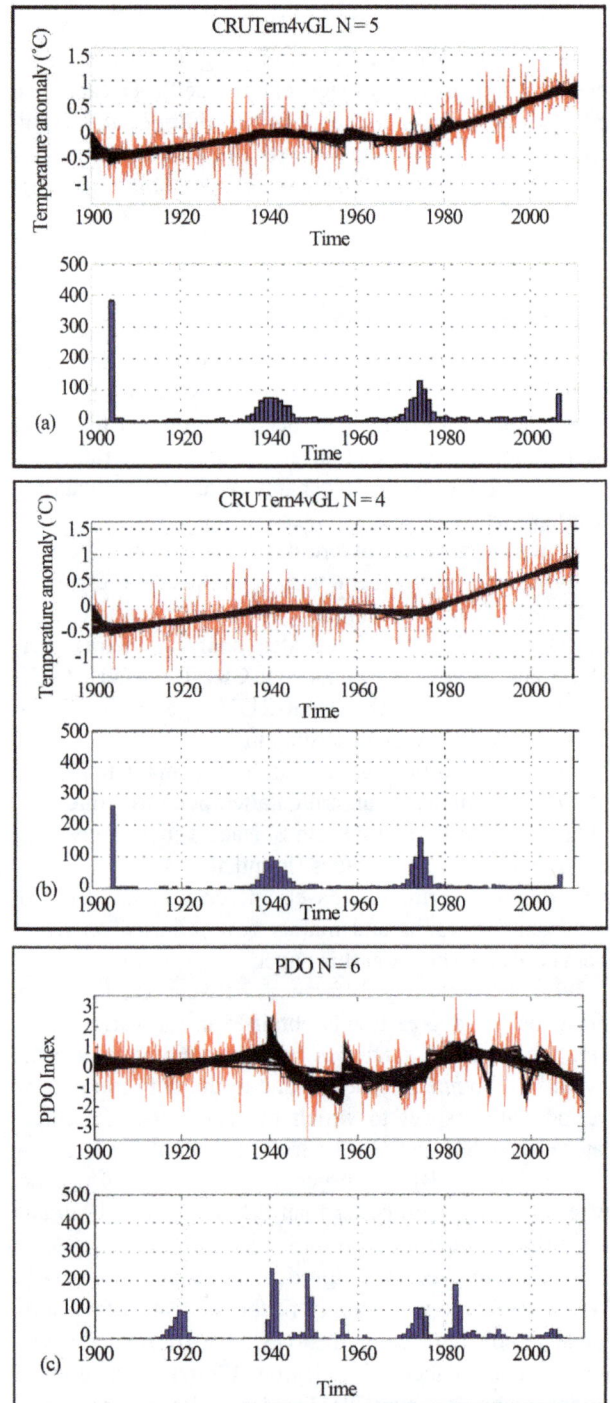

Figure 2. (a) **CRUTem4vGL monthly global temperature anomaly averages from 1900 to 2010 (red curve) and 500 fits of curves made of 5 linear segments (4 nodes) determined with the simulated annealing method described in the text. Below, histogram of node distribution; (b) Same with 4 segments and 3 nodes; (c) Same for the Pacific Decadal Oscillation Index (PDO) with 6 segments and 5 nodes.**

underlines the same nodes, but with more spread in dates. As the number of segments is increased to 10, node distributions become more uniform, as segments are used to adjust to shorter (sub-decadal) fluctuations in the data.

The differences in location of the nodes introduce some curvature in the "cloud" of 500 fitted broken lines, but the linear segments between the nodes are quite clear (**Figures 2(a)** and **(b)**). We confirm our preliminary hypothesis that the 1900-2010 data can be fit well by such a series of linear segments and that the data argue in favor of at least two singular events around 1940 and 1974. The slopes of the three main segments are 1.4 K/100 yr for 1904-1940 (36 years), −0.9 K/100 yr for 1940-1974 (34 years) and 3.1 K/100 yr for 1974-2006 (32 years). This oscillation appears to be better represented by a sequence of linear segments with rather abrupt slope changes at the ~1904, ~1940 and ~1974 nodes (adjusted R-squared > 0.62), rather than a sinusoidal or polynomial fit (see previous section). Climate shifts close to these dates have independently been identified by [11], using methods from the dynamics of synchronized chaotic systems (see Section 5).

In order to get a further idea of the robustness of inferences that can be made from **Figure 2**, we have resumed the same computations as in **Figure 2(a)** (5 segments, 4 nodes) for all 21 datasets that can be obtained from the CRU site
http://www.cru.uea.ac.uk/cru/data/temperature/#datter
(*i.e.* land air temperature anomalies CRUTem3, CRUTem3v, CRUTem4 and CRUTem4v, sea surface temperature anomalies HadSST2 and combined land and marine temperatures HadCRUT3 and HadCRU3v, each with hemispheric means for the northern and southern hemispheres and combined global series), and we have stacked all corresponding histograms, resulting in **Figure 3**. There is a prominent bimodal cluster of nodes at 1938-1940 and 1945-1946 and a single mode cluster at 1975-1976, and two lesser bimodal clusters around 1904/1908 and 2006/2010. Altogether, most data series are well fit by a sequence of 3 linear segments between the early 1900s and the early 2000s, with suggestions for an earlier and a later segment that remain to be confirmed.

4. Comparing Trends in Global Temperature and the PDO and AMO Oceanic Indices

We have applied the same method to an oceanic proxy, the Pacific Decadal Oscillation (PDO is the leading principal component of monthly sea surface temperature anomalies in the North Pacific Ocean, poleward of 20°N). The optimal number of segments is 6 (**Figure 2(c)**). Two prominent doublets of nodes separated by ~33 years are seen, one at ~1940 and ~1948, the other at ~1973 and ~1982. A less prominent node is located at ~1919 (29 years

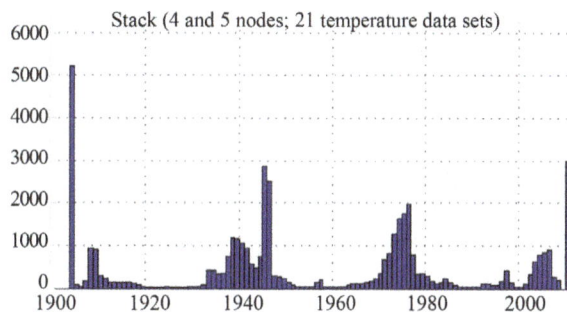

Figure 3. A stack of histograms calculated in the manner of Figure 2(a) with 4 nodes and 5 segments for 21 data sets from site
http://www.cru.uea.ac.uk/cru/data/temperature/#datter: land air temperature anomalies CRUTEM3, CRUTEM3v, CRUTEM4 and CRUTEM4v, sea surface temperature anomalies HadSST2 and combined land and marine temperatures HadCRUT3 and HadCRU3v, each with hemispheric means for the northern (NH) and southern (SH) hemispheres and combined global series (GL). v stands for "variance adjusted".

before the 1948 cluster).

The fitting exercises above suggest a comparison, which is illustrated in **Figures 4(b)** and **(c)**, where the monthly values of PDO and global temperature anomaly are shown on the same time scale. Periods when the multi-decadal slopes of linear segments fitted to the temperature data, as outlined in **Figures 2** and **3**, were positive and negative have been colored, respectively, in pale red and blue. There appears to be a strong correlation between the *sign of the slope of T* (*i.e.* its multi-decadal time derivative) and the dominant *sign of PDO*. There is good correspondence between the ~1940 and ~1974 nodes of CRUTem4vGL and the main changes in PDO sign, with PDO shifting from mainly positive (between ~1930 and ~1940) to mainly negative (between ~1950 and ~1976), then back to mainly positive (between ~1976 and ~2000).

The Atlantic Multidecadal Oscillation index (AMO characterizes the average sea surface temperature over the northern Atlantic; it is defined as the SST averaged over 0°N - 60°N, 0°W - 80°W, minus SST averaged over 60°S - 60°N) also correlates well with global mean surface temperature anomalies (as for instance already pointed out by [8]), and therefore also with the dominant multi-decadal sign changes of PDO, as seen in **Figure 4(d)**: a decreasing linear segment prior to ~1920, an increasing one from ~1920 to ~1940, a decreasing one from ~1950 to ~1975, an increasing one after ~1975, generating a strong "~60-yr oscillation".

5. Discussion

The above analysis suggests the possibility of causal links between some multi-decadal (~60-yr) changes in

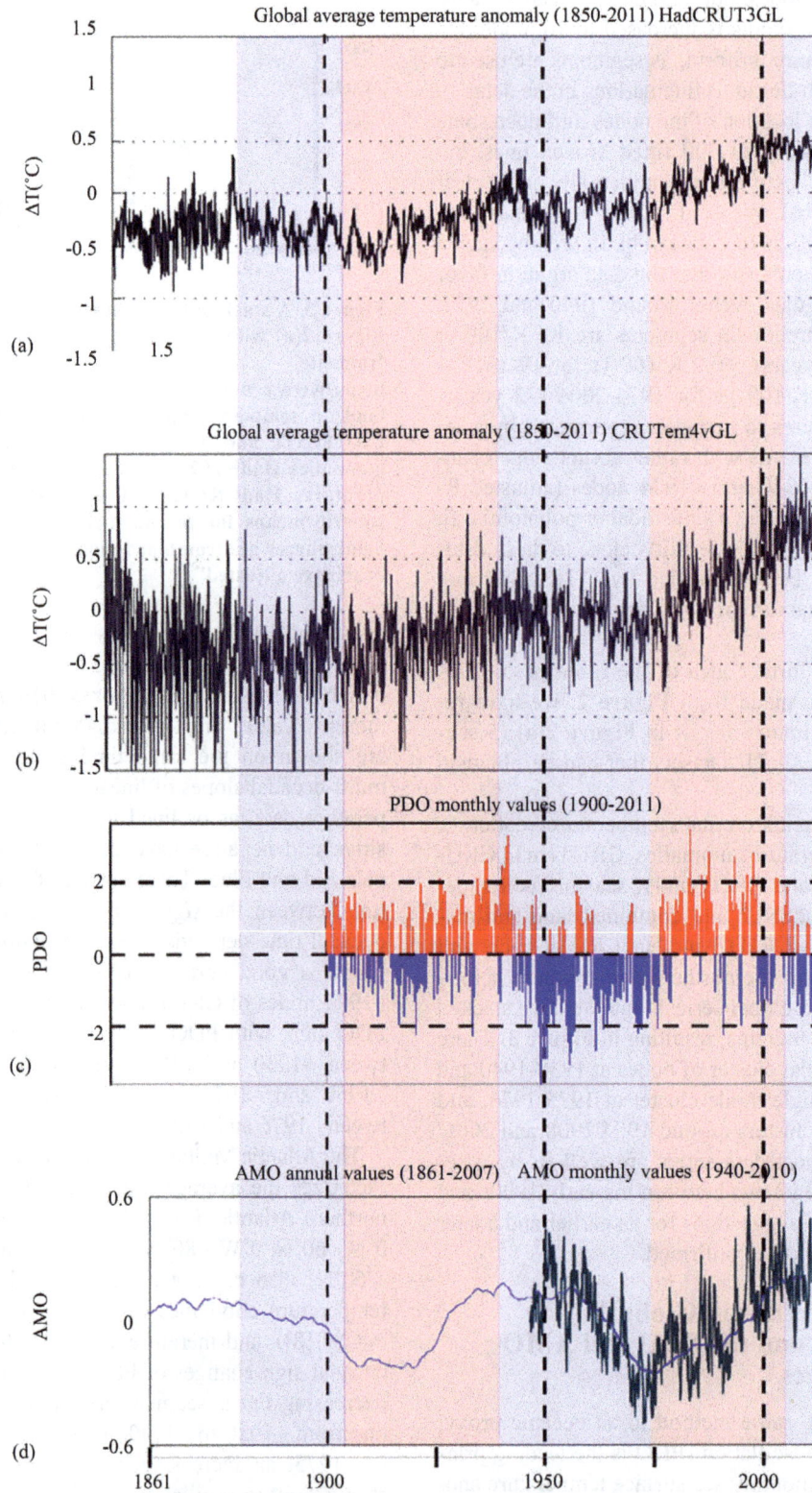

Figure 4. (a) Monthly global average temperature anomaly HadCRUT3GL 1850-2011; (b) Monthly global average temperature anomaly CRUTem4vGL 1850-2011; (c) Monthly values of PDO index 1900-2011 (positive in red, negative in blue); (d)
Monthly values of AMO index 1900-2010 and annual values 1861-2007. Vertical bands shaded in pale red (respectively blue)
emphasize the correlation between periods of linearly increasing (resp. decreasing) temperature, dominantly positive (resp.
negative) PDO index and increasing (resp. decreasing) AMO (see text).

oceanic indices PDO (and AMO) and in global surface temperature since the early 20th century, more precisely a causal connection between changes in the modes of oceanic circulation and changes in multi-decadal (linear) trends of global surface temperature. Given the respective masses, impedances and time constants involved, it is reasonable to argue that the oceanic system forces the atmospheric system on these time scales rather than the opposite.

Tsonis et al. [6] have proposed a dynamical system approach of major climate shifts, such as the ones identified by the nodes at ~1940 and ~1974. These authors define a network of oceanic indices (PDO, the El Nino Southern Oscillation ENSO, the North Atlantic Oscillation NAO and the North Pacific Oscillation NPO). They find that over the period 1900-2000, this network synchronized several times. [6] define the coupling strength between indices and find that "where the synchronous state was followed by a steady increase in the coupling strength (⋯), the previous synchronous state was destroyed, after which a new climate state emerged". The dates of the events they find are near 1910, 1940 and 1975, i.e. essentially identical (given data variance and uncertainties in node locations) to the dates we have identified for CRUTem4vGL temperature anomalies and also most other temperature data sets (Figure 3).

[6] find the same features in models of "systems with nonlinear coupled oscillators, caused by bifurcations as the coupling parameter changes". They cannot conclude (nor can we) whether changes are triggered by some external forcing or are generated within the chaotic system itself. Their figure 2(d) illustrates the shifts in the global-SST ENSO index: this is in essence the same "box-car" correlation that can be observed in our Figure 4 between T and PDO; in the present paper, we believe we provide stronger and more quantitative observational evidence of the reality of these shifts and when they occur.

A recent (~2006) shift of the temperature anomaly slope is suggested by Figures 2 and 3. In the early 2000s, the PDO turned back to a negative-dominated mode (Figure 4(c)) and the slope of the multidecadal linear trend in AMO may have become flat, as had happened towards the end of the ~1920-1940 episode (Figure 4(d)). Also, the succession of ~30 yr long quasi-linear segments in the temperature anomaly curve (Figure 4(b): possibly 1860-1880 (?) and 1880-1910 (?), then 1910-1940, 1940-1970 and 1970-2000) could itself suggest that the ~60 yr oscillation, whatever its source, may continue as a ~2000-2030 segment.

We point out the puzzling fact that an earlier CRU global temperature dataset, HadCRUT3GL, displayed the alternating segments better than the latest version CRUTem4vGL (Figures 4(a) vs (b)). The annual variability

of the monthly averages increased by a factor of 2 in the recent revision. So did the amplitude of the temperature anomaly since 1975, that changed from 0.5 K to about 1.0 K. The segment with negative slope prior to ~1910-1920, and more importantly the segment with flat or even negative slope after 2000, are particularly conspicuous on HadCRUT3GL (Figure 4(a)). [8] discuss the fact that the four available global temperature anomaly curves start diverging in the past two decades (their Figure 8): there is close agreement from 1900 to the late 1980s, but after that GISS and HadCRU display a plateau, whereas the NOAA and BEST curves continue to rise to values that are 0.25 K above the GISS and HadCRU plateau. It is awkward that it is the most recent part of the data compilation that displays this divergence. The plateau after the late 1990s has been the topic of much recent discussion; the previous HadCRUT3 data display this plateau much more clearly than CRUTem4vGL that seems to have suppressed it.

The shift in dominant sign of PDO (Figure 4(c)), the flattening in AMO (Figure 4(d)), and the plateau in several global temperature anomaly data (e.g. Figure 4(a)) taken together would support the hypothesis of a regime change having occurred near the end of the 20th century or early in the 21st century. A ~2000 regime change would complement the series of events at ~30-yr intervals (~1910-1920, ~1940, ~1970, ~2000). This change, if real, could imply that a ~15 year long period with no further warming lies ahead. Klyashtorin and Lyubushin [2], de Jager and Duhau [18], Scafetta [3] and Russell et al. [19] are among several authors who have suggested decades of future cooling: these authors argue for external forcings due to solar or planetary effects (see also the recent paper by [20]), when the mechanism of [6] envisions the chaotic result of interaction between oceanic non-linear dynamical systems. Tsonis et al. [6] conclude that the 1970 to 2000 warming may not be (wholly) due to the radiative effect of greenhouse gases overcoming shortwave reflection effects due to aerosols and that the climate may indeed have shifted to a different state.

Our analysis strongly argues for the presence of a ~60-yr oscillation in the climate system (at least over the limited time interval—100 to 150 years—covered by reliable instrumental observations). More precisely, global temperature data can be interpreted as a series of linear segments interrupted by rather fast changes in slope, i.e. abrupt changes in regimes. Each episode or regime maintains itself for approximately 30 years. We favor the oceanic system as a driver of atmospheric temperatures on these multi-decadal time scales. It remains to be seen whether abrupt changes in climate mode are a result of internal chaotic dynamics of the ocean system only or could be forced by external factors.

6. Acknowledgements

We acknowledge financial support from IPGP as part of the IEPT RAS-IPGP cooperation. IPGP Contribution NS 3391.

REFERENCES

[1]   C. Essex, R. McKitrick and B. Andresen, "Does a Global Temperature Exist?" *Journal of Non-Equilibrium Thermodynamics*, Vol. 32, No. 1, 2007, pp. 1-27.

[2]   L. B. Klyashtorin and A. A. Lyubushin, "On the Coherence between Dynamics of the World Fuel Consumption and Global Temperature Anomaly," *Energy and Environment*, Vol. 14, No. 6, 2003, pp. 773-782.

[3]   N. Scafetta, "Empirical Evidence for a Celestial Origin of the Climate Oscillations and Its Implications," *Journal of Atmospheric and Solar-Terrestrial Physics*, Vol. 72, No. 13, 2010, pp. 951-970.

[4]   J. L. Lean and D. H. Rind, "How Natural and Anthropogenic Influences Alter Global and Regional Surface Temperatures: 1889 to 2006," *Geophysical Research Letters*, Vol. 35, No. 18, 2008, Article ID: L18701.

[5]   J. L. Lean and D. H. Rind, "How Will Earth's Surface Temperature Change in Future Decades?" *Geophysical Research Letters*, Vol. 36, No. 15, 2009, Article ID: L15708.

[6]   A. A. Tsonis, K. Swanson and S. Kravtsov, "A New Dynamical Mechanism for Major Climate Shifts," *Geophysical Research Letters*, Vol. 34, 2007, Article ID: L13705.

[7]   P. D. Jones, D. H. Lister, T. J. Osborn, C. Harpham, M. Salmon and C. P. Morice, "Hemispheric and Large-Scale Land Surface Air Temperature Variations: An Extensive Revision and an Update to 2010," *Journal of Geophysical Research*, Vol. 16, No. 1, 2012, pp. 206-223.

[8]   R. Rohde, J. Curry, D. Groom, R. Jacobsen, R. A. Muller, S. Perlmutter, A. Rosenfeld, C. Wickham and J. Wurtele, "Berkeley Earth Temperature Averaging Process," 2011. http://berkeleyearth.org/available-resources/

[9]   H. Theil, "Economic Forecasts and Policy," North Holland, Amsterdam, 1961.

[10] H. J. Lüdecke, R. Link and F. K. Ewert, "How Natural Is the Recent Centennial Warming? An Analysis of 2249 Surface Temperature Records," *International Journal of Modern Physics*, Vol. 22, No. 10, 2011, pp. 1139-1159.

[11] K. L. Swanson and A. A. Tsonis, "Has the Climate Recently Shifted?" *Geophysical Research Letters*, Vol. 36, 2009, Article ID: L06711.

[12] I. A. Eckley, P. Fearnhead and R. Killick, "Analysis of Change-Point Models," In: D. Barber, A. Taylan Cemgil and S. Chiappa, Eds., *Bayesian Time Series Models*, Cambridge University Press, Cambridge, 2011.

[13] K. Vasko and H. Toivonen, "Estimating the Number of Segments in Time Series Data Using Permutation Tests," *The 2002 IEEE International Conference on Data Mining (ICDM'02)*, Maebashi City, December 2002, pp. 466-473.

[14] S. Kirkpatrick, C. D. Gelatt Jr. and M. P. Vecchi, "Optimization by Simulated Annealing," *Science*, Vol. 220, No. 4598, 1983, pp. 671-680.

[15] G. Bhanot, "The Metropolis Algorithm," *Reports on Progress in Physics*, Vol. 51, No. 3, 1988, pp. 429-457.

[16] W. H. Press, S. A. Teukolsky, W. T. Vetterling and B. P. Flannery, "Section 10.12. Simulated Annealing Methods," In: *Numerical Recipes: The Art of Scientific Computing*, 3rd Edition, Cambridge University Press, New York, 2007.

[17] D. Gibert and J.-L. Le Mouël, "Inversion of Polar Motion Data: Chandler Wobble, Phase Jumps, and Geomagnetic Jerks," *Journal of Geophysical Research*, Vol. 113, 2008, Article ID: B10405.

[18] C. de Jager and S. Duhau, "Forecasting the Parameters of Sunspot Cycle 24 and Beyond," *Journal of Atmospheric and Solar-Terrestrial Physics*, Vol. 71, No. 2, 2009, pp. 239-245.

[19] C. T. Russell, J. G. Luhmann and L. K. Jian, "How Unprecedented a Solar Minimum?" *Reviews of Geophysics*, Vol. 48, 2004, Article ID: RG2004.

[20] J. A. Abreu, J. Beer, A. Ferriz-Mas, K. G. McCracken and F. Steinhilber, "Is There a Planetary Influence on Solar Activity?" *Astronomy & Astrophysics*, Vol. 548, 2012, Article ID: A88.

Appendix

In order to generate broken-line models that fit the data, we have constructed an algorithm that uses two imbricated loops.

The first, internal, loop uses the Metropolis algorithm (e.g. [16]), that generates a suite of models distributed according to a given probability law p. At iteration n the model $m(n)$ has probability $p(n)$. This is perturbed (see below) in order to test a new model $m(test)$ with probability $p(test)$. The new model is accepted if $p(test) \geq p(n)$ and becomes $m(n + 1)$, replacing $m(n)$; if $p(test) < p(n)$ the model is accepted with probability $p(test)/p(n)$.

The perturbation from one step to the next is done in the following way. If N is the (given) number of segments in the broken line to be fitted to the data, there are $N - 1$ free nodes (the ends being assumed fixed in abscissa, $i.e.$ time, not in ordinate). In an individual calculation, the positions of the $N - 1$ nodes are first drawn by assuming each node to be uniformly distributed on the full data interval. The $N + 1$ ordinates that define the best fit broken line (model $m(1)$) to the data are then calculated and the misfit evaluated. One of the $N - 1$ nodes is then selected at random and replaced by a new node with uniform probability over the data interval, and the fitting process is carried out again. This is repeated a large number of times, generating a large suite of models distributed following probability p. There are many models where p is large, few where p is small. Residuals to the fit are modeled as a generalized Gaussian statistics, from which the model posterior probability p is derived.

The second, external loop is that of simulated annealing. The Metropolis algorithm is run with a control parameter α (see below) that slowly transforms the a posteriori probability distribution $\exp(\log(p)/\alpha)$ from a uniform law (in principle $\alpha = \infty$) to the desired a posteriori law ($\alpha = 1$). It allows convergence of the models towards regions of maximum probability. The acceptance criterion for a new model becomes $[p(test)/p(n)]^{1/\alpha}$.

In practice, in this paper, the process was iterated 300 times (inner loop), for each value of the control parameter α. The process was repeated starting with the final model of the previous (inner) loop, with a new, decreased value of the control parameter 0.98α (external loop). This was done starting with $\alpha = 0.1$ and decreasing it to $\alpha = 0.001$, $i.e.$ about 350 steps (values of α less than 1 are due to the fact that probabilities are not normalized). This is what allows the model to escape from secondary minima. The process described above provides one "optimized" model fit, which is based on some 100,000 calculations. The whole process is repeated 500 times, with a new drawing of the $N - 1$ free nodes each time, to yield finally 500 "optimized" broken-line fits with $N - 1$ free nodes. Histograms of the node dates can then easily be calculated from this set.

18

Observed Changes in Long-Term Climatic Conditions and Inner-Regional Differences in Urban Regions of the Baltic Sea Coast

Michael Richter[1*], Sonja Deppisch[1], Hans Von Storch[2]
[1]Urban Planning and Regional Development, Hafen City University, Hamburg, Germany
[2]Institute for Coastal Cesearch, Helmholtz-Zentrum, Geesthacht, Germany

ABSTRACT

This paper presents research outcomes from an investigation into climate change and urban impacts on climate development in urban regions of the Baltic Sea coast. The cities considered were Rostock and Stockholm, and their surrounding regions. The objectives were: 1) to determine whether significant changes in temperature and precipitation have occurred and, if so, to what extent; and 2) to establish whether there is a noticeable urban heat island effect in Stockholm and the medium-sized city of Rostock. Climatic trends were detected by linear regression and the Mann-Kendall test. Different precipitation trends were detected over the whole period of observation. Average annual temperatures increased significantly in both case studies, particularly from the 1970s with highest trends in winter and lowest in autumn (Rostock) and summer (Stockholm). Although changes in temperature extremes were detected for both regions, no overall long-term trend for precipitation extremes was observed. The average temperature in the city of Rostock (Stockholm) was approximately 0.3°C to 0.6°C (1.2°C) higher than in the surrounding rural areas had seasonal variations, with maxima in the warm season. The main outcomes were that significant changes in climatic conditions, particularly temperature patterns, have been occurring in the case study regions since the 1980s, and that there is a considerable urban heat island effect in both Stockholm and Rostock. This could encourage urban planners to consider specific climatic conditions and small-scale climatic influences also in relatively small coastal urban conglomerates in mid latitudes which can follow from land use changes.

Keywords: Urban Heat Island; Climate Change; Urban Climate; Baltic Sea; Extreme Events

1. Introduction

1.1. Climate Change in the Baltic Sea Region

The global climate is changing. The warming trend for the entire globe was 0.04°C per decade from 1850 to 2005. Warming has accelerated since the 1980s to a temperature increase of approximately 0.17°C per decade [1]. The warming trend in the Baltic Sea Region, at 0.10/0.07°C (north/south) per decade, exceeds the global trend. This may be because the globe consists mainly of thermally inert ocean, leading to the more rapid warming of land masses [2]. In this context, the Baltic Sea Region refers to the catchment area of the Baltic Sea, located in north-eastern Europe (see **Figure 1**). In the period of 1871 to 2004, the most obvious seasonal trends occurred in spring, with increases of 0.15°C and 0.11°C per decade in the north and south, respectively. The least warming occurred in summer: 0.06°C/0.03°C per decade [2]. Temperature increases of about 1°C in the 20th century have been reported in the drainage basin of the Baltic Sea [2].

Globally, minimum air temperatures rose to a greater extent than maximum air temperatures, meaning that the diurnal temperature range decreased [2]. Reference [3] analysed data concerning the thermal growing season in the Baltic region, and found that it had extended by about one week since 1951. This phenomenon follows a growing east-west gradient between Denmark and Finland, and higher intra-annual variability in the west [3]. Global trends in precipitation behaved differently, depending on latitude. In northern areas above 30°N, rainfall increased in the 20th century, while it decreased in tropical areas between 10°N and 10°S [1]. Significant increases in

---

*Corresponding author.

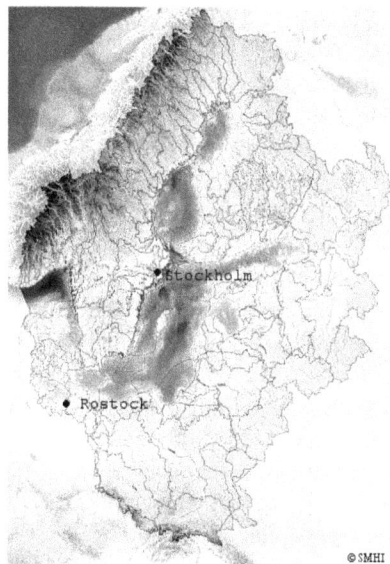

Figure 1. Map of the Baltic Sea region in Northern Europe with its drainage basin outline and the location of the cities of Rostock and Stockholm.

rainfall could be found, for example, in the eastern areas of North and South America, Northern Europe and Central and North Asia. By contrast, the Sahel regions, the Mediterranean, South Asia and southern Africa became drier [1]. According to data by [4], more precipitation was measured in the watershed of the Baltic Sea in the period of 1976 to 2000 than from 1951 to 1975. In the Baltic region, rainfall increased most visibly in Sweden and on the east coast in the winter and spring seasons [2]. In summer there was increased rainfall in the north, whilst less precipitation fell in the southern part of the region. Overall, virtually no long-term trend is visible since rainfall varies both spatially and temporally [2].

1.2. Urban Heat Island and Climate Change

It is a well-known fact that atmospheric conditions in cities differ from those in the surrounding countryside [5]. The urban heat island (UHI) effect describes the phenomenon of differing temperature and other climatic factors between urban areas and their immediate surrounding rural areas. In many cities, this effect is more like a "heat archipel" than an island [6], depending on the urban structure. In addition, there are changes in rainfall, cloud cover, wind conditions and air pollution. These differences are primarily generated in urban areas due to large-scale building structures, sealing and emissions from transport and industry. For a more detailed summary of the reasons for special urban climatic conditions see for example [7]. The average annual air temperature in large cities is 1°C to 2°C higher than in the immediate surrounding area. In exceptional cases, differences of more than 10°C can be determined in mega cities at certain

times [8]. Due to the urban climate, inter-annual temperature variability tends to decrease [9]. Stable stratification and low cloud cover promote the formation of an UHI (see [10]); high wind speeds reduce its intensity. Different dependencies of the UHI intensity on wind speed were found by various authors [8,10-12] for example, that the UHI intensity depends on the inverse square root of wind speed [8,12]. Reference [8] also discovered a logarithmic relation between the population of a city and the maximum UHI intensity. Following Oke's relation, a medium-sized city such as Rostock, with approximately 200,000 inhabitants, should have a maximum UHI intensity of about 6°C to 7°C. The urban climate effect will impact on an increasing number of people worldwide in the 21st century because over half of the world's population was city-dwellers in 2010, and this tendency is growing [13]. Regional climate models have revealed that more severe consequences can be expected concerning future climate change in metropolitan areas than in their surrounding areas [14]. It is assumed that the number of summer days ($T_{max} \geq 25°C$) and hot days ($T_{max} \geq 30°C$) will increase in many regions [14], and hence the areas where people suffer from thermal stress [15]. Here, heat waves play an important role and are very likely to become more common and severe in Europe [16]. If societies are not prepared for such changes, morbidity and the mortality rate will increase due to the consequences of a higher heat load [15], an example of which occurred in France in the summer of 2003 [17]. In Germany, 3500 people died as a consequence of the 2003 heat wave [18]. Across Europe, there were about 22,000 to 45,000 heat-related deaths within two weeks of August [19]. These consequences of climate change may, however, be reduced by introducing appropriate adaptation measures to cities. On the other hand, cold stress and mortality is likely to decrease especially in mid-lattitudes [6] and in cities due to the urban climate effects [20].

1.3. Aim of the Paper

To evaluate past and current changes in climatic conditions in the case study regions, this paper presents the results of analyses of measured meteorological parameters from different time scales and locations. Annual, seasonal, monthly and daily series were analysed for the case study regions to investigate long-term trends for temperature and precipitation. The effects of UHI were investigated by comparing the measured data of inner-city weather stations with nearby suburban/rural stations with a time resolution of hours and days.

The main research questions were:

How has the climate (temperature, precipitation) changed in the case study regions and are there regional differences in the case study regions, for example, between coastal and inland areas?

Is there a noticeable UHI effect in the case study regions, and under which specific weather conditions do the largest temperature differences occur?

2. Analysis of the Data-Study Cases Rostock and Stockholm

2.1. Study Areas

Rostock, a city on the German Baltic Sea coast (see **Figures 1** and **2**), is the largest city in the Federal State of Mecklenburg-Vorpommern (Germany). The city is the commercial centre of Mecklenburg-Vorpommern, and had approximately 202,000 inhabitants in 2010 (1990: 250,000). One feature of the city is the mouth of the river Warnow, called Breitling. With an area of about 1 km^2, the Breitling is located in the middle of the city, with the seaport. The urban centre is located at the southern end

of the Warnow delta. The areas with highest degree of sealing are to the west of the Warnow, where most inhabitants live. Rostock is relatively flat; its highest elevation (approximately 50 m above sea level) is in the southeast. After data from German Meteorological Service (DWD), the average annual temperature is 8.4°C; the average annual precipitation is 590 mm (1961-1990, station Rostock-Warnemünde).

The city of Stockholm is the capital of Sweden, located between the Baltic Sea in the west and Lake Malären in the east (see **Figures 1** and **2**). With an area of 187 km^2, the city is slightly larger than Rostock, but its population density is much higher (4.552 per km^2). Stockholm has about 850,000 inhabitants. After data from Swedish Meteorological and Hydrological Agency (SMHI), the average annual temperature is 6.6°C and the average annual precipitation 540 mm (1961-1990, station Stockholm-Observatorielunden).

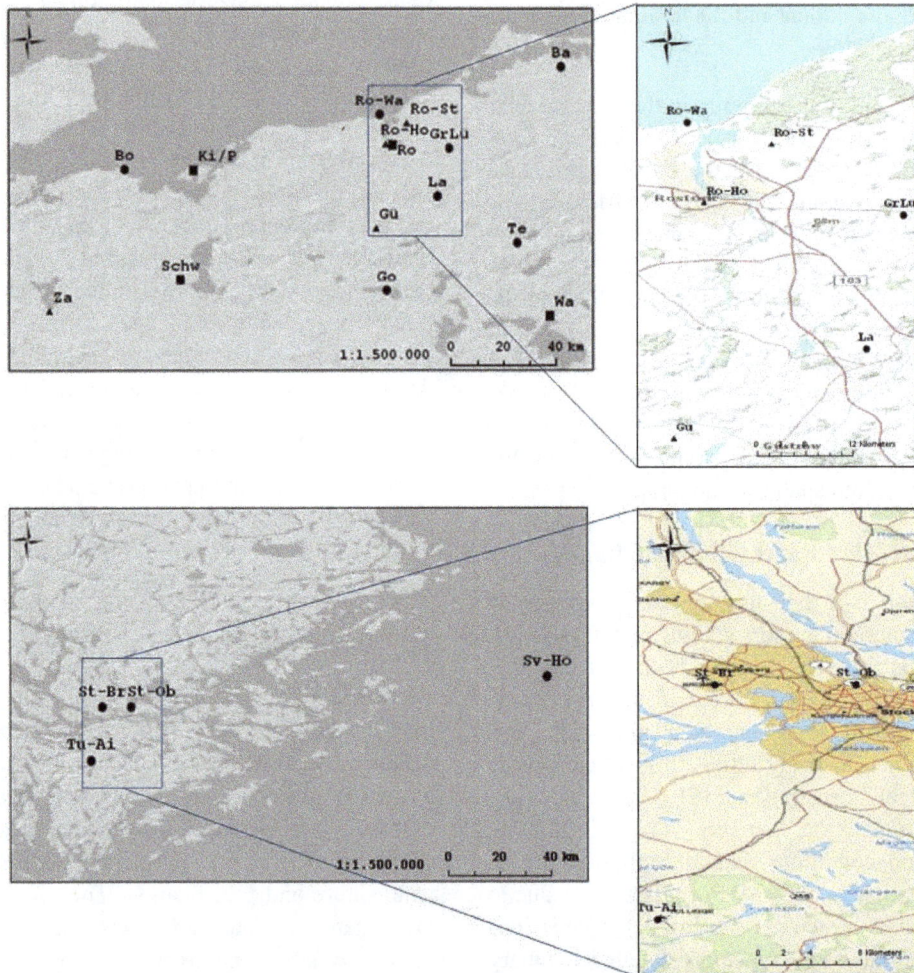

Figure 2. Location of measurement sites for the case study regions Rostock (54°05'20''N, 12°08'24''E) and Stockholm (59°19'30''N, 18°3'0''E). Station names: Za—Zarrentin, Bo—Boltenhagen, Schw—Schwerin, Ki/P—Kirchdorf/Poehl, Go—Goldberg, Gü—Gülzow, Ro-Wa—Rostock-Warnemünde, Ro-Ho—Rostock-Holbeinplatz, Ro—Rostock, Ro-St—Rostock-Stuthof, GrLü—Groß Lüsewitz, La—Laage, Ba—Barth, Te—Teterow, Wa—Waren, Tu-Ai—Tullinge Airport, St-Br—Stockholm-Bromma, St-Ob—Stockholm-Observatorielunden, Sv-Ho—Svenska-Högarna.

2.2. Measurement Sites, Data Collection and Preparation

Homogenised monthly data of temperature and precipitation from DWD were used to analyse the long-term climatic trends in the region of Rostock. All of the chosen stations were located in Mecklenburg-Vorpommern, within 100 km of Rostock (see **Figure 2**). Trends for extreme values were examined using daily data concerning precipitation and maximum and minimum temperatures from various stations. This data was also supplied by DWD. In addition, data from the air monitoring network of Mecklenburg-Vorpommern was supplied by the State Office for Environment, Nature Conservation and Geology to examine the urban effects on temperature, in particular. The stations were close to the city centre (Rostock-Holbeinplatz) located near a busy road, or at rural sites near Rostock (Rostock-Stuthof, Gülzow) surrounded by agricultural areas or grassland. The station Rostock-Warnemünde is located directly behind the dunes of the Baltic Sea Coast. The UHI intensity was calculated from the difference between the temperatures recorded at the weather station near the city centre (Rostock-Holbeinplatz) and those at the rural stations (Rostock-Stuthof, Rostock-Warnemünde, Gülzow). Data for the Stockholm region (see **Figure 2**) was provided by the Swedish Meteorological and Hydrological Agency (SMHI). This data comprised the long-term daily series of Stockholm-Observatorielunden 1756-2009 [21,22], from which effects such as the UHI effect (see [23]) were removed. Temperature extremes were calculated using the 3-hour average data from Stockholm-Bromma and Svenska-Högarna (1961-2009) and by calculating the daily maximum and minimum temperature. Urban effects on temperature were computed using the weather station at Stockholm-Bromma airport as the urban station and Tullinge Airport as the rural station. Stockholm-Bromma was used as urban station instead of Stockholm-Observatorielunden because of free availability of data from 3-hourly measurements. The time series for evaluating climatic changes were divided into different normal periods of 30 years' length. Due to the expected different results for both climatic changes and the UHI effect at several times of the year, the time series were divided further into seasons by months (DJF/MAM/JJA/SON).

2.3. Statistical Methods

To check the data supplied by the air monitoring network of Mecklenburg-Vorpommern, outliers were identified and removed by checking the data series graphically and looking for implausible differences between measured temperature values. This data was statistically reviewed and compared with data from DWD by calculating descriptive statistics, correlation coefficients and cross-correlations on a daily and hourly basis between the time series. Various statistical methods were applied to determine and test the significance of long-term trends. Trends concerning annual or seasonal temperature and precipitation were calculated using linear regression with least-square fitting to obtain a linear trend per time interval. The significance of these trends was tested using the Mann-Kendall test [24]. Due to the possible serial correlation of the time series, the Mann-Kendall test may become liberal, falsely rejecting the null hypothesis [25]. For this reason, all time series were filtered by a "prewhitening" approach [25] to eliminate the problems caused by serial correlation. These statistical analyses were conducted using the ANCLIM software by Petr Štěpánek [26] and SPSS Statistics. Gliding linear trends of annual mean temperatures were estimated by linear regression using moving 31-year windows. The trends in extremes were calculated using the RClimDex Package for R. After a simple quality control of the data (replacement of missing & unreasonable values, identification of outliers), the programme computes all 27 core indices recommended by the CCl/CLIVAR Expert Team for Climate Change Detection Monitoring and Indices (ETCCDMI), as well as other temperature and precipitation indices with user-defined thresholds [27].

3. Results

This section is divided into two main parts. In the first part, a description is given of the long-term climatic changes in temperature, precipitation and indices of extremes in the case study regions. In the second part, the urban impacts on temperature and the dependence of the UHI on wind speed and cloud cover are explored.

3.1. Long-Term Changes

Temperature changes in the Rostock region were examined with monthly data from the stations at Schwerin, Rostock-Warnemünde (**Figure 3**) and Teterow. Representative for the Rostock region, **Figures 3-5** show results from the Rostock-Warnemünde station. For the whole periods under study, there are positive trends focusing on an annual base of about 0.25°C/decade (**Figure 4(a)**) (significant ($p < 0.05$) for Rostock-Warnemünde, Schwerin). The largest increases in average annual temperature occurred in the period from 1976 to 2005, with linear trends of 0.5°C to 0.62°C/decade. When focusing on seasonal trends, it becomes apparent that autumn (SON) warmed least considerably at all stations. The strongest increases in seasonal temperature occurred in winter between 1976 and 2005, at Warnemünde with a trend of 0.79°C/decade (**Figure 4(a)**). Temperatures in spring (MAM) and summer (JJA) increased by nearly 0.7°C/decade at Warnemünde. There seems to be a stronger

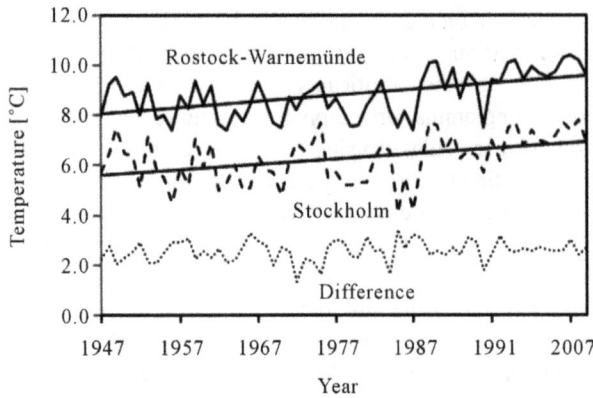

Figure 3. Annual average temperature (1947-2009) at Rostock-Warnemünde with linear trend from 1951 to 2005 and Stockholm-Observatorielunden with linear trend from 1947 to 2009 and plotted differences between these annual temperatures.

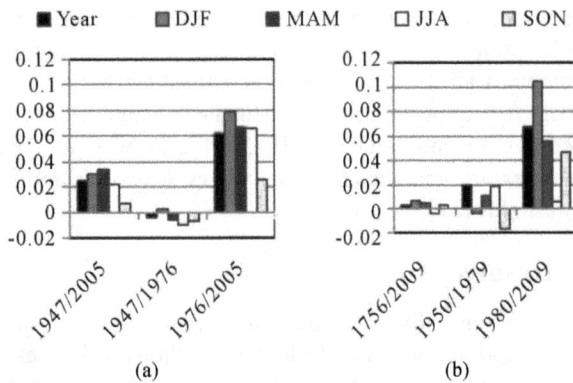

Figure 4. Linear trends [°C per year] of observed annual temperatures at Rostock-Warnemünde weather station (a) and Stockholm-Observatorielunden (b). Trends are for different seasons or for the whole year and, due to different measurement periods and data availability, for different periods.

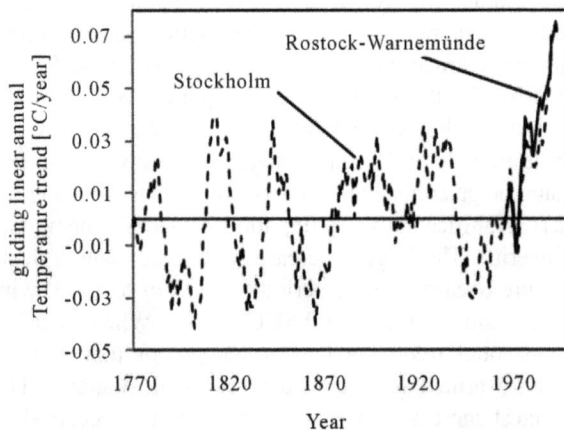

Figure 5. Gliding linear trends of the annual mean temperature, estimated by 31-year windows, of weather stations Rostock-Warnemünde and Stockholm-Observatorielunden for the time between 1951 and 2009.

increase in annual average temperatures since the late 1980s for all series (see **Figure 3**, for instance). For Stockholm, the long-term series from 1756 to 2009 [21, 22] was used to examine trends in average annual temperatures. These results (**Figures 3** and **4(b)**) also indicate an increase in average annual temperature in the late 1980s. During the period from 1756 to 2009, slight increases (0.03°C to 0.07°C/decade) occurred throughout the year and for all seasons with the exception of summer, where a decrease of 0.04°C/decade was found. All trends were significant ($p < 0.05$) for this period. No significant trends were detected for the period from the 1950s (1950-1979). These differences ranged from −0.17°C/decade in autumn to 0.2°C/decade for the entire year. For the period after 1980, trends range from 0.06°C/decade in summer to 1.05°C/decade in winter (**Figure 4(b)**). For the entire year, the trend is significant with 0.67°C/decade ($p < 0.05$). Gliding linear annual mean temperature trends, estimated by linear fit in moving 31-year windows (**Figure 5**), showed temporal variations in the rate of change. The longer time series for Stockholm showed variations ranging from −0.4 to +0.4°C/decade until the 1980s. From this point in time (1985 for Rostock and 1989 for Stockholm), the trend values exceeded this range. After 1972/1973, the rate was positive and accelerated over the entire period, with a few interruptions. The maximum linear 31-year trend was achieved in both Rostock and Stockholm, at 0.75°C/decade, in 1993 (period 1978-2003). Gliding linear trends were similar over the entire period from 1951 to 2009. After 1972, the warming trend at Rostock-Warnemünde was almost always higher.

The time series of precipitation were filtered by a "prewhitening" approach [25] to eliminate the problems caused by serial correlation and linear trends were calculated with 10-year running means and corresponding significance was detected with the Mann-Kendall-Test.

For any period, there was an overall positive trend for all seasons and for the whole year for all stations, as was the case for temperature. The linear 31-year gliding mean precipitation trend at Rostock weather station (Ro in **Figure 2**) was negative until 1975 (**Figure 6**), after which the trend was positive up to 2009. There was a maximum increase in 1986 (1971-2001) at 6.9 mm/a, and decreasing but positive trends thereafter.

Table 1 shows linear trends of observed 10-year running means of annual and seasonal precipitation from 1951 to 2005. No significant trends were detected by the Mann-Kendall Test for the period from 1951 to 2005 ($p \leq 0.05$). For Schwerin, Rostock and Kirchdorf, precipitation increased by 0.32 mm/a to 0.49 mm/a. The largest increases in precipitation occurred in winter and autumn; by contrast, decreases or small increases were mainly recorded in spring and summer. The annual linear precipitation trend for Waren was negative for the whole

(a)

(b)

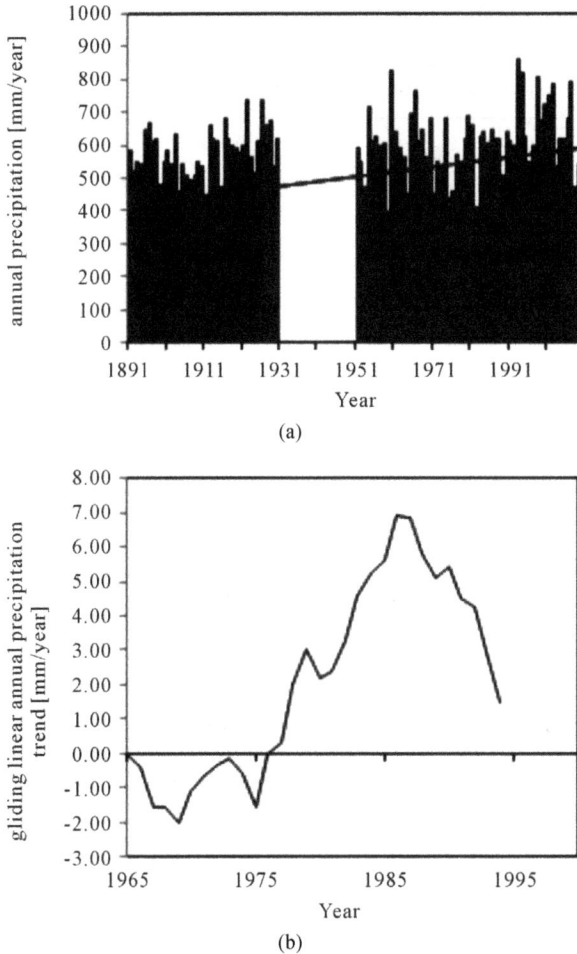

Figure 6. Annual precipitation at Rostock weather station with linear trend (a) and 31-year gliding linear trend (b) for 1951 to 2005.

Table 1. Linear trends [mm per year] of observed 10-year running means of annual precipitation from 1951 to 2005. Trends are for different seasons and for the whole year.

Station	Year	DJF	MAM	JJA	SON
Schwerin	0.48	0.52	0.135	−0.67	0.57
Rostock	0.49	0.41	0.07	−0.09	0.63
Kirchdorf	0.32	0.15	−0.003	0.29	0.32
Waren	−1.5	−0.16	−0.39	−0.86	−0.51

year and for all seasons.

Climate indices for the weather stations were calculated from daily values of maximum temperature, minimum temperature and precipitation. **Table 2** shows values with significant ($p < 0.05$) increasing or decreasing trends. It can be seen that the length of the thermal growing season (GSL) increased significantly for all stations. The number of frost days per year (FD0) decreased, and the number of summer days (SU25) increased sig-

nificantly at most locations.

Other indices show only a few significant decreases or increases at various sites. Schwerin is the only station with an increase in extreme precipitation indices, R20 mm (number of very heavy precipitation days, annual count of days when $p > 20$ mm) and R95 pp (very wet days, annual total precipitation when $p > 95$th percentile).

Indices for the Stockholm region were computed using data from Stockholm-Bromma and Svenska-Högarna (**Table 3**).

Here, the number of ice days per year (ID0) and the number of frost days per year (FD0) decreased significantly, and the monthly minimum value of daily minimum temperature (TNn) and the monthly maximum value of daily minimum temperature (TNx) increased significantly at both sites from 1961 to 2009. Cold extremes (FD0, ID0) decreased by about 0.5 to 1 day per year. Additionally, the number of summer days (SU25), the length of the growing season (GSL) and the monthly maximum and minimum value of daily maximum temperature (TXx, TXn) increased at Svenska Högarna.

3.2. Urban Impacts

For the Rostock region, data with a time step of one hour were available for Rostock-Warnemünde (HRO_Warnemuende), Rostock-Holbeinplatz (HRO_Holbeinplatz), Rostock-Stuthof (HRO_Stuthof) and Guelzow (see **Figure 2**). **Table 4** shows descriptive statistics of the hourly temperature measurements. The station near the city centre of Rostock (HRO_Holbeinplatz) had the highest mean temperature (10.2°C) during the period from 2001 to 2009.

Rostock-Holbeinplatz and Rostock-Warnemünde had considerably higher absolute minimum temperatures (−13.3°C and −13.9°C). The mean temperature difference between the station in the inner city (HRO-Holbeinplatz) and rural stations near the city, or the mean UHI effect, is between 0.3°C and 0.6°C. The same values are produced when focusing on daily average temperatures (**Table 5**). Here, the mean temperature difference between HRO-Holbeinplatz and Gross Lüsewitz weather station (Gr_ Luesewitz, rural station, distance about 15 km) was even 1.4°C. The average diurnal cycle (**Figure 7**) shows that at night, HRO_Warnemünde had the highest temperatures, which were about 1°C higher than at HRO_Stuthof and Guelzow at 5:00 hrs. HRO_Holbeinplatz was warmer than these stations, but about 0.5°C colder than at HRO_ Warnemuende at night. At 9:00 hrs the mean temperatures were the same; temperatures at HRO_Holbeinplatz increased more rapidly than at the other locations until 14:00/15:00 hrs, when the average differences between the nearest stations—HRO_Stuthof and HRO_Warnemünde—were about 1°C to 2°C. From late afternoon onwards, temperatures decreased until they again reached

Table 2. Significant (p ≤ 0.05) changes in climatic core indces (following [27]) in days/year or °C/year for various sites in and around Rostock.

Climate indices	Warne-Münde	Teterow	Schwe-Rin	Gross Lüse-Witz	Goldberg	Boltenhagen	Barth
DTR		0.018				−0.009	0.013
FD0	−0.385		−0.146	−0.568	−0.432	−0.49	
ID0		−0.241				−0.2	
GSL	0.91	0.668	0.335	0.901	0.901	0.758	0.79
R20 mm			0.009				
R95 pp			0.35				
SU25	0.169	0.329		0.268	0.209		0.16
TNn	0.05				0.083	0.064	
TNx	0.028		0.011		0.002	0.022	
TXx		0.058		0.051			
TR20	0.015						

Table 3. Significant (p ≤ 0.05) changes in climatic core indices (following [27]) in days/year or °C/year for various sites in and around Stockholm.

Climate indices	Stockholm-Bromma	Svenska Högarna
SU25		0.056
ID0	−0.485	−0.667
FD0	−0.531	−0.985
GSL		0.499
TXx		0.078
TXn		0.084
TNx	0.028	0.048
TNn	0.092	0.096

Table 4. Descriptive statistics for stations including temperature range, absolute minimum and maximum temperature, mean temperature and standard deviation for hourly data measured 2001-2009 in °C.

Station	Range	Min	Max	Mean	Std dev
HRO_Holbeinplatz	49.3	−13.9	35.4	10.2	7.6
HRO_Stuthof	54.2	−19.2	35.0	9.6	7.4
Guelzow	54.4	−18.4	36.0	9.9	7.7
HRO_Warnemuende	46.3	−13.3	33.0	9.9	7.0

Table 5. Descriptive statistics for stations including temperature range, absolute minimum and maximum temperature, mean temperature and standard deviation for daily data measured 2001-2009 in °C.

Station	Range	Min	Max	Mean	Std dev
HRO_Holbeinplatz	38.8	−11.0	27.8	10.2	7.1
HRO_Stuthof	37.7	−11.4	26.3	9.6	6.8
HRO_Warnemünde	36.2	−10.7	25.5	9.9	6.8
Laage	38.7	−12.2	26.5	9.3	7.1
Gr_Luesewitz	37.8	−12.4	25.4	8.8	6.9
Guelzow	38.7	−11.6	27.1	9.9	7.1

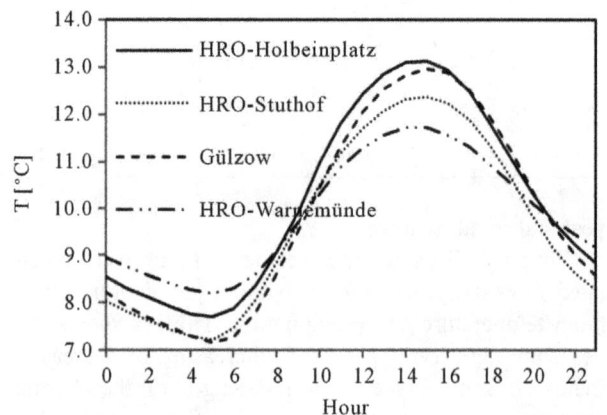

Figure 7. Average diurnal temperature cycles for different stations 2001-2009.

Observed Changes in Long-Term Climatic Conditions and Inner-Regional Differences in Urban Regions of the Baltic Sea Coast

173

a point when they were nearly the same, at 20:00/21:00 hrs.

This pattern remained the same throughout the year. However, temperature differences between HRO_Holbeinplatz and HRO_Warnemünde were lower at night (<0.6°C) and higher in the early afternoon (up to 2.5°C), particularly in spring (MAM) and summer (JJA, **Figure 8(b)**). The differences in temperature over the day in winter (DJF, **Figure 8(a)**) and autumn (SON) were less considerable. When comparing the temperature differences at HRO_Holbeinplatz and HRO_Stuthof, the highest temperature difference was always in the early afternoon. In summer there was another UHI maximum at midnight and two minima at 7:00 hrs and 19:00 hrs. Minima of temperature differences between HRO_Holbeinplatz and Guelzow occurred in the late afternoon in summer and winter; the summer maximum was in the early morning (5:00/6:00 hrs) and the winter maximum was at midday. **Figure 9** shows that the monthly mean UHI was highest in May, with a difference of over 1°C for

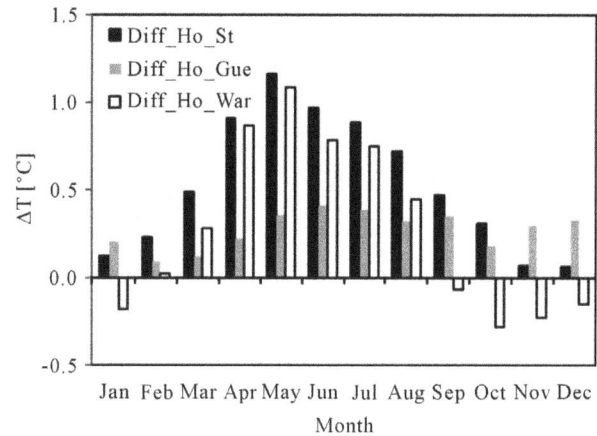

Figure 9. Average monthly UHI intensity for 2001 to 2009. UHI intensities were computed from each difference of monthly averages between Rostock-Holbeinplatz (Ho) and Rostock-Stuthof, Rostock-Warnemünde and Gülzow weather stations.

HRO_Warnemünde and HRO_Stuthof, and lowest in winter. When HRO_Holbeinplatz was compared with the coastal station HRO_Warnemünde, average temperatures near the coast were higher in autumn and winter. There is no such clear seasonal cycle for the difference between HRO_Holbeinplatz and Gülzow but two maxima in summer and early winter and two minima in autumn and at the end of winter. The maximum UHI intensity over the observed period was 8.5°C for HRO_Stuthof and 8.3°C for Guelzow for hourly measurements and 4.4°C and 5.7°C, respectively, for daily mean temperatures. Data from two stations was compared for the Stockholm region. Stockholm-Bromma is a weather station at Stockholm-Bromma Airport to the northwest of the city centre; Tullinge Airport is a more rural station south of Stockholm. Three-hourly temperature data was available for the period from 1996 to 2009 for both sites, which were then compared. In this period, Stockholm-Bromma had an average temperature of 7.4°C and Tullinge Airport 6.2°C. This would be a mean UHI of 1.2°C, comparable, for example, to Hamburg [12] and London [28]. The maximum UHI intensity for 3-hourly measurements and daily means (in brackets) is 12.9°C (8.3°C), the maximum Urban cool island (UCI) intensity is −5.4°C (−2.4°C), respectively. As in Rostock, the diurnal cycle of the UHI changed throughout the year. As can be seen in **Figure 10**, the winter UHI was only weak, averaging at between 0.4°C at midday and 0.8°C before midnight. There were significant diurnal variations of the UHI intensity during the other seasons. The average UHI was at its maximum at around midnight, and was highest in summer, at about 2.3°C. The average UHI intensities decreased from 3:00 hrs to 6:00 hrs; the minimum level was at around 9:00 hrs, after which UHI intensities increased again until around midnight.

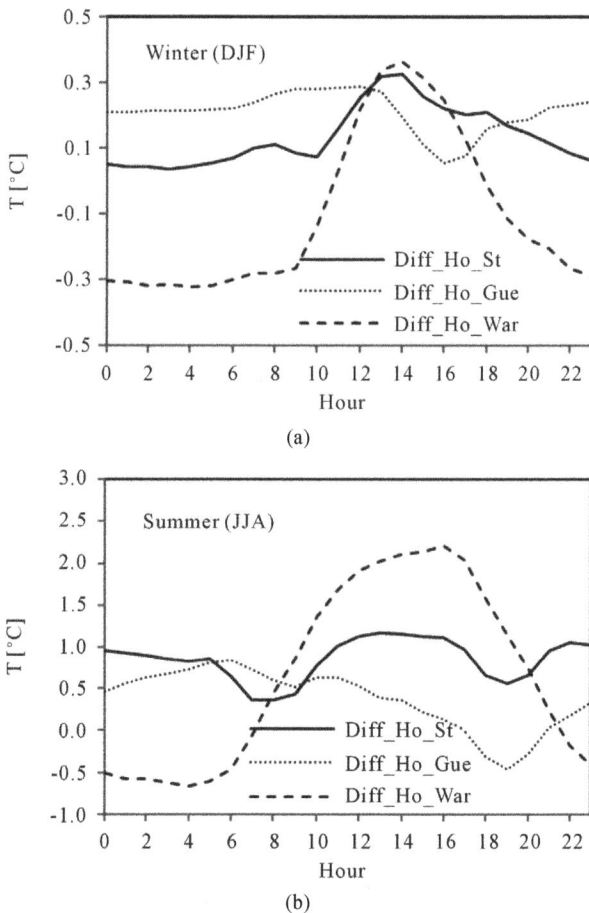

(a)

(b)

Figure 8. Average diurnal cycle of UHI intensity for winter months ((a), DJF) and summer months ((b), JJA) for 2001-2009. UHI intensities were computed from each difference between Rostock-Holbeinplatz (Ho) and Rostock-Stuthof, Rostock-Warnemünde and Gülzow weather stations.

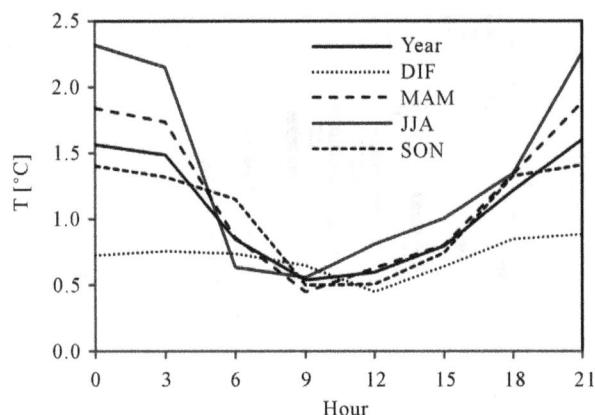

Figure 10. Average diurnal cycle of UHI intensity for the whole year, winter months (DJF), spring months (MAM), summer months (JJA) and autumn months (SON) for 1996 to 2009. UHI intensities were computed from the difference between averages at Stockholm-Bromma and Tullinge-Airport.

The dependence of the UHI intensity on different meteorological conditions has been reported in various studies [10,12,29]. Here it was examined whether there was a significant relationship between the daily mean UHI intensity (daily mean ΔT between Rostock-Holbeinplatz and Rostock-Stuthof) in Rostock and the conditions of wind speed and cloud cover. Rostock-Warnemünde weather (synoptic station) station was used as the reference station for the meteorological conditions due to the reliability of its data. Because of the eventual disruptive influences of f.e. land-sea breeze circulation on wind speed measurement, the analysis was performed with daily wind speed data of other synoptic stations near Rostock (Schwerin, Goldberg), too. A linear regression analysis with the least square method was carried out for all data and seasons with Equation (1):

$$UHI = a*x+b \qquad (1)$$

where x is the daily mean of wind speed or cloud cover. For wind speed, the parameters are given in **Table 6**. All regressions are significant ($p < 0.05$). Parameters for the other stations were very similar in order of magnitude and seasonality and are not shown here. The coefficient of determination r^2 ranged from 0.11 in winter to 0.20 in autumn. The value b determines the theoretical daily average UHI intensity when the average daily wind speed is 0. This ranges from 0.42°C in winter to 1.53°C in spring. The stronger determined variance of UHI intensity can be explained by cloud cover (**Table 7**, all significant for $p < 0.05$). The r^2 values are mostly higher than those for wind speed except for winter and autumn. For cloud cover, the strongest dependence of UHI intensity is during spring and summer; the least dependence is in winter. The theoretical UHI intensity when there was no cloud cover is 0.3°C in winter and 1.81°C in summer.

Table 6. Parameters of linear regression between daily average UHI intensity (temperature difference between Rostock-Holbeinplatz and Rostock-Stuthof) and wind speed at Rostock-Warnemünde.

	a	b	r^2
Year	−0.11	1.08	0.16
DJF	−0.05	0.42	0.11
MAM	−0.15	1.53	0.19
JJA	−0.12	1.40	0.16
SON	−0.01	0.74	0.20

Table 7. Parameters of linear regression between daily average UHI intensity (temperature difference between Rostock-Holbeinplatz and Rostock-Stuthof) and cloud cover (in %) at Rostock-Warnemünde.

	a	b	r^2
Year	−0.02	1.36	0.22
DJF	0.00	0.30	0.02
MAM	−0.02	1.63	0.25
JJA	−0.02	1.81	0.27
SON	−0.01	0.83	0.14

4. Discussion

Trend analyses for long-term precipitation and temperature on an annual and seasonal basis revealed significant changes in climatic conditions in both case study regions, Rostock and Stockholm. All of the examined time series of temperature measurements showed sustained accelerated warming since the 1980s compared to earlier periods. This is in line with increased warming on a global scale [30] and in the Baltic Sea catchment [2] since the 1980s. Both study areas warmed most in winter months (Rostock: 0.7°C to 0.79°C/decade, Stockholm: 1.05°C/decade). This is in line with the BACC-Report [2], which stated that the warming trends are higher in northern latitudes. Differences exist in the least warming periods. The Rostock region warmed least in autumn (−0.1°C to 0.26°C/decade); for the Stockholm region, the lowest warming trend (0.06°C/decade) was in summer. Our analysis of the precipitation time series revealed no clear increasing or decreasing trend in the Rostock region, but a strong variability in annual precipitation quantities. The long-term trends were partly significant, but only in the order of magnitude of 1‰ of annual precipitation quantities. The shorter 31-year gliding linear trend at Rostock weather station was up to the order of magnitude of 1% of the annual precipitation quantity. Three of the four stations have negative trends in summer; as expected, the two most continental stations, Schwerin and Waren, had the greatest negative trend. GSL (growing season length) changed significantly for most stations in the Rostock region at nearly one day per year for the whole measured period (in most cases from 1947 to 2009). This trend can be assumed to be stronger when focusing on the last 30 years only, which is the case for all of the calculated indices. The number of ice days and frost days per year decreased for all stations in both regions, albeit not always significantly. In the northernmost region, Stockholm, the decrease is up to one day higher per year. The number of summer days per year increased in all cases (most of them significantly), at 0.056 to 0.329 days per year, by a smaller amount than for cold extremes. The number of tropical nights only increased significantly at Rostock-Warnemünde. This may be due to the generally rare occurrence of these events at other stations. Extremes such as summer days and tropical nights can become more important when considering UHI intensities, which are added to the high temperatures in the regions and are not considered in these climatic trends. Both case studies showed considerable UHI intensities, especially in the warm period (spring and summer). The UHI intensity was at its minimum in winter. The maximum UHI intensities slightly exceeded the expectations concluded from Oke's relationship between maximum UHI intensity and number of inhabitants. [7] generalised UHI intensities by reviewing empirical studies. He found that they were greatest in summer, during anticyclonic conditions, at night, and that they increased with decreasing wind speed, cloud cover and increasing city size or population. There were differences in time of occurrence of the maximum UHI intensity between the case studies. The maximum temperature difference at Stockholm occurred in the middle of the night and the minimum at midday. In Rostock, however, two maxima occurred: the larger in the early afternoon and the smaller at night. With its location near the Baltic Sea Coast, a land-sea breeze circulation could be expected which could suppress the UHI in the afternoon. In this special case of Rostock and the location of the measurement station Rostock-Holbeinplatz this seemed not to be the case. Similar to Rostock, maximum heat island intensities approaching 8°C on sunny midsummer days were reported for Saskatoon, Canada, a similar sized city with 200,000 inhabitants [31] and, for example, Helsinki [32] and Lodz [33]. The mean UHI of 1.2°C at Stockholm is comparable, for example, to Hamburg [12] and London [28], cities with many more inhabitants than Stockholm. This confirms the hypothesis that UHI intensities tend to increase from low to high latitudes [34]. The warmer temperatures at Rostock-Warnemünde compared to Rostock-Stuthof could be due to the influence of the Baltic Sea temperature, since this station is located directly at the coast. This hypothesis is supported by the diurnal temperature cycle of Rostock-Warnemünde, which has a significantly lower temperature range. The expectations of the relationship between UHI intensity and large-scale meteorological conditions were confirmed for the Rostock region. There was a negative dependence of UHI intensity on both wind speed and cloud cover. This has already been shown by various studies [8,12,35]. With $r^2 = 0.22$, the determined variance for cloud cover for the whole year was similar to Hamburg [35]. The r^2 values have contrasting characteristics through the seasons. Whereas the dependence of UHI intensity on cloud cover is highest in summer and spring, for wind speed it is highest in spring and autumn. This is important when considering the expected climatic changes accompanied by an increase in summer high-pressure situations with low wind speed and cloud cover [36]. This could possibly lead to the accumulation of situations with high UHI intensities and to diverse impacts induced by these intensities, such as thermal stress to humans and to urban ecosystems, and increased air pollution. Finally, this study could only show results for special locations and limited time periods due to its concentration on a few weather stations at sites in the two chosen cities and the set observation times. Further weather stations at different locations with different degrees of sealing, plant cover, canopy forms, building structures, etc. would be required to generate a better understanding of the UHI and its accompanying effects. For example, Rostock-Holbeinplatz weather station is not located at the heart of the city, which has even more sealed areas and buildings, and less vegetation cover. Hence, the maximum UHI intensity in the inner city of Rostock could be even higher. With longer time series of meteorological parameters in cities, it would even be possible to evaluate ongoing effects of climate change on the urban climate, such as whether cities are affected to a greater extent by temperature changes than their surrounding countryside.

5. Conclusion

This research confirms the findings of various other studies which determined that climate changes (for this study, temperature and precipitation) in the case study regions and the speed of change have increased since the 1980s. No uniform trend was detected for precipitation, although the annual amount of precipitation increased in most cases, and there is less precipitation in summer and more in winter. The changes in climatic conditions were proved for the case study regions of the Baltic Sea coast, Rostock and Stockholm. The changes in temperature are most noticeable in winter, and particularly affect the cold extremes, expressed in decreasing ice and frost days. In the northernmost region, Stockholm, warming in winter months is significantly higher; for Rostock, this is the case in summer. Given that the observed temperature trends (since the 1980s) will continue, average annual temperatures could rise by up to 1°C in 20 years. This increase in temperature could be problematic for urban regions, which already have average temperatures and maximum UHI intensities that are several degrees Celsius higher (e.g. Rostock 8.5°C) at night and over the day. In the investigated temperate climatic regions with expected land-sea breeze circulations, however, it does not yet seem to constitute a serious problem. Nevertheless, there are benefits for cities due to the UHI like less energy consumption for heating and less cold stress for people. Both case studies had significant UHI effects with maximum intensities in summer and at night in Stockholm and in the afternoon in Rostock. For Rostock it was proved that the UHI intensity increases with decreasing wind speed and cloud cover, whereas cloud cover seemed to have a greater influence. With a possible increase in heat wave occurrence and intensity due to climate change, urban regions must make precautions to ensure they can care for the health of their inhabitants, for example. With meteorological observations in cities covering stations at different sites and for longer time periods it would be possible to examine the causalities between a changing (warmer) climate and the (spatial distribution of the) occurrence and intensity of the UHI. For the first time, we were able to show that even relatively small urban conglomerates affected by marine weather in mid latitudes, such as Rostock and Stockholm, with 200,000 to 850,000 inhabitants, are subject to a significant UHI effect. Lower UHI intensities can particularly be expected for Rostock, which has only about 200,000 inhabitants and a pronounced land-sea-breeze circulation. This may encourage urban and regional planners to consider specific climatic conditions in urban areas and small-scale climatic influences which can follow from land use changes. For longlasting infrastructure, there may be long-term negative consequences due to the changing climate. However, this may not be the case in the next few years, under certain circumstances. With greater knowledge of the spatial and temporal distribution of the UHI in the case studies, it would be possible to develop specific UHI-mitigation and climate change adaptation measures, which could become important in the future when heat waves are expected to occur more often. This would require a larger observational network or high-resolution model simulations.

6. Acknowledgements

The authors would like to thank the Swedish Meteorological and Hydrological Institute (SMHI), the German Meteorological Service (DWD) and the State Office for Environment, Nature Conservation and Geology of Mecklenburg-Vorpommern for providing meteorological data. They are also grateful to the research group plan B: altic for its ongoing discussions and to the German Federal Ministry of Research and Education for funding the research work from its Social-Ecological Research Programme (FKZ 01UU0909).

REFERENCES

[1]   Intergovernmental Panel on Climate Change, "Climate Change 2007: The Physical Science Basis: Summary for Policy Makers," Paris, 2007.

[2]   BACC, "Regional Climate Studies. Assessment of Climate Change for the Baltic Sea Basin," Springer, Berlin, 2008.

[3]   H. W. Linderholm, A. Walther and D. Chen, "Twentieth-Century Trends in the Thermal Growing Season in the Greater Baltic Area," *Climatic Change*, Vol. 87, No. 3-4, 2008, pp. 405-419.

[4]   C. Beck, J. Grieser and B. Rudolph, "A New Monthly Precipitation Climatology for the Global Land Areas for the Period 1951 to 2000," *Climate Status Report*, 2004.

[5]   A. Kratzer and Das Stadtklima, "Vieweg," Braunschweig, 1937.

[6]   T. R. Oke and F. Hanell, "The Form of the Urban Heat Island in Hamilton, Canada," *WMO Technical Note*, No. 254, Geneva, 1970.

[7]   A. J. Arnfield, "Two Decades of Urban Climate Research: A Review of Turbulence, Exchanges of Energy and Water, and the Urban Heat Island," *International Journal of Climatology*, Vol. 23, No. 1, 2003, pp. 1-26.

[8]   T. R. Oke, "City Size and the Urban Heat Island," *Atmospheric Environment*, Vol. 7, No. 8, 1973, pp. 769-779.

[9]   I. Camilloni and V. Barros, "On the Urban Heat Island Effect Dependence on Temperature Trends," *Climatic Change*, Vol. 37, No. 4, 1997, pp. 665-681.

[10] C. J. G. Morris, I. Simmonds and N. Plummer, "Quantification of the Influences of Wind and Cloud on the Nocturnal Urban Heat Island of a Large City," *Journal of Applied Meteorology*, Vol. 40, No. 2, 2001, pp. 169-182.

[11] Y. H. Kim and J. J. Baik, "Daily Maximum Urban Heat Island Intensity in Large Cities of Korea," *Theoretical and Applied Climatology*, Vol. 79, No. 3-4, 2004, pp. 151-164.

[12] K. H. Schlünzen, P. Hoffmann, G. Rosenhagen and W. Riecke, "Long-Term Changes and Regional Differences in Temperature and Precipitation in the Metropolitan Area of Hamburg," *International Journal of Climatology*, Vol. 30, No. 8, 2010, pp. 1121-1136.

[13] B. Crossette, "State of the World Population 2010," United Nations Population Fund, New York, 2010.

[14] B. Früh, P. Becker, T. Deutschländer, J. D. Hessel, M. Kossmann, I. Mieskes, J. Namyslo, M. Roos, U. Sievers, T. Steigerwald, H. Turau and U. Wienert, "Estimation of

Climate-Change Impacts on the Urban Heat Load Using an Urban Climate Model and Regional Climate Projections," *Journal of Applied Meteorology and Climatology*, Vol. 50, No. 1, 2011, pp. 167-184.

[15] C. Souch and C. Grimmond, "Applied Climatology: 'Heat waves'," *Progress in Physical Geography*, Vol. 28, No. 4, 2004, pp. 599-606.

[16] Intergovernmental Panel on Climate Change, "Climate Change 2007: The Physical Science Basis. Contribution of Working Group I to the Fourth Assessment Report of the IPCC," Paris, 2007.

[17] L. Filleul, S. Cassadou, S. Médina, P. Fabres, A. Lefranc, D. Eilstein, A. le Tertre, L. Pascal, B. Chardon, M. Blanchard, C. Declercq, J. F. Jusot, H. Prouvost and M. Ledrans, "The Relation between Temperature, Ozone, and Mortality in Nine French Cities during the Heat Wave of 2003," *Environmental Health Perspectives*, Vol. 114, No. 9, 2006, pp. 1344-1347.

[18] M. Rückversicherungsgesellschaft, "Topics: Jahresrückblick Naturkatastrophen," Munich, 2003.

[19] J. A. Patz, D. Campbell-Lendrum, T. Holloway and J. A. Foley, "Impact of Regional Climate Change on Human Health," *Nature*, Vol. 438, No. 7066, 2005, pp. 310-317.

[20] S. Thorsson, F. Lindberg, J. Björklund, B. Holmer and D. Rayner, "Potential Changes in Outdoor Thermal Comfort Conditions in Gothenburg, Sweden Due to Climate Change: The Influence of Urban Geometry," *International Journal of Climatology*, Vol. 31, No. 2, 2011, pp. 324-335.

[21] A. Moberg, H. Alexandersson, H. Bergström and P. D. Jones, "Were Southern Swedish Temperatures before 1860 as Warm as Measured?" *International Journal of Climatology*, Vol. 23, No. 12, 2003, pp. 1495-1521.

[22] A. Moberg, H. Bergström, J. Ruiz Krigsmand and O. Svanered, "Daily Air Temperature and Pressure Series for Stockholm (1756-1998)," *Climatic Change*, Vol. 53, No. 1-3, 2002, pp. 171-212.

[23] A. Moberg and H. Bergström, "Homogenization of Swedish Temperature Data. Part III: The Long Temperature Records from Uppsala and Stockholm," *International Journal of Climatology*, Vol. 17, No. 1, 1997, pp. 667-699.

[24] R. Sneyers, "On the Statistical Analysis of Series of Observations," *WMO Technical Note*, No. 415, Geneva, 1990.

[25] A. Kulkarni and H. von Storch, "Monte Carlo Experiments on the Effect of Serial Correlation on the Mann-Kendall Test of Trends," *Meteorologische Zeitschrift N.F.*, Vol. 4, No. 2, 1995, pp. 82-85.

[26] P. Štěpánek, "AnClim—Software for Time Series Analysis," 2008.

[27] X. Zhang and F. Yang, "RClimDex (1.0) User Manual," Downsview, 2004.

[28] R. Watkins, J. Palmer, M. Kolokotroni and P. Littlefair, "The Balance of the Annual Heating and Cooling Demand within the London Urban Heat Island," *Building Services Engineering Research and Technology*, Vol. 23, No. 4, 2002, pp. 207-213.

[29] J. P. Montávez, A. Rodríguez and J. I. Jiménez, "A Study of the Urban Heat Island of Granada," *International Journal of Climatology*, Vol. 20, No. 8, 2000, pp. 899-911.

[30] Intergovernmental Panel on Climate Change, "Climate change 2007: Synthesis Report," Paris, 2007.

[31] E. Ripley, O. Archibold and D. Bretell, "Temporal and Spatial Temperature Patterns in Saskatoon," *Weather*, Vol. 51, No. 12, 1996, pp. 398-403.

[32] U. Wienert, "Untersuchungen zur Breiten—Und Klimazonenabhängigkeit der Urbanen Wärmeinsel—Eine Statistische Analyse," *Essener Ökologische Schriften*, No. 16, 2002.

[33] K. Fortuniak, K. Klysik and J. Wibig, "Urban-Rural Contrasts of Meteorological Parameters in Lodz," *Theoretical and Applied Climatology*, Vol. 84, No. 1-3, 2006, pp. 91-101.

[34] U. Wienert and W. Kuttler, "The Dependence of the Urban Heat Island Intensity on Latitude—A Statistical Approach," *Metereologische Zeitschrift*, Vol. 14, No. 5, 2005, pp. 677-686.

[35] P. Hoffmann, O. Krueger and K. H. Schlünzen, "A Statistical Model for the Urban Heat Island and Its Application to a Climate Change Scenario," *International Journal of Climatology*, Vol. 32, No. 8, 2012, pp. 1238-1248.

[36] D. J. Jacob and D. A. Winner, "Effect of Climate Change on Air Quality," *Atmospheric Environment*, Vol. 43, No. 1, 2009, pp. 51-63.

19

Performance Evaluation of Four Commercial Optical Particle Counters

Franco Belosi, Gianni Santachiara, Franco Prodi
Institute of Atmospheric Sciences and Climate (ISAC), National Research Council, Bologna, Italy

ABSTRACT

The performances of four optical particles counters, Aerosol Spectrometer (Grimm 1.108), Enviro Check (Grimm 1.107), DustMonit and ParticleScan, were evaluated in laboratory tests employing monodisperse aerosol particles. The study focused on how commercial instruments perform during routine measurements respect to OPC scientific understanding, because it is important for users of such instruments to be aware of their limitations. Measurements were performed using aerosol generated by a Monodisperse Aerosol Generator (MAGE), which produced carnauba wax particles of diameter (1.00 ± 0.08) μm and (1.40 ± 0.15) μm, and monodisperse Polystyrene Latex (PSL) aerosol with nominal diameter of 1.0 μm. The results show comparable total particle number concentrations for all the counters, when the count of the first size channel (0.3 - 0.4 μm) for the 1.108 Grimm counter was left out. In the said channel the Grimm counter 1.108 always showed much higher particle counts than those inferred from the tested aerosols. The overcount was proved by the fact that the aerosol sampled in each test on a Nuclepore filter showed no particles in the 0.3 - 0.4 μm range when examined under Scanning Electronic Microscope (SEM). The presence of an artefact produced by the counter was assumed as a likely explanation. For all the counters, the Count Median Diameters (CMDs) of aerosol size distributions, were far below the expected value for the aerosol used. The nearest CMD values to the expected ones were shown by the Grimm 1.107 counter.

Keywords: Optical Particle Counter; Air quality; Aerosol Size Distribution

1. Introduction

Single particle light scattering is one of the most widely used techniques for measuring the particle number size distribution in the range from 0.2 μm up to several microns. Single Optical Particle Counters (OPCs) measure the light elastically scattered from a single particle illuminated by a well defined light source while it is passing through the sensing volume of the instrument. The scattered light intensity is utilized as a measure of the particle's size. The Mie theory is used to predict the light scattering intensity of an electromagnetic wave by a homogenous spherical particle. Consequently, the relative signal response of an OPC can be determined by knowing the characteristics of the light source, the detector, and the physical configuration of the sensitive volume, and the properties (mainly size, refractive index and shape) of the particle [1]. Several efforts have been made to test experimental OPC's response against Mie theory, using polystyrene latex [2-4].

In recent years, measurement techniques based on aerosol light scattering have received greater attention because of the possibility of deploying the aerosol size distribution as a proxy of the health related fractions like PM10, PM2.5 and PM1 [5]. The increasing importance of adverse health effects of air pollution has driven research towards the development of real time measurement techniques, which can couple meteorological parameters and particulate matter characteristics (mainly concentration and size distribution), thus providing an improved understanding of the relationship between sources and effects of mitigation actions that could be taken. In order to estimate the aerosol mass concentration from the OPC size distribution measurements, as well as the influence of particle shape and refractive index, particle density must also be taken into account. Since the latter is generally not known over the size distribution spectrum, only empirical conversion matrix factors can be used. Therefore, only sampling site-dependent correlations can be obtained [6].

In the present study, four commercial OPCs were tested in parallel with monodisperse generated aerosols, in order to compare their particle counting efficiencies and particle size classification capabilities.

The aim of the present study is to caution against careless use of the counters in monitoring situations. The study was promoted by the Italian Aerosol Association (IAS) in the context of the working group on "PMx aerosols".

2. Experimental Set-Up

The experimental work was carried out at the Institute of Atmospheric Sciences and Climate (ISAC) of the Italian National Research Council (CNR) in Bologna during summer 2010. Monodisperse particles were generated by means of MAGE (Monodisperse Aerosol Generator, Lavoro & Ambiente, Bologna, Italy). MAGE is a condensation-type generator capable of producing monodisperse particles in a wide size range. A stream of nuclei (mainly NaCl nuclei) is exposed to the vapour of a low-volatile liquid at an elevated temperature, and the controlled heterogeneous condensation of the vapour onto the nuclei results in the generation of monodisperse aerosol [7,8]. MAGE was used with Carnauba Wax (CW) at 270°C and 300°C oven temperatures in order to obtain two different monodisperse aerosol distributions. Particles size was ascertained by aspirating the aerosol on a Nuclepore filter (Whatman, 0.20 µm pore size) at the outlet of the MAGE and observing the filter under SEM. The measured diameter were (1.00 ± 0.08) µm at the lower temperature and (1.40 ± 0.15) µm at the higher temperature value. **Figure 1** shows a typical SEM image of the aerosol generated by the MAGE at 270°C, sampled directly from the generator outlet. It can be seen that the particles are of spherical shape. Their density is supposed to be the same as the CW bulk density (about 990 $Kg \cdot m^{-3}$), while the refractive index is 1.45°C at 20°C.

Since the particle number concentration was very high at the MAGE outlet (of the order of 10^{12} particles·m^{-3}), two diluters were positioned in series, after which all OPCs sampled from a mixing volume. The diluters were based on the Venturi effect, as shown in the paper of Yoon *et al.* [9].

A test was also carried out with Polystyrene Latex particles (PSL, Agar Scientific) with nominal diameter of

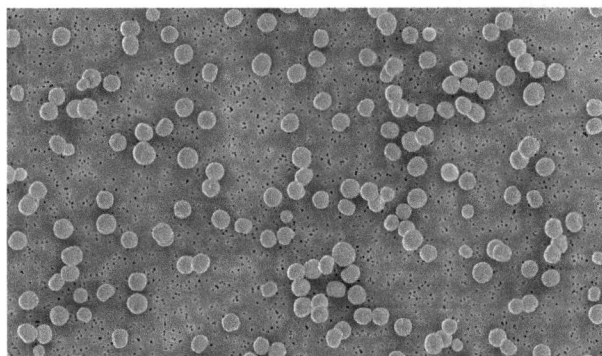

Figure 1. Monodisperse carnauba wax particles obtained with the MAGE at the working temperature of 270°C.

1.0 µm and refractive index of 1.59. The PSL particles were nebulized with a six-jet TSI atomizer model 9306, aerosol generator, and were sampled without dilution.

The generated aerosols were not strictly monodisperse due to various causes. As a matter of fact some NaCl nuclei could not have been nucleated by the CW vapour. Based on the concentration of the NaCl solution and the size of the droplets generated by the Collison, the nuclei should have a diameter of about 60 nm [10]. In addition the generation of PSL particles involves even residue particles which are below 100 nm [11]. Therefore the OPC counters should not be able to detect both the NaCl nuclei and PSL residue particles. A possible coagulation of the aerosol at the exit of the MAGE due to the high particle concentration was avoided by diluting the aerosol.

The OPCs tested were the following: DustMonit (Con. Tec Engineering srl, Milan) with sampling flow rate of 1 lpm, size separation range of 0.3 - 10 µm and 8 channels; Grimm 1.107 (Grimm Aerosol Technik, GmbH) with sampling flow rate 1.2 lpm, size separation range of 0.2 - 32 µm and 31 channels; Grimm 1.108 (Grimm Aerosol Technik, GmbH) with sampling flow rate 1.2 lpm, size separation range of 0.3 - 20 µm and 15 channels; ParticleScan CR (ParticleScan, Advanced Particle Counters. IQAir) with sampling flow rate 2.8 lpm, size separation range of 0.3 - 5 µm and 6 channels. Each OPC name is shortened to the following initials: DM (DustMonit), EN (Grimm 1.107), GR (Grimm 1.108) and PS (ParticleScan). When comparing the OPCs, the particle number concentration of the EN counter was considered only for sizes larger than 0.3 µm (first two channels excluded). The tested OPCs have a laser wavelengths in the range from 675 nm (Grimm 1.107) to 780 nm (Grimm 1.108). In accordance with the manufacture's recommendations, Grimm OPCs are calibrated with polystyrene latex, while for DustMonit and ParticleScan the calibration procedures are not available.

The OPC performances were compared by measuring the size distribution of monodisperse aerosol particles simultaneously sampled from a 1 liter mixing volume. Before each test session the background particle number concentration in the laboratory was measured and verified to be less than 1% of the particle number concentration generated in the comparison experiments. In order to compare the findings from each OPC, the particle number concentration in each size channel (dN) was normalized by considering the channel width (dN/dDp). During each run (for both particle diameters), an aerosol sample generated by MAGE was sampled on Nuclepore filter media for SEM observations. Only tests with a total particle number concentration less than 2×10^6 l^{-1} were considered, as this is the maximum particle number concentration recommended by the respective manufacturers to avoid coincidence errors.

In addition, in order to confirm the results of the comparison, the GR counter was subsequently sent to the manufacturer for the yearly revision (calibration checks and adjustments). On its return, a further test was performed against a Las-X (PMS, Inc) active cavity laser.

3. Results and Discussion

The results are presented below, separately for the total particle number concentration and the particle size classification.

3.1. Total Particle Number Concentration Measurements

1) *Carnauba wax particles*

Figure 2 shows the ratio of the total particle number concentration measured by each OPC and the average total particle number concentration (M), obtained by averaging the total counts of the single OPC, for the four tests performed with monodisperse particles at (1.00 ± 0.08) μm (MAGE at 270°C). The bars show one standard deviation. The histogram shows the results considering the OPC channels from 0.3 μm and from 0.5 μm separately.

Figure 3 shows the results obtained with the particle size of (1.40 ± 0.15) μm (MAGE at 300°C).

Compared to the other counters the results show a higher total particle number concentration in the case of GR, when the full size spectrum is considered. The agreement among the OPCs improves when only channels starting from 0.5 μm are considered.

2) *Polystyrene latex particles*

A test with PSL particles of 1.0 μm diameter was carried out to investigate any possible artifacts in the gener-

Particle diameter: (1.00±0.08) μm

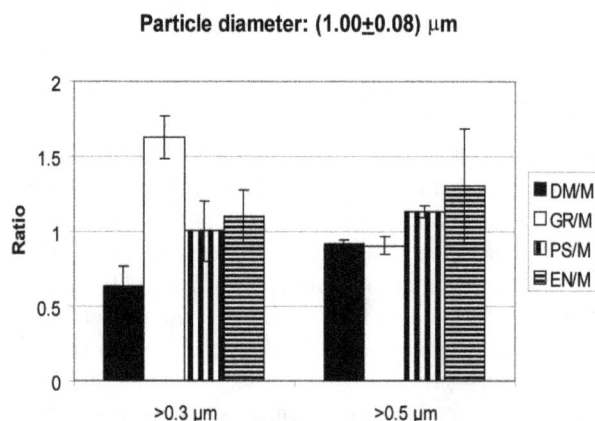

Figure 2. Ratio of the total particle number concentration measured by each OPC and the average total particle number concentration. Particle diameter: (1.00 ± 0.08) μm. Bars show one standard deviation. Results are shown considering the OPC channels from 0.3 μm and from 0.5 μm separately.

Particle diameter: (1.40±0.15) μm

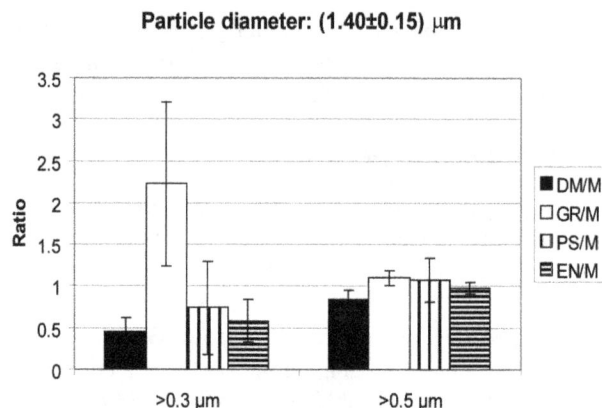

Figure 3. Ratio of the total particle number concentration measured by each OPC and the average total particle number concentration. Particle diameter: (1.40 ± 0.15) μm. Bars show one standard deviation. Results are shown considering the OPC channels from 0.3 μm and from 0.5 μm separately.

ated CW aerosol particles. The results, shown in **Figure 4**, are comparable to those of the CW particle tests, showing a higher particle number concentration of the GR sampler in the first two size channels. It can be concluded that the GR counter counts more particles than the other samplers, in the first two size channels (0.3 - 0.4 μm and 0.4 - 0.5 μm).

The behaviour of the GR counter conflicts with the results of Heim *et al.* [4], who found a decrease in the counting efficiency of the OPCs towards the first size channels (lower particle diameters).

3.2. Particle Size Distribution Measurements

Figures 5 and **6** show the particle size distributions obtained with PS, GR, DM and EN optical particle counters, by measuring generated CW particles. The sampled particle diameter obtained from SEM observation were (1.00 ± 0.08) μm and (1.40 ± 0.15) μm, respectively.

The results show that EN, and to a greater extend GR, measured higher particle number concentrations in smaller size channels, as compared to PS and DM.

Table 1 shows the CMD and the Geometric Standard Deviation (GSD) for each counter. The CMD and GSD were obtained by reporting cumulative particle number distributions on log-probability plots. The CMD from all counters turned out to be much lower than expected according to the tested particles.

By excluding the count of the first channel (size 0.3 - 0.4 μm) and sampling aerosol of size (1.00 ± 0.08) μm and (1.40 ± 0.15) μm, the CMD for the EN counter was 0.73 μm and 0.98 μm, and for the GR, 0.52 μm and 0.68 μm, respectively. In the case of PSL particle tests, the GR counter showed the highest particle number concentration in the first size channel. By excluding this channel, a CMD of 0.56 μm was obtained, instead of the

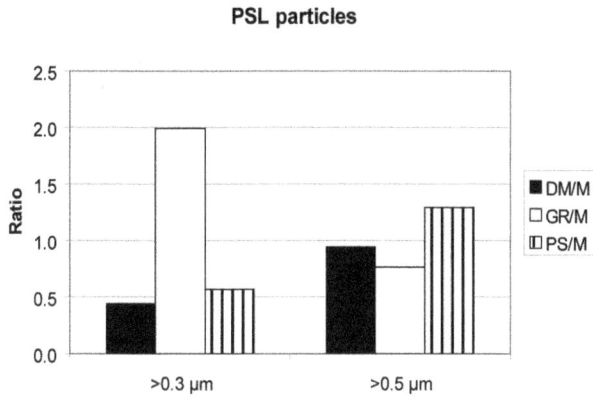

Figure 4. Ratio of the total particle number concentration measured by each OPC and the average total particle number concentration. PSL particles. Results are shown considering the OPC channels from 0.3 μm and from 0.5 μm separately.

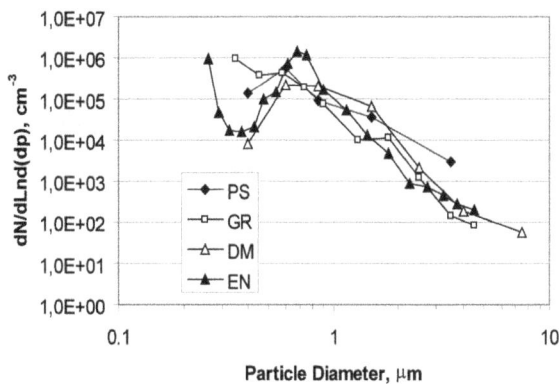

Figure 5. Particle size distributions obtained for CW particles. Particle diameter: (1.00 ± 0.08) μm.

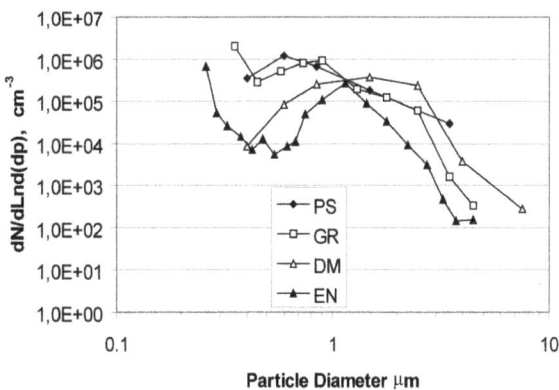

Figure 6. Particle size distributions obtained for CW particles. Particle diameter: (1.40 ± 0.15) μm.

1.0 μm sphere diameter. The CMD underestimation by GR is in agreement with the results obtained by Heim *et al.* [4], which showed, albeit for the Grimm 1.109 OPC, a ower detected PSL particle diameter in the size range from around 0.8 μm to approximately 2 μm. After the comparison tests the GR optical particle counter was sent

Table 1. Averaged CMD and GSD obtained during the tests with different monodisperse particle class ranges: (a) Particle diameter (1.00 ± 0.08) μm; (b) Particle diameter (1.40 ± 0.15) μm; (c) PSL (1.0 μm). For the EN counter only size channels larger than 0.3 μm were considered.

Optical Counter	(a)		(b)		(c)	
	CMD	GSD	CMD	GSD	CMD	GSD
GR	0.35	1.69	0.50	1.60	<0.30	/
EN	0.70	1.26	0.90	1.33	ND	ND
PS	0.50	1.39	0.60	1.40	0.44	1.53
DM	0.67	1.39	1.03	1.52	0.55	1.62

l to the manufacturer for the yearly calibration and maintenance. On its return a further comparison against a Las-X (PMS, Inc) active cavity laser was performed. **Figure 7** shows the results for GR-A before maintenance (same curve as **Figure 5**) and after (GR-B). It can be seen that the GR better classifies the particles size than it did before the calibration, but the over count in the first size channel is still present.

The reported results evidence that OPC users are not able to check whether the instruments they employ are working well or not. As a matter of fact the software installed in all OPCs takes into account only the lens cleaning effect, and therefore the instrument self-test procedure can indicate "no errors", even when the size classification capabilities of the sampler is compromised.

In addition, since these instruments could be used to assess the environmental aerosol fractions (PM1, PM2.5 and PM10), a comparison between particle number and particle mass concentrations was undertaken. This could be done only for EN and DM counters, because GR and PS counters do not give particle number and mass distribution simultaneously. The particle mass distribution was computed from the particle number size distribution for all tests and aerosols by considering spherical and unity density particles.

Figure 8 shows the ratio between the measured PMx fractions given by the two counters (size-segregated according to the environmental particle fractions) and the computed mass concentrations (9 tests considered).

Table 2 summarizes the ratio between measured and computed PMx fractions. Since this ratio is related to the particle density, the computed and measured mass concentrations become equivalent by assuming an average particle density of about 1.9 for EN (excluding PM1 fraction) and about 2.7 for DM. The comparison with data published by Tuch *et al.* [12] gave a value of around 1.6 g·cm^{-3}; Morawska *et al.* [13] gave a derived density ranging from 1.2 to 1.8 g·cm^{-3} (comparison between TEOM and SMPS), while Burkart *et al.* [14] reported an instrument-specific factor of 2.8 for GR 1.108 to calculate mass concentration from particle number size distribution.

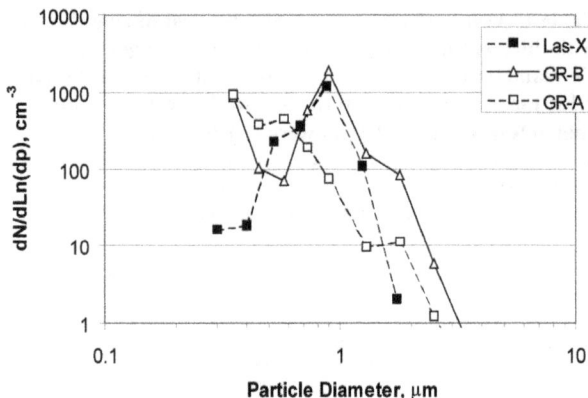

Figure 7. Particle size distribution obtained with GR and LAS-X. Particle diameter: (1.00 ± 0.08) μm. GR-A means before maintenance and GR-B after maintenance.

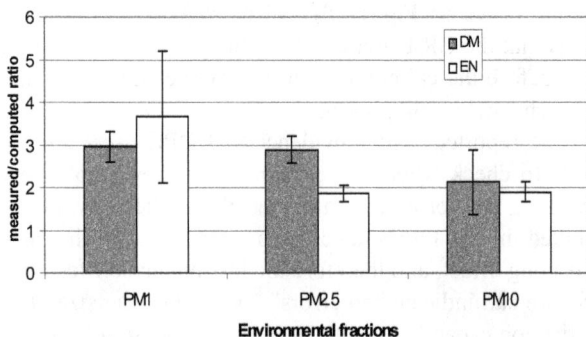

Figure 8. Ratio between measured and computed PMx fractions for the EN and DM counters. Bars show one standard deviation.

Table 2. Averaged particle number mass ratio obtained during the tests with different monodisperse particle class ranges for EN and DM. For EN only size channels larger than 0.3 μm were considered.

Optical Counter	PM1		PM2.5		PM10	
	Ratio	St.Dev.	Ratio	St.Dev.	Ratio	St.Dev.
EN	3.66	1.54	1.86	0.19	1.90	0.23
DM	2.97	0.36	2.89	0.32	2.12	0.74

4. Summary and Conclusions

The results show a comparable total particle concentration number for all the counters, if the count of the first size channel 0.3 - 0.4 μm, for the Grimm 1.108 counter is excluded. The GR counter always showed much higher particle counts for this channel. The over count was proven by the fact that the aerosol sampled in each test on a Nuclepore filter showed no particles in the 0.3 - 0.4 μm range when examined under SEM. The presence of an artefact produced by the counter was supposed to be a likely explanation.

The CMDs of the aerosol size distributions were far below the expected value for all the counters, even taking

into account the low size resolution of the DM and the PS instruments. The GR counter showed improbable CMDs values, and only by excluding the first size channel it was possible to obtain comparable CMDs with that of other counters. The nearest CMD value to the expected one was obtained by the Grimm 1.107 counter.

Since these counters could be used to assess the environmental aerosol fractions (PM1, PM2.5 and PM10), a comparison for EN and DM counters was performed between the particle mass concentration derived from the particle size distribution and the particle mass shown by the counters. The obtained ratio, which is related to an averaged particle density, is consistent with published data. GR and PS do not allow to get simultaneous measurements of aerosol number and mass concentrations.

In view of their widespread use in air quality programs, also in unattended mode, a check-kit for the instrument should be developed and provided to the users in order to guarantee reliability of the measured data. For example, a microscope slide with deposited calibrated particles could be supplied for periodical transfer into the optical chamber, thus allowing the comparison of the instrument's output with reference values.

5. Acknowledgements

The Authors wish to thank Prof. M. Causà (Univ. of Naples), C. Giglioni (ConTec Engineering) and Prof. P. Prati (Univ. of Genova) for making their respective counters available for the tests. Special thanks to C. Colombi and V. Gianelle (ARPA-Milano) for the useful discussions, and to the Italian Aerosol Association (IAS).

REFERENCES

[1]   J. Gebhart, "Optical Direct-Reading Techniques: Light Intensity Systems," In: P. A. Baron and K. Willeke, Eds., *Aerosol Measurement*, Van Nostrand Reinhold, New York, 1993, pp. 313-344.

[2]   B. T. Chen, Y. S. Cheng and H. C. Yeh, "Experimental Response of Two Optical Particle Counters", *Journal of Aerosol Science*, Vol. 15, No. 4, 1984, pp. 457-464.

[3]   H. Y. Wen and G. Kasper, "Counting Efficiencies of Six Commercial Particle Counters," *Journal of Aerosol Science*, Vol. 17, No. 6, 1986, pp. 947-961.

[4]   M. Heim, B. J. Mullins, H. Umhauer and G. Kasper, "Performance Evaluation of Three Optical Particle Counters with an Efficient 'Multimodal' Calibration Method," *Journal of Aerosol Science*, Vol. 39, No. 12, 2008, pp. 1019-1031.

[5]   A. Tittarelli, A. Borgini, M. Bertoldi, E. De Saeger, A. Ruprecht, R. Stefanoni, G. Tagliabue, P. Contiero and P. Crosignani, "Estimation of Particle Mass Concentration in Ambient Air Using a Particle Counter," *Atmospheric*

Environment, Vol. 42, No. 36, 2008, pp. 8543-8548.

[6]   H. Grimm and D. J. Eatough, "Aerosol Measurement: The Use of Optical Light Scattering for the Determination of Particulate Size Distribution, and Particulate Mass, including Semi-Volatile Fraction," *Journal of Air Waste Management Association*, Vol. 59, No. 1, 2009, pp. 101-107.

[7]   V. Prodi, "A Condensation Aerosol Generator for Solid Monodisperse Particles," In: T. T. Mercer, P. E. Morrow and W. Stoeber, Eds., *Assessment of Airborne Particles*, Charles C Thomas Publisher, New York, 1972, pp. 169-181.

[8]   V. Prodi and W. Mölter, "Temperature Characteristics of a Monodisperse Aerosol Condensation Generator, Aerosols: Formation and Reactivity," *Proceedings of the 2nd International Aerosol Conference*, Berlin, 22-26 September 1986, pp. 1065-1068.

[9]   Y. J. Yoon, S. Cheevers, S. G. Jenings and C. D. O'Dowd, "Performance of a Venturi Dilution Chamber for Sampling 3 - 20 nm Particles," *Journal of Aerosol Science*, Vol. 36, No. 4, 2005, pp. 535-540.

[10] K. D. Horton, R. D. Miller and J. P. Mitchell, "Characterization of a Condensation-Type Monodisperse Aerosol Generator (MAGE)," *Journal of Aerosol Science*, Vol. 22, No. 3, 1991, pp. 347-363.

[11] K. T. Whitby and B. Y. H. Liu, "Polystyrene Aerosols—Electrical Charge and Residue Size Distribution," A*tmospheric Environment*, Vol. 2, No. 2, 1968, pp. 103-116.

[12] Th. Tuch, E. Tamm, J. Heinrich, J. Heyder, P. Brand, Ch. Roth, H. E. Wichmann, J. Pekkanen and W. G. Kreyling, "Comparison of Two Particle-Size Spectrometers for Ambient Aerosol Measurements," *Atmospheric Environment*, Vol. 34, No. 1, 2000, pp. 139-149.

[13] L. Moraswka, G. Johnson, Z. D. Ristovski and V. Agranovski, "Relation between Particle Mass and Number for Submicrometer Airborne Particles," *Atmospheric Environment*, Vol. 33, No. 13, 1999, pp. 1983-1990.

[14] J. Burkart, G. Steiner, G. Reischel, H. Moshammer, M. Neuberger and R. Hitzenberger, "Characterizing the Performance of Two Optical Particle Counters (Grimm OPC1.108 and OPC1.109) under Urban Aerosol Conditions," *Journal of Aerosol Science*, Vol. 41, No. 10, 2010, pp. 953-962.

20

GOSAT CH$_4$ and CO$_2$, MODIS Evapotranspiration on the Northern Hemisphere June and July 2009, 2010 and 2011

Reginald R. Muskett

Geophysical Institute, University of Alaska Fairbanks, Fairbanks, USA

ABSTRACT

The Greenhouse gases Observing Satellite (GOSAT) affords an ability to assess and monitor CH$_4$ and CO$_2$ near-surface atmospheric concentrations globally on monthly scales pertaining to biogeochemical cycles and anthropogenic emissions. In addition to GOSAT our investigation incorporates global-monthly estimates of evapotranspiration (ET) from the Moderate Resolution Spectroradiometer (MODIS) and fire/wildfire locations for correspondence and comparison. We restrict the investigation to the months of June and July in years 2009, 2010 and 2011. After processing and assessment on the northern hemisphere we focus on two regions in Eurasia for interrogation: 40° to 80°E by 50° to 58°N and 100° to 140°E by 50° to 58°N. The regions allow for contrasting regional settings, an agricultural-industrial-urban west-region to a boreal-steppe discontinuous permafrost zone palsa and thaw lake east-region. Joint probability density functions allow us to identify significant modes, the highest probable values of background levels of CH$_4$ and CO$_2$ to ET and develop regressions for correlated relationships. We found that background levels of CH$_4$, CO$_2$ and ET were not affected by the wildfires of 2010. Regressions indicate significant inverse relationships of CH$_4$ and CO$_2$ to ET in the west-region and no significant relationships in the east-region. The east-region shows significantly higher background levels of CH$_4$, CO$_2$ and ET owing to the heterogeneity of ecosystems, hydrology, physical processes and terrain in the discontinuous permafrost zone of the central Siberian Plateau.

Keywords: GOSAT; CH$_4$; CO$_2$; MODIS; Evapotranspiration; Wildfire

1. Introduction

Addressing the uncertainties in near-surface atmospheric CH$_4$ and CO$_2$ sources and sinks is a task that space-based remote sensing has engaged in recent years to complement and extend ground-based sensing methods. GOSAT a joint project by the Japan National Institute of Environmental Studies, the Japan Ministry of Environment and the Japan Aerospace Exploration Agency (JAXA) employs dual-channel Fourier transform spectrometry to measure dry-air concentrations of CH$_4$ and CO$_2$ globally since launch in early 2009 [1].

CH$_4$ and CO$_2$ are constituents of the Earth's atmosphere. Their concentrations derive from physical processes of long-lived biogeochemical cycles on lands and in oceans. Additionally variations in their concentration owe to anthropogenic activities since wide spread industrialization in the 19th century. Their affects on the Earth's near-surface energy budget, *i.e.* greenhouse gases warming, have been speculated, postulated and debated extensively [2-5].

Our purpose is to investigate correspondences of background levels of CH$_4$, CO$_2$ and ET through joint probability density functions and develop regressions for correlated relationships. We compose and co-geolocate the datasets on the northern hemisphere, and then focus on two regions of interest in Eurasia during June and July 2009, 2010 and 2011. We test a hypothesis that the wildfires in the summer of 2010 affected background levels of CH$_4$, CO$_2$ and ET on a regional and hemispheric basis.

2. Data and Methods

2.1. GOSAT

Launched and operating since January 2009 the Greenhouse gases Observing Satellite (GOSAT) has been providing global estimates of the column average dry air mole fractions of CH$_4$ and CO$_2$ [1]. The GOSAT sensing instrument is the Thermal and Near Infrared Sensor for Carbon Observations (TANSO) Fourier Transform Spectrometer (FTS). TANSO-FTS exploits absorption band characteristics of backscatter sunlight from surface and atmosphere in two channels: a spectral interferometer. The spectral resolution of TANSO-FTS is 0.3 cm. The nadir footprint radius is about 5 km. A second instrument the TANSO-Cloud and Aerosol Imager (CAI) observes surface and atmosphere reflectance in the shortwave and near infrared to provide cloud screening flags at several hundred points within a FTS footprint. Vertical profiles of pressure, temperature and humidity together with surface pressure and wind speed are taken from European Center for Medium-Range Weather Forecasts. Surface elevation data (within the FTS footprint) are taken from the GTOPO30 database.

The retrieval algorithm uses a radiative transfer inverse method based on Phillips-Tikhonov regularization and the L-curve method to estimate retrieval parameters to simultaneously derive near surface and atmosphere column average concentrations of CH$_4$ and CO$_2$ [1]. The algorithm uses radiances in 4 spectral windows cover the O$_2$A-band, weakly absorbing CO$_2$ band, CH$_4$ band and strongly absorbing CO$_2$ band. Retrieval parameters are in 12-layer vertical profiles of CH$_4$ and CO$_2$ column number and total number density of H$_2$O (interfering absorber) and scattering-spectral shift parameters.

During April 2009 through July 2010 validation campaigns were performed using six Total Carbon Column Observing Network (TCCON) sites that use ground-based FTS: two in Australia, Europe and North America. Bias in CH$_4$ is $-0.3\% \pm 0.26\%$ and in CO$_2$ is $-0.05\% \pm 0.37\%$, on average.

GOSAT TANSO-FTS by design has a critical dependence on solar angle and cloud (contamination by meteorological cloud water vapor) that reduces the number of reliable CH$_4$ and CO$_2$ retrievals in the high latitudes. Therefore we limit our time period of interest to June and July 2009, 2010 and 2011 to obtain the maximum number of retrievals. From the files we extract geolocation information (latitude-longitude relative to the WGS-84 reference ellipsoid), concentration of CH$_4$ and CO$_2$ in ppb and ppm, respectively, and the standard error values.

June and July retrievals across the northern hemisphere 40°N latitude and higher are irregular grids which become sparse at 70°N. We apply a least squares Green's Function interpolation such that the original concentration values per geolocation are constraints to produce regular equal-area projection grids at 5 km intervals [6]. The resultant CH$_4$ and CO$_2$ mixing ratios (concentrations) and standard error grids are shown in **Figure 1**. Gray-areas indicate no data. The oceans are masked; generally too few data for reliable processing. Values on Greenland are mostly artifacts from too few data at times.

2.2. MODIS

Proto-Flight Model and Flight Model 1 of the Moderate

Resolution Imaging Spectroradiometer (MODIS) operate onboard the NASA-Terra launched in December 1999 and NASA-Aqua launched in May 2002, respectively [7]. The orbits of Terra and Aqua are near polar and sun synchronous. They acquire reflectance and radiance (atmosphere and surface) in the optical through thermal portion of the electromagnetic spectrum in 36 bands. Spatial (nadir) resolutions are from 250 m, 500 m and 1 km. Terra acquires its data with equator crossing times of 10:30 (daytime) with repeat pass 22:30 (nighttime). Aqua acquires its data with equator crossing times of 13:30 (daytime) with repeat pass 01:30 (nighttime). Both instruments have a viewable swath width of 2330 km. Temporal resolutions of the MODIS products are 1-day, 8-day, 16-day, 1-month, 4-month and 1-year. Detailed information can be found online at the Land Processes Distributed Active Archive Center USGS website[1].

2.3. MODIS Fire Information for Resource Management System (FIRMS)

MODIS FIRMS produces daily near-real time and science quality fire location hotspots data from standard MODIS MOD14 (Terra) and MYD14 (Aqua) Fire and Thermal Anomalies products, Level 3 version 5 [8]. The fire detection algorithm of MOD14/MYD14 (version 2.4 as of Oct. 2006) uses radiances in thermal 4 µm (two channels) and 11 µm (one channel) for fire detection relative to background brightness temperature, 0.65, 0.68 and 2.1 µm water-body (glint false alarm rejection) and 11 and 12 µm for fire/cloud masks [9]. Daily FIRMS products of hotspot locations give the detection thermal brightness temperature in Kelvin, probability of detection quality flag and geolocations given by latitude and longitude relative to the WGS-84 reference ellipsoid.

Validation campaigns have been carried out globally since 2000 using co-located thermal images from ASTER, Landsat 5/7 and Geostationary Operational Environment Satellites (GOES) for example [10]. On a global basis estimated commission errors vary with scan angle from 1.5% up to 8%. Commission errors in dense vegetation equatorial regions can approach 35% [11].

Our processing of the MODIS FIRMS data takes the daily detection files to produce June and July month fire-spot sums at 1 km resolution on the northern hemisphere, **Figure 2**. This gives us a monthly landscape and regional fire-spot geolocations for comparison to the GOSAT and other MODIS monthly datasets.

2.4. MODIS Evapotranspiration

The Global Terrestrial Evapotranspiration Data Set project of the Numerical Terradynamic Simulation Group at the University of Montana Collage of Forestry produces

---

[1]https://lpdaac.usgs.gov/products/modis_overview.

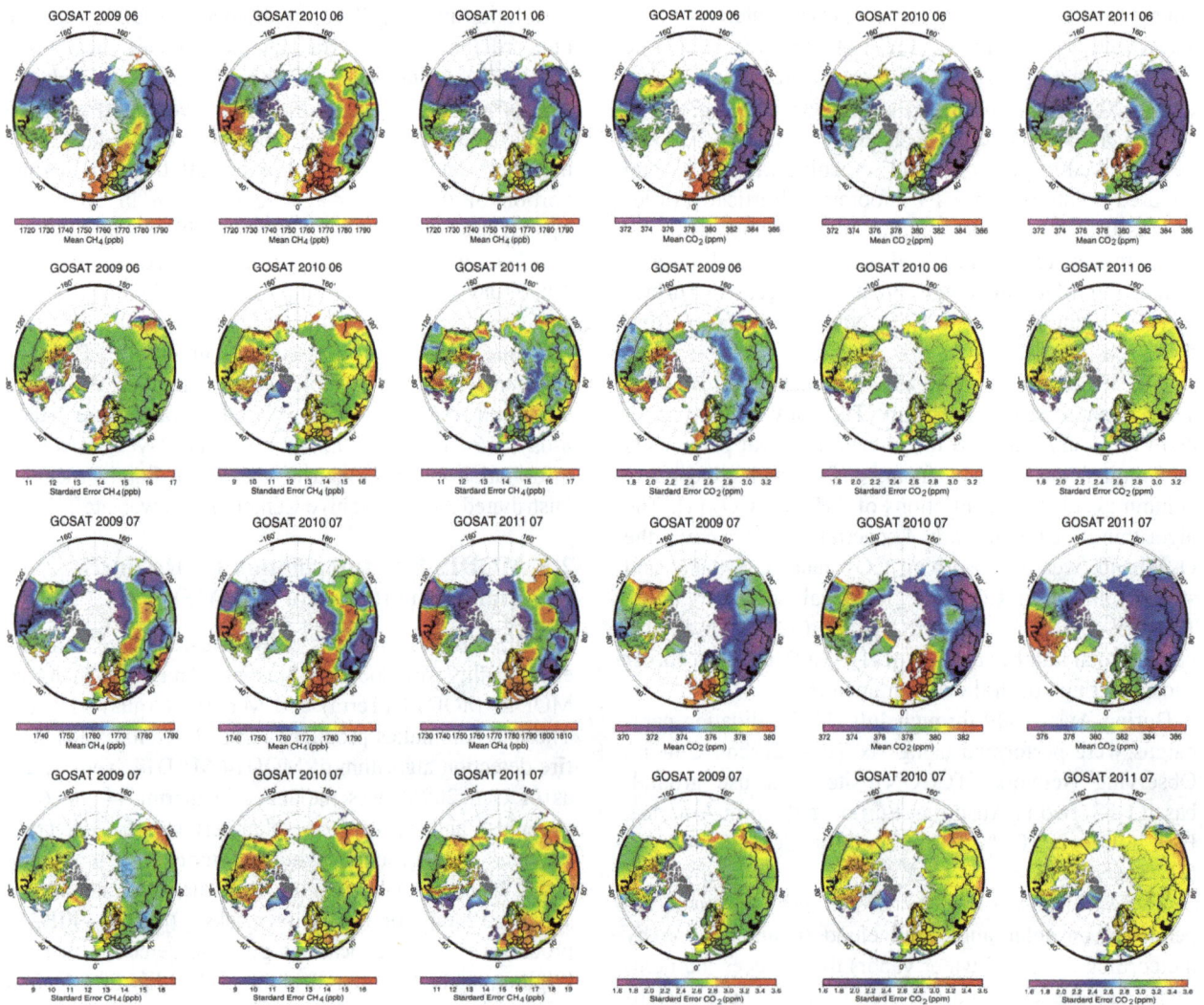

Figure 1. GOSAT mean CH$_4$ and CO$_2$ mixing ratio (concentrations) and standard error (ppm) fields during June and July 2009-2011.

the MODIS MOD16 product Global Terrestrial Evapotranspiration for the NASA[2]. Files contain derived evapotranspiration (ET, mm/mo) and include latent heat flux (LE), potential ET and potential LE with quality control flags. The time period covered begins January 2000 and continues through the current year with about one to two month processing lag. Spatial resolution is 1 km with temporal resolution at 8-day, 1-month and 1-year intervals. Geolocation information is relative to the WGS-84 reference ellipsoid.

Since 2012 MOD16 ET is derived algorithmically based on the Penman-Monteith equation [12]. The algorithm derived ET includes surface evapotranspiration from vegetation stomata, evaporation from moist and wet soil, evaporation from open water bodies and evaporation of rain water on vegetation. During winter months,

evaporation from surface snow is included in ET. Cloud contamination estimates derive from albedo and fraction of photosynthetic available radiation/leaf area index (FPAR/LAI) MODIS inputs. Validations have been performed using available tower eddy covariance observations at FLUXNET stations. Correlation of tower measurements with the current version of the algorithm shows R = 0.86 with the distribution centered on the 1:1 line. Errors in tower measured eddy covariance can themselves be as great as 30%. This arises from problems in the energy balance closure at tower sites and scaling issues when estimating landscape scale ET from specific and sparse site measurements.

Our processing of the monthly MOD16 files extracts ET and geolocation values into equal-area projection grids. We co-geolocate the GOSAT and MODIS grids on the WGS-84 reference ellipsoid. This is to ensure that for any geolocation there are paired ET rates, CH$_4$ and CO$_2$

---

[2]http://www.ntsg.umt.edu/project/mod16.

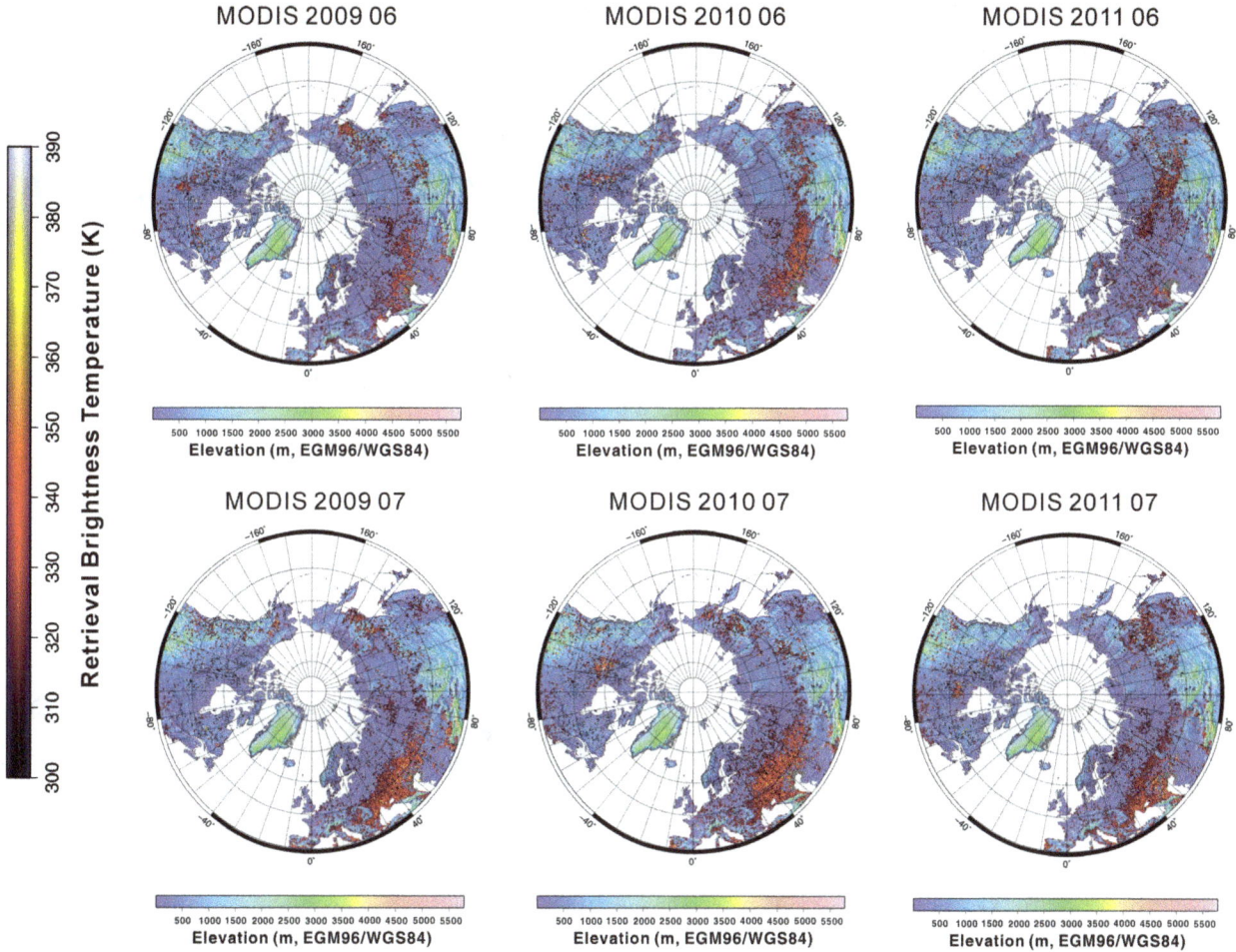

Figure 2. MODIS FIRMS fire hotspot retrievals (K, 1 km resolution) during June and July 2009-2011 projected on the ACE2 DEM (1 km resolution).

concentrations. The MODIS ET grids are shown in **Figure 3**.

2.5. Regions of Interest in Eurasia

Our investigation focuses on two regions of Eurasia: west-region 40° to 80°E by 50° to 58°N and east-region 100° to 140°E by 50° to 58°N. The west-region includes agricultural, industrial and urban area of Moscow and drained peatlands. The east-region is within the southern portions of the central Siberian Plateau and including heterogeneous terrains of the discontinuous permafrost zone, palsa and thermokarst lakes, and southern boreal and steppe ecosystems.

Wildfires are visually spectacular and physically dynamic events. We use these regions to investigate the wildfire events in June-July 2010 relative to 2009 and 2011 for affecting regional CH$_4$ and CO$_2$ levels, *i.e.* background concentration levels. We state this as a hypothesis to test: Did regional background levels of CH$_4$, CO$_2$ and ET change in response to the June-July wildfires in Russia.

3. Results

Within the east- and west-regions we extract up to 299,800 co-geolocated CH$_4$ and CO$_2$ concentrations and ET rates in months June and July 2009, 2010 and 2011 and extract fire/wildfire hotspots. We then compute joint probability density functions of CH$_4$ with ET and CO$_2$ with ET. The joint probability density functions allow us to isolate the dominant mode, the most likely co-geolocated levels. We then compute least squares regression of the modes of CH$_4$ and ET with CO$_2$ with ET. We further plot the co-geolocation standard deviations of the dominant modes. This allows for evaluation of the test hypothesis.

3.1. Fire/Wildfire Hotspots

Table 1 shows the counts of MODIS FIRMS fire/wildfire hotspots (1 km) in the east- and west-regions for June and July 2009, 2010 and 2011.

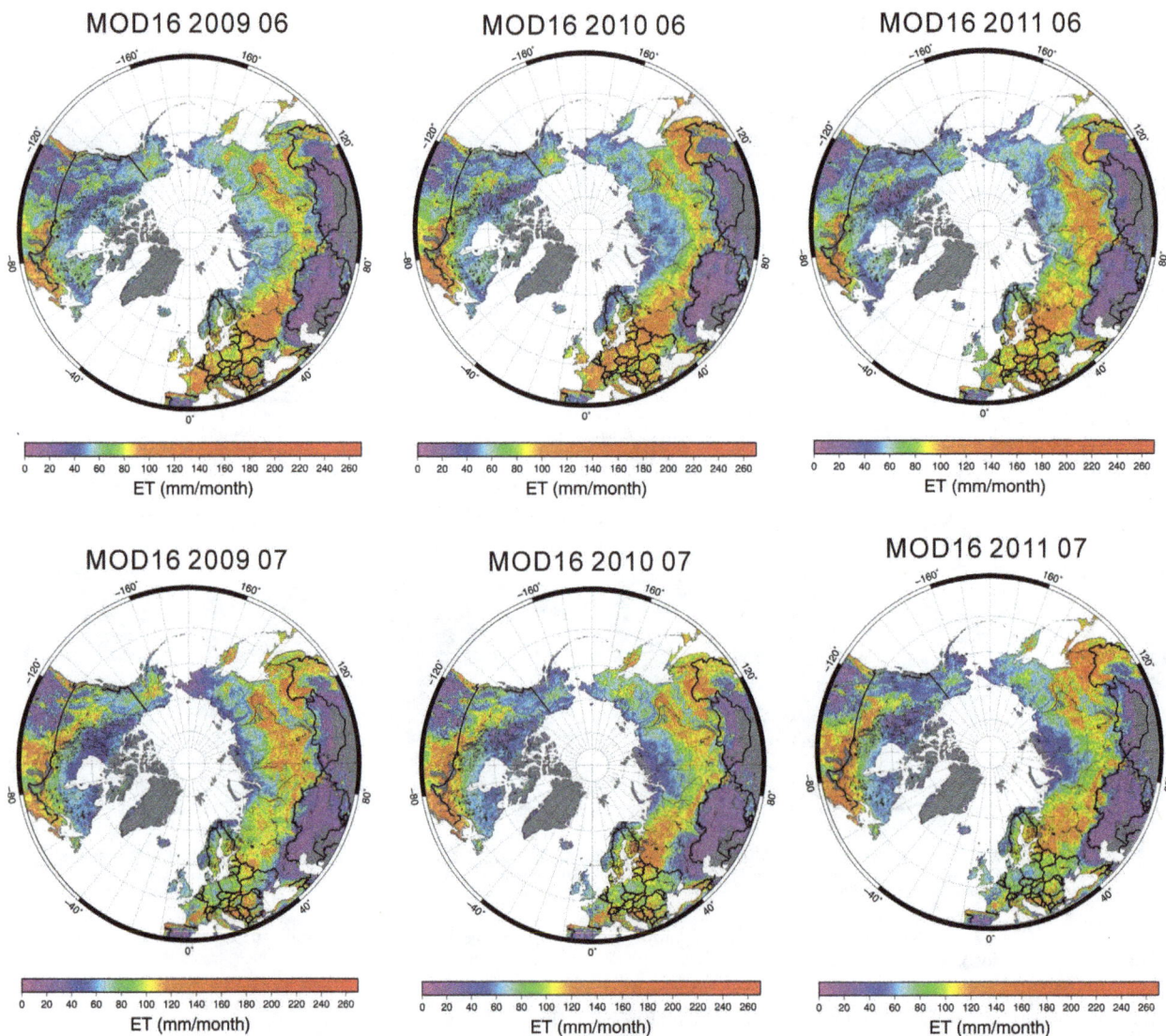

Figure 3. MODIS Evapotranspiration (mm/month) fields during June and July 2009-2011.

Table 1. Counts of fire/wildfire hotspots from MODIS FIRMS for regions 40° to 80°E by 50° to 58°N (West) and 100° to 140°E by 50° to 58°N (East).

2009			
June		July	
West-Region	East-Region	West-Region	East-Region
2701	2705	5495	572
2010			
June		July	
West-Region	East-Region	West-Region	East-Region
8560	3915	18,022	1324
2011			
June		July	
West-Region	East-Region	West-Region	East-Region
992	12,679	3155	29,986

July counts in the west-region are greater by a factor of two above counts in June. In the east-region, July counts are lower by more than a factor of two than counts in June for years 2009 and 2010. In year 2011 the east-region July count is greater by a factor of two compared to the June count.

3.2. Joint Probability Density Functions

Figures 4 shows the joint probability density functions of CH_4 with ET and CO_2 with ET. The distribution change by month and year as well as the dominant modes, which represent values of co-geolocated CH_4 with ET and CO_2 with ET that are the most likely.

On inspection the dominant modes correspond to regional background levels of CH_4, CO_2 and ET. Dominant modes do not correspond to any co-geolocated event or group of events such as a wildfire hotspot or industrial

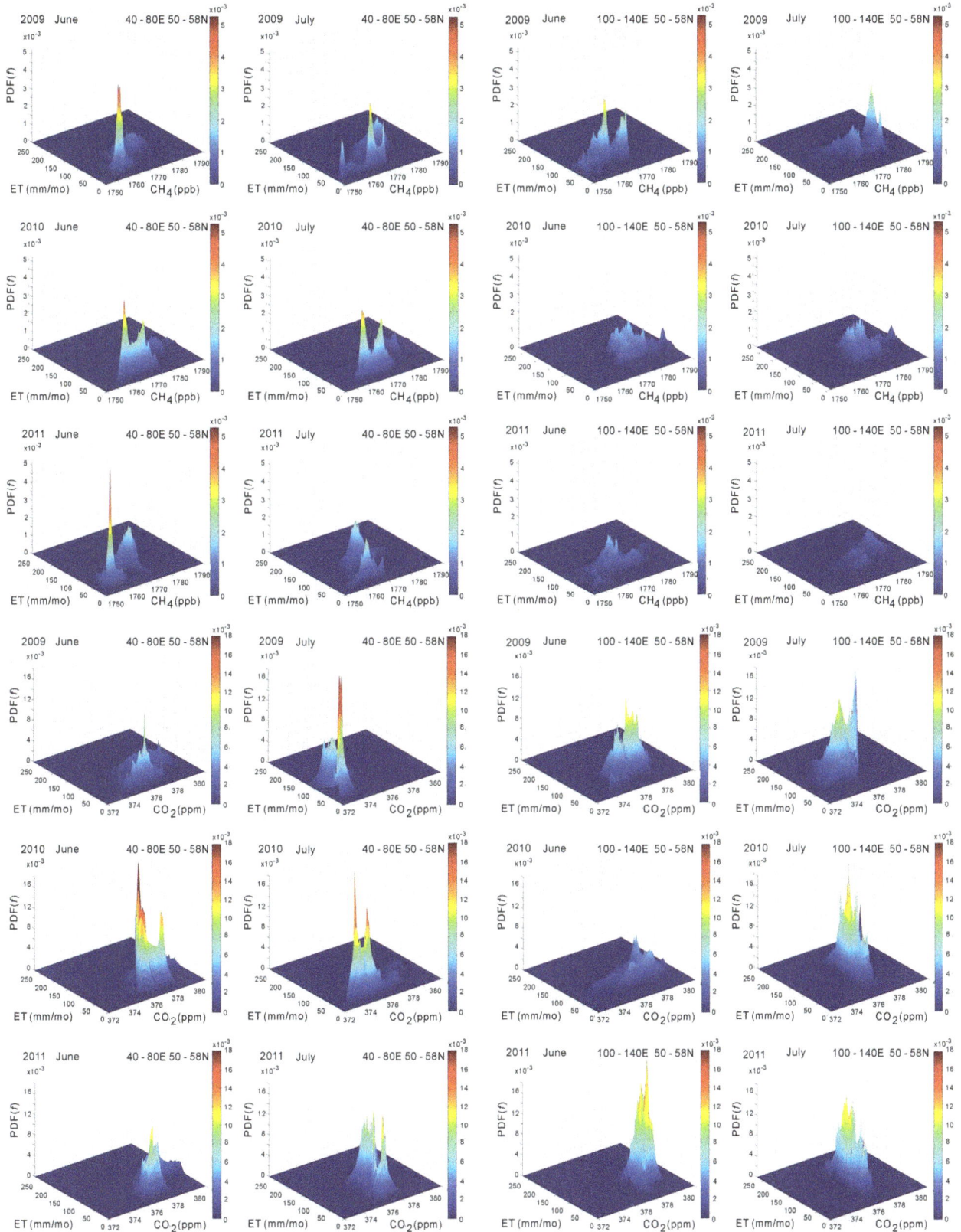

Figure 4. Joint probability density functions of MOD16 evapotranspiration (mm/month) and GOSAT CH$_4$ concentrations (ppb) and CO$_2$ concentrations (ppm), June 2009-2011 and July 2009-2011 in the region 40° to 80°E by 50° to 58°N and 100° to 140°E by 50° to 58°N.

site (such as a coal-fire powered electrical generation facility, or a city in general). Events lend themselves to the spread of the probability distribution and have much lower probability, *i.e.* less significance.

3.3. Regressions of Dominant Modes

We next isolate the dominant modes in their values of CH_4 ET and CO_2 ET and plot these in **Figure 5**. **Table 2** also gives the values with their mode standard deviations. Modes occurring in the regions of interest are identified by symbol and the month is identified by color.

The west-region, which includes Moscow and locations of the 2010 fires show strong R^2 values. The regressions indicate inverse relationships of CH_4 to ET and CO_2 to ET in time of month and year. This equates low CH_4 and CO_2 concentrations to high ET rates.

The east-region, which includes much of Siberia and includes boreal and steppe ecosystems, locations wildfire activity and very numerous peatlands-wetlands and palsa and thermokarst lakes show very weak R^2 values. Significant forward or inverse relationships of CH_4 to ET and CO_2 to ET are not indicated.

The spread of the dominant modes of CH_4 ET and CO_2 ET are much greater for the east-region than for the west-region. This is an expression of the heterogeneity of eco-hydrologic systems, the discontinuous permafrost zone and the varied physical processes and terrain governing CH_4 and CO_2 concentrations and ET rates.

3.4. Standard Deviations of Dominant Modes

Figure 6 shows the standard deviation comparison plot

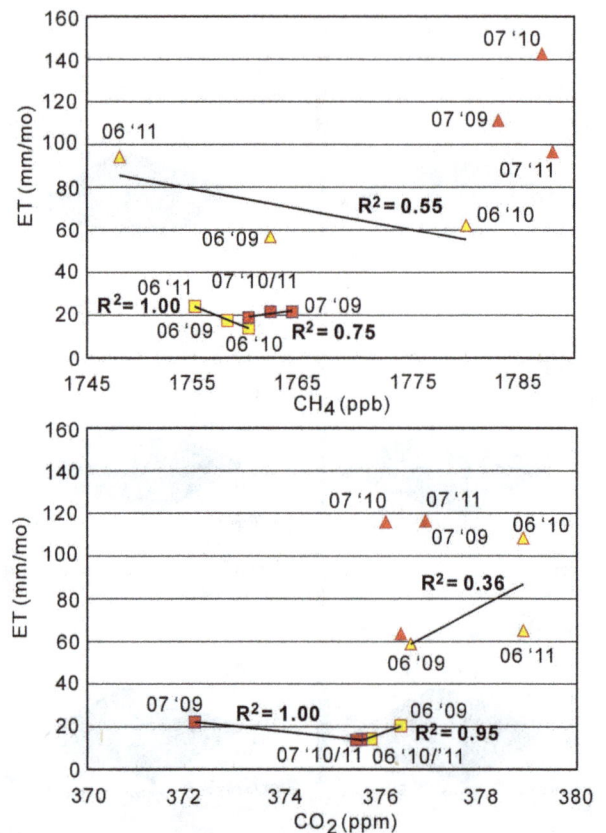

Figure 5. Regressions of dominant modes from joint probability density functions of MODIS Evapotranspiration (ET, mm/mo) with GOSAT CH_4 (ppb) and CO_2 (ppm) in regions 40° - 80°E by 50° - 58°N (Square) and 100° - 140°E by 50° - 58°N (Triangle) during June (yellow) and July (red), 2009, 2010 and 2011.

Table 2. Dominant modes and standard deviations of joint probability density functions of GOSAT CH_4 and CO_2 with MODIS Evapotranspiration (ET) during June and July 2009, 2010 and 2011 in west and east regions.

	CH_4 (ppb)	ET (mm/mo)	CO_2 (ppm)	ET (mm/mo)
		West Region		
2009 June	1758 ± 1.81	17.5 ± 2.69	376.4 ± 0.96	20.4 ± 3.65
July	1764 ± 5.31	21.6 ± 3.12	372.2 ± 0.51	22.2 ± 3.39
2010 June	1760 ± 4.41	14.0 ± 2.22	375.8 ± 0.98	14.3 ± 2.29
July	1760 ± 4.32	18.8 ± 2.96	375.6 ± 0.74	14.0 ± 1.61
2011 June	1755 ± 4.71	24.1 ± 3.76	375.6 ± 0.51	14.2 ± 0.81
July	1762 ± 4.20	21.6 ± 3.34	375.5 ± 0.44	13.8 ± 0.97
		East Region		
2009 June	1762 ± 6.07	56.7 ± 9.07	376.6 ± 1.03	58.8 ± 9.78
July	1783 ± 7.72	111.3 ± 18.18	376.4 ± 0.99	63.6 ± 9.51
2010 June	1780 ± 6.79	62.2 ± 9.79	378.9 ± 1.93	108.4 ± 17.04
July	1788 ± 6.52	96.6 ± 18.09	376.1 ± 1.03	115.9 ± 18.63
2011 June	1748 ± 5.12	94.2 ± 16.0	378.9 ± 0.66	65.1 ± 11.17
July	1787 ± 9.66	142.5 ± 21.79	376.9 ± 0.45	116.4 ± 20.28

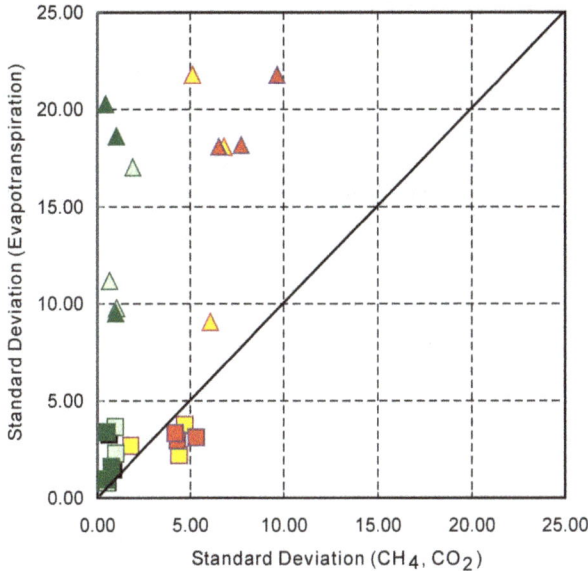

Figure 6. Standard deviation plot of dominant modes of GOSAT-derived CH$_4$, CO$_2$ to MODIS-derived Evapotranspiration in June (yellow, light green), July (red, dark green) of 2009, 2010 and 2011 in regions 40° - 80°E by 50° - 58°N (Square) and 100° - 140°E by 50° - 58°N (Triangle).

of the co-geolocated joint probability density function dominant modes. The standard deviations of the dominant modes in the west-region have restricted ranges and plot close to the 1:1 line near the origin. The east-region, while having a similar range in standard deviation CH$_4$ and CO$_2$ has a much larger range in standard deviation of ET.

This indicates that physical processes governing CH$_4$, CO$_2$ concentrations and ET rates in the west-region are different than those of the east-region. The background levels of CH$_4$ and CO$_2$ concentrations remain near constant, with some variation over June-July of 2009, 2010 and 2011. Evidence supports an inverse relationship of CH$_4$ and CO$_2$ concentrations to ET rates in the west-region.

4. Discussion

In the summer of 2010 numerous media outlets including companies and governments in Europe and North America reported on the visually spectacular fires in southwest Russia particularly those southeast of Moscow[3]. There were many reported impacts to local air quality, and problems associated with the summer "heat wave" that accompanied the fires of that year [13].

Our investigation focus is on comparison and interrogation of GOSAT and MODIS datasets at both fine scale (landscape) and course scale (regional to hemispheric). Our analysis shows correspondence of increase-regions

of CH$_4$, CO$_2$ concentrations and ET rates during June and July 2009, 2010 and 2011. These increase-regions have correspondence to active-fire (wildfire, including forest and steppe and peatlands-wetlands including palsa and thermokarst lakes) during the same months and years. Fire/wildfire counts in June and July 2009, 2010 and 2011 show variations of a factor of two. The west-region in general has higher fire/wildfire counts in July 2010. The highest count of fire/wildfire occurs in July 2011 in the east-region. CH$_4$, CO$_2$ concentrations and ET rates are significantly higher in the east-region (Siberia) relative to the west-region (agricultural, industrial and urban) and during June and July in 2010, on average. Standard deviations of CH$_4$ and CO$_2$ concentrations occupy a narrow range in both regions of interest. Standard deviations of ET rates are much broader in the east-region relative to the west-region. This and the higher values of CH$_4$, CO$_2$ and ET point to heterogeneous ecosystems, hydrology including palsa and thermokarst lake processes, discontinuous permafrost zone, terrain of the east-region relative to the west-region. The east-region is well known for its long history of lightning ignited wildfires within larch-dominated communities of the central Siberian Plateau [14].

On average, background levels of CH$_4$, CO$_2$ concentrations and ET rates are not significantly perturbed during June-July 2010 relative to 2009 and 2011. This points to the buffering capacity of the lower atmosphere and terrestrial ecosystems [15-17]. Therefore, evidence does not at this time support the test hypothesis. In conclusion the fires and wildfires of summer 2010 did not significantly affect regional or hemispheric and multi-year background levels of CH$_4$, CO$_2$ concentrations and ET rates.

5. Conclusion

We investigate the correspondence of GOSAT CH$_4$ and CO$_2$ near-surface concentrations and standard errors to MODIS ET rates and fire/wildfire hotspots on the northern hemisphere during June-July 2009, 2010 and 2011. Our month-year time frame is constrained by the sun-angle and cloud-free requirements of the Fourier Transform Spectrometer, spectral interferometer TANSO-FTS sensor onboard GOSAT. After processing our datasets to 5 km resolution we focus on two regions in Eurasia for analysis: west-region 40° to 80°E by 50° to 58°N and east-region 100° to 140°E by 50° to 58°N. Joint probability density functions identify significant dominant modes of CH$_4$ and CO$_2$ concentrations to ET rates. Evidence indicates that background levels of CH$_4$ and CO$_2$ are specific to each region and that these levels are near constant across the month-by-year observations. The west-region (agricultural, industrial and urban) shows strongly

---

[3]http://www.jpl.nasa.gov/news/news.php?release=2010-261.

correlated inverse relationships of CH_4 and CO_2 concentrations to ET rates. CH_4, CO_2 concentrations and ET rates in the east-region have significantly greater levels than those in the west-region. In the east-region (Siberia) evidence shows no relationships (forward or inverse) of CH_4 and CO_2 concentrations to ET rates. This is likely an expression of the heterogeneity of ecosystems, hydrology including numerous palsa and thermokarst lakes and varied physical processes and terrain of the discontinuous permafrost zone of the central Siberian Plateau.

6. Acknowledgements

R. R. Muskett thanks Hiroshi Watanabe, Center for Global Environmental Research and National Institute for Environmental Studies and the GOSAT Project for encouragement. R. R. Muskett thanks Arctic Region Supercomputing Center for computational facilities assistance. R. R. Muskett was supported through grants to Vladimir E. Romanovsky Geophysical Institute University of Alaska Fairbanks from the USGS Alaska Climate Science Center in coordination with the Arctic and Western Alaska Landscape Conservation Cooperative and the Scenarios Network for Alaska and Arctic Planning and the National Science Foundation Arctic Observing Network. Datasets used in this research are available through the Geophysical Institute Permafrost Laboratory website and server[4] with anonymous ftp available on request. The Generic Mapping Tools and MATLAB are used in computational processing of this research.

REFERENCES

[1]   A. Butz, S. Guerlet, O. Hasekamp, D. Schepers, A. Galli, I. Aben, C. Frankenberg, J.-M. Hartmann, H. Tran, A. Kuze, G. Keppel-Aleks, G. Toon, D. Wunch, P. Wennberg, N. Deutscher, D. Griffith, R. Macatangay, J. Messerschmidt, J. Notholt and T. Warneke, "Toward Accurate CO_2 and CH_4 Observations from GOSAT," *Geophysical Research Letters*, Vol. 38, No. 14, 2011, Article ID L14812.

[2]   S. Arrhenius, "On the Influence of Carbonic Acid in the Air upon the Temperature of the Ground," *Philosophical Magazine and the Journal of Science*, Vol. 41, No. 251, 1896, pp. 237-276.

[3]   M. I. Hoffert and C. Covey, "Deriving Global Climate Sensitivity from Palaeoclimate Reconstructions," *Nature*, Vol. 360, 1992, pp. 573-576.

[4]   J. T. Houghton, Y. Ding, D. J. Griggs, M. Noguer, P. J. van der Linden, X. Dai, K. Maskell and C. A. Johnson, "Climate Change 2001: Scientific Basis," Cambridge University Press, New York, 2001.

[5]   G. Kramm and R. Dlugi, "Scrutinizing the Atmospheric Greenhouse Effect and Its Climatic Impact," *Natural Sci-*

ence, Vol. 3, No. 12, 2011, pp. 971-998.

[6]   P. Wessel, "A General-Purpose Green's Function-Based Interpolator," *Computers & Geoscience*, Vol. 35, No. 6, 2009, pp. 1247-1254.

[7]   R. R. Muskett, "MODIS-Derived Arctic Land-Surface Temperature Trends," *Atmospheric and Climate Sciences*, Vol. 3, No. 1, 2013, pp. 55-60.

[8]   C. O. Justice, L. Giglio, D. Roy, L. Boschetti, I. Csiszar, D. Davies, S. Korontzi, W. Schroeder, K. O'Neal and J. Morisette, "MODIS-Derived Global Fire Products," In: B. Ramachandran, C. O. Justice and M. J. Abrams, Eds., *Land Remote Sensing and Global Environmental Change*, Springer, New York, Vol. 11, 2011, pp. 661-679.

[9]   C. Justice, L. Giglio, L. Boschetti, D. Roy, I. Csiszar, J. Morisette and Y. Kaufman, "MODIS Fire Products: Algorithm Technical Background Document," Ver. 2.3, University of Maryland, 2006. http://modis-fire.umd.edu/AF_usermanual.html

[10]   I. A. Csiszar, J. T. Morisette and L. Giglio, "Validation of Active Fire Detection from Moderate-Resolution Satellite Sensors: The MODIS Example in Northern Eurasia," *IEEE Transactions on Geoscience and Remote Sensing*, Vol. 44, No. 7, 2006, pp. 1757-1764.

[11]   S. Wilfrid, E. Prins, L. Giglio, I. Csiszar, C. Schmidt, J. Morisette and D. Morton, "Validation of GOES and MODIS Active Fire Detection Products Using ASTER and ETM+ Data," *Remote Sensing of Environment*, Vol. 112, No. 5, 2008, pp. 2711-2726.

[12]   Q. Mu, M. Zhao and S. W. Running, "Improvements to a MODIS Global Terrestrial Evapotranspiration Algorithm," *Remote Sensing of Environment*, Vol. 115, No. 8, 2011, pp. 1781-1800.

[13]   I. B. Konovalov, M. Beekmann, I. N. Kuznetsova, A. Yurova and A. M. Zvyagintsev, "Atmospheric Impacts of the 2010 Russian Wildfires: Integrating Modelling and Measurements of an Extreme Air Pollution Episode in the Moscow Region," *Atmospheric Chemistry and Physics*, Vol. 11, No. 19, 2011, pp. 10031-10056.

[14]   V. I. Kharuk, K. J. Ranson, M. L. Dvinskaya and S. T. Im, "Wildfires in Northern Siberian Larch Dominated Communities," *Environmental Research Letters*, Vol. 6, 2011, Article ID: 045208.

[15]   O.-Y. Kwon and J. L. Schnoor, "Simple Global Carbon Model: The Atmosphere-Terrestrial Biosphere-Ocean Interaction," *Global Biogeochemical Cycles*, Vol. 8, No. 3, 1994, pp. 295-305.

[16]   M. Pagani, K. Caldeira, R. Berner and D. Beerling, "The Role of Terrestrial Plants in Limiting CO_2 Decline for 24 Million Years," *Nature*, Vol. 460, 2009, pp. 85-88.

[17]   D. Taraborrelli, M. G. Lawrence1, J. N. Crowley, T. J. Dillon, S. Gromov, C. B. M. Groß, L. Vereecken and J. Lelieveld, "Hydroxyl Radical Buffered by Isoprene Oxidation over Tropical Forests," *Nature Geoscience*, Vol. 5, 2012, pp. 190-193.

---

[4]http://permafrost.gi.alaska.edu and www.permafrostwatch.org.

21

Characterizing PM$_{2.5}$ Pollution of a Subtropical Metropolitan Area in China

Guojin Sun[1,3], Lin Yao[2], Li Jiao[2], Yao Shi[1], Qingyu Zhang[1], Mengna Tao[1], Guorong Shan[1], Yi He[1*]

[1]Department of Chemical and Biological Engineering, Zhejiang University, Hangzhou, China
[2]Environmental Monitoring Center, Hangzhou, China
[3]Zhejiang Environmental Protection Agency, Hangzhou, China

ABSTRACT

The chemical and physical characteristics of PM$_{2.5}$, especially their temporal and geographical variations, have been explored in metropolitan Hangzhou area (China) by a field campaign from September 2010 to July 2011. Annual average concentrations of PM$_{2.5}$ and PM$_{10}$ during non-raining days were 106 - 131 µg·m^{-3} and 127 - 158 µg·m^{-3}, respectively, at three stations in urban breathing zones, while corresponding concentrations of PM$_{2.5}$ and PM$_{10}$ at an urban background station (16 m above ground level in a park) were 78 and 104 µg·m^{-3}, respectively. For comparison, the annual average PM$_{10}$ concentration at a suburban station (5 m AGL) was 93 µg·m^{-3}. Detailed chemical analyses were also conducted for all samples collected during the campaign. We found that toxic metals (Cd, As, Pb, Zn, Mo, Cu, Hg) were highly enriched in the breathing zones due to anthropogenic activities, while soluble ions (SO_4^{2-}, NO_3^-, NH_4^+) and total carbon accounted for majority of PM$_{2.5}$ mass. Unlike most areas in China where sulfate was several times of nitrate in fine PM, nitrate was as important as sulfate and highly correlated with ammonium during the campaign. Thus, a historical shift from sulfate-dominant fine PM to nitrate-dominant fine PM was documented.

Keywords: PM Pollution; PM$_{2.5}$ Composition; Breathing Zone; Air Pollution Measurements; Personal Exposure

1. Introduction

PM$_{2.5}$ is an important air pollutant. Due to its relatively long suspension time in the air and special optical properties, PM$_{2.5}$ is responsible for the formation of regional haze [1-3]. Depending on its position in the air, PM$_{2.5}$ may be responsible for warming or cooling of surface [4]. When inhaled, PM$_{2.5}$ may cause severe health problems, such as asthma, lung cancer, and heart disease [5-10], though the exact component of PM$_{2.5}$ responsible for certain adverse health effect is uncertain [8,9].

To protect the public from PM$_{2.5}$ pollution, significant efforts have been conducted worldwide [11-15]. For example, US EPA has started to regulate ambient PM$_{2.5}$ pollution since 1998, and recently tightened its ambient air quality standards for 24-hour and annual averages to 35 and 15 µg·m^{-3} [16], respectively. As PM$_{2.5}$ pollution started to draw public attention in China, China Ministry of Environmental Protection recently released its first ambient air quality standard for PM$_{2.5}$, to be effective in 2016; the standard sets the same PM$_{2.5}$ limits for recreational areas as US EPA, but allows non-recreational areas to have higher limits, namely, 70 and 35 µg·m^{-3} for 24-h

and annual averages, respectively.

While intensive field campaigns have been carried out in China and elsewhere [17-20], the temporal and geographical variations of PM$_{2.5}$ in urban breathing zones (0.5 - 2 m above ground) in China remain poorly documented. To obtain such detailed information about PM$_{2.5}$ pollution, an extensive field campaign was conducted in metropolitan Hangzhou area (HZ), China, during September 2010-July 2011. HZ is experiencing both fast urbanization and growing environmental issues: as of 2010, the population of HZ ranked sixth in China, and the population density was 1214/km^2. The goals of this campaign were: 1) to determine the pollution level of PM$_{2.5}$ in the breathing zone of urban air; 2) to investigate chemical and physical characteristics of PM$_{2.5}$ pollution; 3) to identify areas for further research before next field campaign. Section 2 will describe experimental methods of the campaign, and Section 3 will present major findings. A summary will be presented in Section 4.

2. Experimental Methods

The sampling, analysis, and quality assurance procedures are described below.

---

*Corresponding author.

2.1. Sampling

2.1.1. Sites

HZ is situated in a subtropical area with distinct seasonal weather conditions, and is 50-km away from the western rim of the Pacific Ocean. As HZ is a traditional hotspot for national and international tourists, there are two automatic monitoring stations ("A" and "B", **Table 1**) running since 2005 for $PM_{2.5}$ as well as other air pollutants, independent of this campaign: one ("A") represents suburban recreational conditions, and the other ("B") represents urban mixed conditions. Sampling inlets were at 5 m ("A") and 16 m ("B") above ground level (AGL) at the automatic stations. For this campaign, four additional stations were chosen to represent breathing zone conditions in various urban areas, namely, nearby highway ("C"), a college gate ("D"), a business area with peak hours ("E"), and a business area without peak hours but with dense population ("F"). At these special stations, sampling inlets were at 1 - 1.5 m AGL. **Figure 1** shows the sampling sites for this campaign.

2.1.2. Duration

Sampling was conducted at all stations from September 2010 to July 2011. During the sampling period, seasonal average temperatures in the area were 21°C for fall (September-November 2010), 6.9°C for winter (December 2010-February 2011), 16°C for spring (March-May 2011), and 29°C for summer (June-August 2011). As it was raining frequently in the sampling region, $PM_{2.5}$ samples were only collected at sites "A"-"E" concurrently during

non-raining days. On each sampling day, samples were collected for 18 - 24 hours; in each season, 7 - 10 days were selected for concurrent sampling. In addition, $PM_{2.5}$ samples were also collected at sites "B" and "F" during January-June 2008.

2.1.3. Samplers

At sites "A", "B", "F", air pollutants were continuously measured for $PM_{2.5}$, O_3, CO, NO_2, CH_4, and NMHC, using automatic monitoring equipment (Air Point, Austria; DURAG, Germany; SYNSPEC, Holland). At sites specially selected for this campaign, $PM_{2.5}$ samplers were the model TH-150C, manufactured by Wuhan Tianhong Instruments Co., Ltd. The operational volume flow rate for $PM_{2.5}$ samplers was set at 100 L·min^{-1}, and the size cut was at 2.5 μm. At each site, three samplers were used to collect three concurrent $PM_{2.5}$ samples for analyzing inorganic elements, soluble ions, and carbon items, respectively. For inorganic elemental analysis, PTEF organic filters (Sumitomo Electric Fine Polymer, Inc.) were used. For organic analyses, quartz filters (Whatman 1851047) were used. The pore size was 0.3 μm for both types of filters, and the collection efficiency at 0.15 μm was 99.97% for organic filters and 99% for quartz filters.

2.2. Analyses

The bulk masses of $PM_{2.5}$ samples were weighed with a balance (0.0001 accuracy), and the speciation was conducted for 25 inorganic elements, 9 ions, and 3 carbon items. Details are elaborated below.

Table 1. PM concentration in a subtropical metropolitan area (HZ, CHINA).

Monitoring sites	Period	Mass concentration(μg/m³)		$PM_{2.5}/PM_{10}$
		PM_{10}	$PM_{2.5}$	
C: nearby highway (1 - 1.5 m AGL)	Winter	185.2	164.4	0.888
	Spring	157.3	138.3	0.879
	Summer	111.6	97.6	0.875
	Autumn	177.6	122.0	0.687
	All Seasons	157.9	130.6	0.827
D: college gate (1 - 1.5 m AGL)	Winter	149.9	119.6	0.798
	Spring	94.7	86.6	0.914
	Summer	87.3	77.8	0.891
	Autumn	191.2	140.3	0.734
	All Seasons	130.8	106.1	0.811
E: business area with peak hours (1 - 1.5 m AGL)	Winter	121.0	100.2	0.828
	Spring	122.4	107.2	0.876
	Summer	113.2	87.7	0.775
	Autumn	149.4	134.9	0.903
	All Seasons	126.5	107.5	0.850
B: urban mixed conditions (16 m AGL)	All Seasons	104	77.5	0.745
A: suburban recreational conditions (5 m AGL)	All Seasons	93	/	/

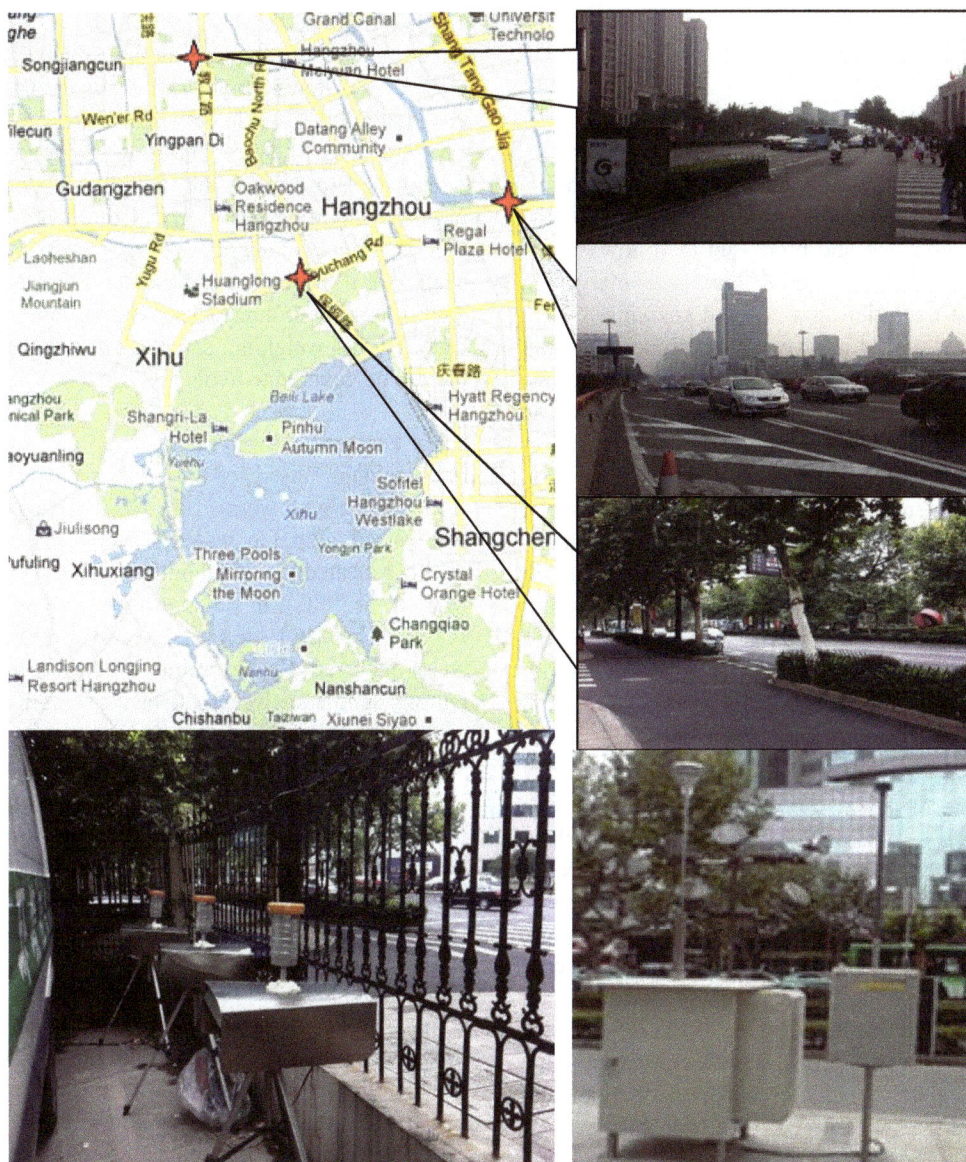

Figure 1. Monitoring stations and sampling setup. The distributions of the sampling locations are marked with red stars in the Google map.

2.2.1. Inorganic Elements

After sampling, organic filters loaded with PM$_{2.5}$ were carefully cut along the inner rim and put into a 50 ml microwave digestion tank. 6 ml HNO$_3$ and 2 ml H$_2$O$_2$ were then added into the tank to dissolve PM$_{2.5}$ at pre-designed procedures. Then, the bulk solution was moved into a 50 mL flask. High-purity water was used to fix the volume and for mixing well. Finally, the solution was used for instrumental analyses.

Instrumental analyses were conducted by Inductively Coupled Plasma (ICP, Optima 7300DV) emission spec-trometers for elements (Li, V, Cr, Co, Ni, Cd), by Atomic Fluorescent Spectrometers (AFS-9230) for ele-ments (Hg, As, Se), and by ICP-Mass Spectrometers (MS, X Series 2) for elements (Al, Ba, Ca, Cu, Fe, K, Mg, Mn,

Mo, Na, P, Pb, S, Si, Ti, Zn).

For the AFS analysis, sample solutions were first heated in acidic media to convert all mercury into Hg^{2+}, and Hg^{2+} was then reduced by KBH$_4$ to form Hg vapor that was drawn into AFS for detection. As and Se were detected similarly.

2.2.2. Soluble Ions

After sampling, a piece of organic filters loaded with PM$_{2.5}$ were put into a 25-ml colorimetric tube. Then, 20.00-ml deionized water was added to and bubbles ex-pelled from the tube. The tube was placed in a supersonic cleaner running for 20 minutes. After stilled for a while, clear solution in top layer was drawn and filtered before being analyzed by Ion Chromatography for water-soluble anions (F^-, Cl^-, NO_3^-, SO_4^{2-}; Dionex IC DX600) and cations (NH_4^+, K^+, Na^+, Ca^{2+}, Mg^{2+}; Dionex ICS-90).

2.2.3. Carbon Analysis

After sampling, quartz filters loaded with $PM_{2.5}$ were put on new and clean aluminum foil, and a representative piece was obtained carefully. A second piece with the same area was obtained and processed as follows to remove carbonate. First, the second piece was merged in concentrated HCl solution in a covered container. Then, the container was placed in a ventilator cabinet for about an hour to remove carbonate in the form of CO_2 and for another hour to let remaining HCl evaporate. The sample was then kept at low temperature and saved for subsequent analyses.

The organic, elemental, and total carbon contents of $PM_{2.5}$ samples were determined with a thermo-photometric carbon analyzer (DRI Model 2001A). A piece of processed sample filter (0.495 cm^2) was put in an environment with pure He gas without O_2, and was heated progressively at 120°C, 250°C, 450°C, and 550°C first to determine organic carbon contents OC1, OC2, OC3, and OC4, respectively. Then, in an environment with 2% O_2 and 98% He, the sample was further heated progressively at 550°C, 700°C, and 800°C to determine elemental carbon contents EC1, EC2, and EC3. The CO_2 produced within each temperature ladder was reduced into CH_4, catalyzed by MnO_2, for being detected with Fire Ionization Detectors. During heating processes, part of organic carbon was converted into black carbon, which hindered clear distinction between organic carbon and elemental carbon. Hence, the reflection intensity of the He-Ne laser light at 633 nm by a monitoring filter was used to gauge the starting temperature of the oxidation of elemental carbon, to ensure science-based distinction between organic carbon and elemental carbon.

2.3. QA/QC Procedures

2.3.1. Sampling

All sampling instruments used for this study were calibrated by the Technology Supervision Bureau of Zhejiang Province within valid periods. Before each sampling, pump volumes of $PM_{2.5}$ samplers were calibrated, and so were sampling volumes of other instruments. During filter sampling, a specialist was responsible for checking conditions of volume-flow meters of sampling pumps, and the project manager randomly checked conditions of flow meters at a frequency. Records were made to ensure the precision of sampling volumes and subsequent calculations of pollutant concentrations. No sample was collected in raining weather, and filter samples were ensured to conserve particulate matter during transportation.

2.3.2. Weighing

Before and after sampling, organic filters were baked at 60°C ± 2°C for 8 hours and blank quartz filters were baked at 500°C for 4 hours. Then, filters were put in dryers for at least 48 hours before weighing. This procedure is expected to remove the disturbance of water, and to provide weighting results comparable with other studies though some compounds such as ammonium nitrate may evaporate if original samples were collected at relatively humid condition, e.g., when relative humidity was over 80%. During weighing, samples were put into glass agar plates and covered to avoid the effect of static electricity.

2.3.3. Elemental Analyses

Each container used for this study was cleaned first, and then washed with warm, 10% HNO_3. Then, the container was rinsed with tap water. Finally, the container was rinsed repeatedly with deionized water (Mill-Q, >18.2 MΩ·cm at 25°C), in order to reduce blank background. For each batch of samples, at least three blank samples were analyzed to determine blank background value. During sample analyses, a standard sample was added for every 10 - 15 samples, to check instrumental stability. When an element showed abnormally high content, analysis was stopped and the inlet system was rinsed with 2% HNO_3 Then, standard solutions were used to calibrate instruments, and samples were diluted before subsequent analyses.

2.3.4. Ionic Analyses

Glass containers were cleaned, and soaked in deionized water for over 24 hours. Then, containers were cleaned with supersonic waves for 30 minutes. Blank background was determined in the same way as for elemental analyses, and so was the use of standard samples.

2.3.5. Carbon Analysis

At the beginning and the end of each day when samples were analyzed, CH_4/CO_2 standard gases were used to calibrate instruments. For every 10 samples, one was randomly chosen for parallel analysis. Two standard samples were analyzed every two weeks. Compared with DRI (USA) instruments which have the same model or use different methods [21-28], the detection precision was <5% for TC and <10% for OC and EC in this study.

3. Results and Discussions

PM mass concentrations and $PM_{2.5}$ compositions are presented below.

3.1. Mass Concentration

Table 1 lists average PM concentrations in the metropolitan breathing zones, at an urban background station, and at a suburban background station during this campaign. It is shown that the average PM$_{2.5}$ concentration in the breathing zone ranged from 106 to 131 µg·m^{-3} annually and from 78 to 164 µg·m^{-3} seasonally, with peak values nearby highway in winter and in business areas in fall. PM$_{2.5}$ accounted for 81% - 85% of PM$_{10}$ annually and 69% - 91% of PM$_{10}$ seasonally. Concentrations of PM$_{2.5}$ and PM$_{10}$ at the metropolitan background station (16 m AGL in a park) were lower than in the breathing zones by (27 - 41)% and (18 - 34)%, respectively. The concentration of PM with diameters between 2.5 - 10 µm was ~25 µg·m^{-3} in urban stations during the campaign.

3.2. Elemental Composition

Table 2 lists concentrations of 25 elements detected in PM$_{2.5}$ samples collected in the breathing zones during the campaign. A number of metals, including toxic heavy metals (Hg, Cd, Cr, Pb, As), were detected, and Fe showed the highest concentration of 2.1 µg·m^{-3} on seasonal average basis. Sulfur concentration ranged 3.0 - 9.7 µg·m^{-3} seasonally, corresponding to 9 - 29 µg·sulfate·m^{-3} if all sulfur element existed as sulfate in PM$_{2.5}$. Total concentration of non-sulfur elements ranged 2.6 - 11 µg·m^{-3} seasonally. Together, elemental concentrations accounted for 7.5 - 18 µg·m^{-3}, or 9% - 23%, of PM$_{2.5}$ concentration seasonally while oxygen, nitrogen, and carbon elements were excluded.

Table 3 lists average enrichment factors (EF) of 25 elements during the campaign. It is shown that the EF value exceeded 1000 for (Cd, Se, S, Mo, Zn), and exceeded 100 for (Pb, Cu, As) in this campaign, while an EF value larger than 10 suggests that the element is an-

thropogenic [29]. Compared with 5 years ago, the EF value decreased for (S, Mo, Zn) and especially for (Se, As), which indicate the decrease in coal-related emissions; meanwhile, the EF value increased for (Pb, Cu, Mn, Fe, V), which suggests the increase in vehicle emissions.

3.3. Ionic Composition

Nine detected soluble ions (F$^-$, Cl$^-$, NO$_3^-$, SO$_4^{2-}$; NH$_4^+$, K$^+$, Na$^+$, Ca^{2+}, Mg^{2+}) contributed an average of 42 µg·m^{-3} to PM$_{2.5}$ concentration measured in breathing zones during this study, and three major ions (SO$_4^{2-}$, NO$_3^-$, NH$_4^+$) accounted for 85% of total ionic mass. On seasonal average basis, concentrations of soluble ions ranged 10 - 86 µg·m^{-3} while the three major ions accounted for 9 - 74 µg·m^{-3} with maximum occurred in winter and minimum in summer. Sulfate concentration ranged from 5.6 to 21 µg·m^{-3} seasonally.

Figure 2 shows seasonal variations of ionic concentrations in the breathing zones during the campaign. All ions showed the lowest concentration in summer, and most ions peaked in winter. Ions (SO$_4^{2-}$, K$^+$, F$^-$, Mg^{2+}) showed high values in fall and winter, and Ca^{2+} peaked in spring. As major ions are secondary in origin and their formation time from corresponding precursors is long enough for regional transport to take place, one could argue that seasonal trends of ionic concentrations might reflect the fact that the study area was cleaner than surrounding regions in the nation, though local emissions and the absolute pollution level are also rather significant. This hypothesis was, however, not supported by further analyses using concurrently measured concentrations of

Table 2. Concentrations of 25 elements (µg/m³) detected in PM$_{2.5}$ samples.

Element	Winter	Spring	Summer	Autumn	Year	Element	Winter	Spring	Summer	Autumn	Year
Al	0.596	1.175	0.293	0.875	0.735	Mn	0.084	0.075	0.039	0.097	0.074
As	0.022	0.012	0.015	0.023	0.018	Mo	0.0028	0.0120	0.0064	0.0028	0.0060
Ba	0.058	0.058	0.036	0.047	0.050	Na	0.504	0.731	0.545	0.282	0.491
Ca	1.934	1.979	1.349	1.530	1.698	Ni	0.0051	0.0269	0.0055	0.0047	0.0105
Cd	0.006	0.0052	0.0032	0.0106	0.0061	P	0.018	0.022	0.018	0.043	0.025
Co	0.0006	0.0008	0.0003	0.0007	0.0006	Pb	0.186	0.127	0.109	0.194	0.154
Cr	0.0064	0.0086	0.0057	0.0053	0.0065	S	5.725	4.465	5.068	5.299	5.139
Cu	0.102	0.069	0.049	0.088	0.077	Se	0.0080	0.0047	0.0035	0.0099	0.0065
Fe	1.303	1.304	0.634	1.525	1.191	Si	0.610	0.5212	0.0000	1.270	0.6002
Hg	0.0001	0.0001	/	0.0003	0.0001	Ti	0.0216	0.0232	0.0054	0.022	0.0179
K	1.571	1.137	0.293	1.972	1.243	V	0.0060	0.0077	0.0054	0.0063	0.0063
Li	0.0017	0.0015	0.0010	0.0015	0.0015	Zn	0.495	0.471	0.343	0.689	0.499
Mg	0.255	0.184	0.096	0.202	0.184	Sum/					
Sum	13.5	12.4	8.9	14.2	12.2	PM$_{2.5}$(%)	10.5	11.2	10.1	10.7	10.6

Table 3. Enrichment factors of 24 elements in PM$_{2.5}$ samples.

Element	Sites				Average of Hangzhou in 2006
	Highway	College Gate	Business Area with Peak Hours	Average	
As	361	683	543	529	1114
Ba	21.3	18.3	15.3	18.3	/
Ca	19.1	26.1	38.2	27.8	18.43
Cd	4746	7874	7959	6860	/
Co	8.6	12.8	8.7	10.0	/
Cr	17.3	28.7	16.8	20.9	42.84
Cu	441	710	611	588	574
Fe	5.5	6.6	5.5	5.9	3.15
Hg	92.3	221	182	165	/
K	7.8	13.8	9.7	10.4	9.13
Li	6.8	12.4	8.3	9.2	/
Mg	2.4	2.7	3.4	2.8	6.96
Mn	23.0	37.9	24.4	28.4	21.62
Mo	1960	2240	1000	1733	/
Na	2.7	5.1	4.2	4.0	6.66
Ni	104	55.1	38.6	65.8	79.65
P	4.9	5.2	4.2	4.7	21.51
Pb	693	1350	832	958	797
S	2794	3850	2710	3118	3332
Se	1994	5675	3545	3738	14496
Si	0.1	0.1	0.2	0.1	0.8
Ti	0.6	0.8	0.8	0.8	13.88
V	11.2	20.9	16.1	16.1	10.88
Zn	742	1608	944	1098	1242

(sulfate, SO$_2$) and (nitrate, NO$_2$) in this campaign. On molar basis, sulfate accounted for 10% - 30% of total sulfur in breathing zones, and nitrate accounted for 1% - 20% of total oxidized nitrogen during this study. Thus, seasonal variations of ionic concentrations may also result from changes in relative humidity near sampling sites during this campaign.

Also shown in **Figure 2** is the ratio of cation/anion in PM$_{2.5}$ samples. The ratio ranged 1.0 - 1.6 with the average of 1.2 in breathing zones, and was 1.1 at a routine automatic monitoring station. Thus, organic anions might be present significantly during the campaign.

As sulfate concentration was measured concurrently with the concentration of elemental sulfur in this study, a comparison between the two concentrations may offer unique insights into the existence of organic sulfur compounds. During this study, the average concentration of S was 5.14 $\mu g \cdot m^{-3}$, and that of sulfate 14.5 $\mu g \cdot m^{-3}$; thus, it is likely that organic sulfur concentration was minor ($\sim$0.3 $\mu g \cdot S \cdot m^{-3}$) compared with inorganic sulfur.

3.4. Carbon Distribution

Concentrations of total carbon (TC), organic carbon (OC), and elemental carbon (EC) in PM$_{2.5}$ were measured to be 32, 21, and 11 $\mu g \cdot m^{-3}$, respectively, during the campaign. Total carbon accounted for 28% of PM$_{2.5}$ concentration in the breathing zones.

Figure 3 shows seasonal concentrations of OC and EC in PM$_{2.5}$ and corresponding ratio of OC/EC at various stations. It is shown that seasonal TC and OC concentrations peaked in fall and showed minimal values in summer, while EC showed much smaller seasonality; thus, OC drove the seasonal trend of TC in the campaign. The seasonal ratio of OC/EC showed minimum value in summer, and peaked in fall in breathing zones. At the suburban background station (5 m AGL), the ratio of OC/EC peaked in summer while the TC concentration was the lowest.

The correlation between OC and EC concentrations is often used to determine whether they come from the

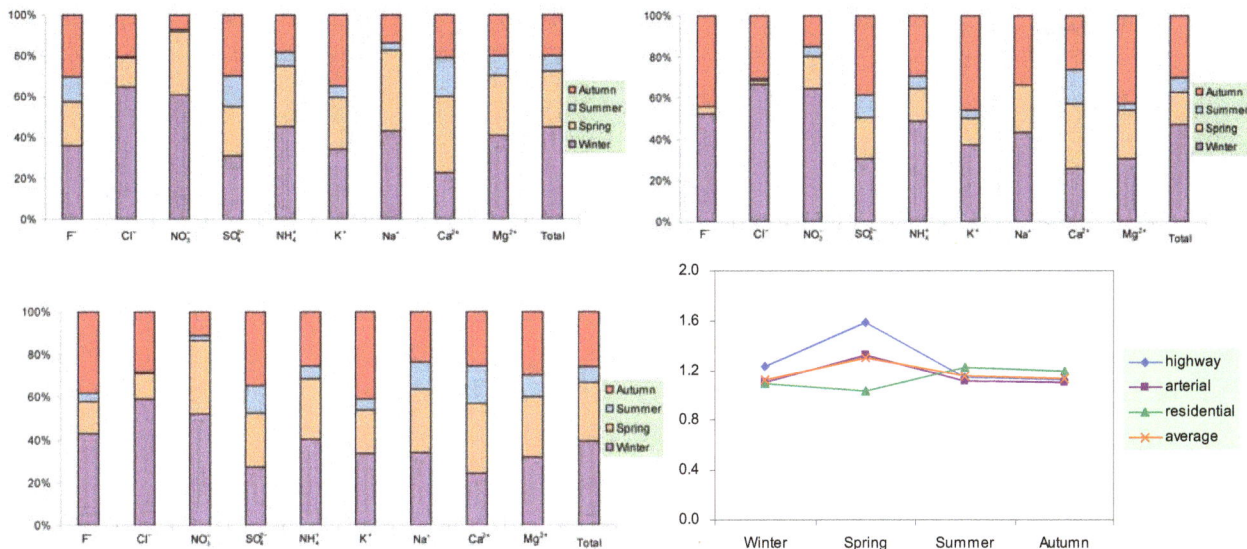

Figure 2. Seasonal variations of soluble ions in PM$_{2.5}$ samples. Top left: highway; top right: arterial; bottom left: residential; bottom right: C/A ratios.

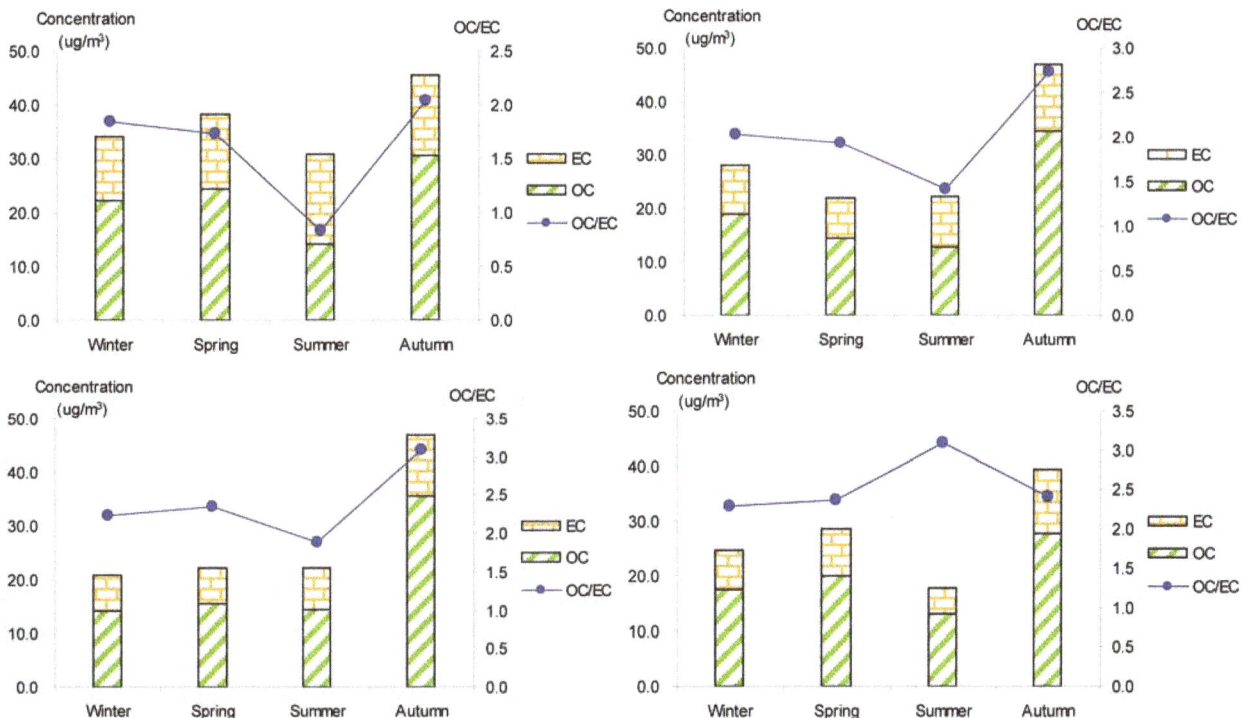

Figure 3. Seasonal average concentrations of OC and EC and their ratios in PM$_{2.5}$ samples. Top left: arterial; top right: highway; bottom left-residential road, bottom right: residential areas.

same origin [30-32], as OC may be either produced by gas to particle conversion processes or directly emitted, while EC is usually directly emitted from fossil fuel combustion and biomass burning [33]. **Figure 4** shows seasonal scatterplots of OC versus EC measurements in breathing zones during the campaign. It is shown that OC and EC concentrations were highly correlated in winter ($R^2 = 0.85$, slope = 2.0) and spring ($R^2 = 0.78$, slope =

1.7). In summer and especially in fall, the OC concentration was poorly correlated with EC. In summer, data points split into two groups: one with a slope about 2, and the other with a slope about 0.5. In fall, data points may be bracketed by two lines with the ratio of OC/EC equals 2 and 4, respectively. The higher ratio of OC/EC was likely due to photochemical aging, and the lower ratio of OC/EC was likely due to fresh emissions or due

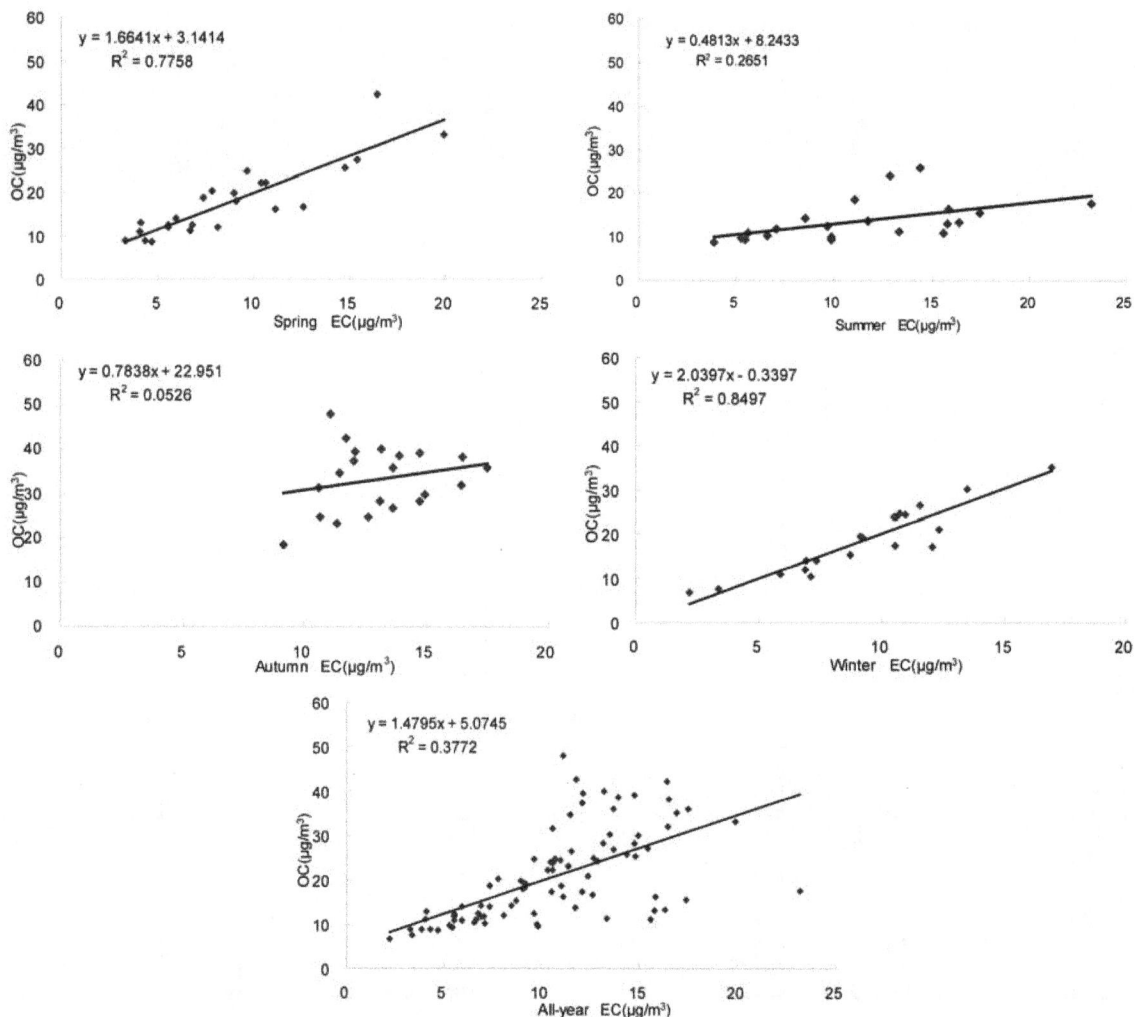

Figure 4. Scatter-plots of PM$_{2.5}$ OC and EC concentrations in four seasons.

to frequent rainfall that might remove more OC than EC from the air besides other factors.

Secondary organic aerosols are important for many scientific reasons, and consist of many compounds which are only partly identified [34]. Yu *et al.* [35] for the first time estimated the spatial distribution of secondary organic carbon (SOC) over USA in summer, 1999. During the study period, major sources of primary OC were agricultural burning, soil dust, paved road dust and non-road diesel, while major sources of EC were non-road diesel, on-road heavy-duty diesel vehicle, agricultural burning and jet fuel combustion in USA. They showed that, on seasonal scale, modeled ratio of primary (OC/EC) changed significantly at specific locations, with values ranging from 0.8 to 3.5 at stations due to various combinations of emission sources. Meanwhile, the contribution of SOC to total OC ranged 50% - 80% based on combined analyses of observational and modeled data. In a subsequent study employing US EPA CMAQ, simulated seasonal average ratio of primary (OC/EC) over various

regions of USA ranged from 2.0 to 3.6, and estimated that SOC contributed less to total OC in 2001 compared with that in 1999 [36,37].

While it is elegant to estimate SOC from modeled ratio of primary (OC/EC) and measured EC, using measured EC as a tracer for primary OC may provide a preliminary estimate for measured SOC, with all possible uncertainties embedded in field data [35]. Using empirical formula (F1) below, SOC may be estimated from OC, EC, and the minimum ratio of OC/EC during a period.

$$SOC = OC - EC \times (OC/EC)_{min} \qquad (F1)$$

On seasonal basis, $(OC/EC)_{min}$ ranged 0.71 - 2.3 in this campaign, and SOC was estimated to range from 1.2 to 9.6 $\mu g \cdot m^{-3}$ in breathing zones using formula (F1).

4. Conclusions

In this work, we observed significant temporal and geographical variations in PM pollution over a subtropical metropolitan area through a multi-institutional field campaign in Hangzhou, China. During the campaign, we captured serious PM$_{2.5}$ pollutions in the breathing zone of the scenic city in all seasons. A number of toxic metals were highly enriched in the breathing zone due to anthropogenic activities. The annual average concentration of PM$_{2.5}$ ranged from 106 to 131 µg·m^{-3} in the metropolitan breathing zones, which were 40% - 70% higher than at typical ambient stations. Speciation of PM$_{2.5}$ was carried out for 25 elements using instrumental analyses by ICP, ICP-MS, and AFS, for 9 soluble ions using IC, and for (TC, OC, EC) using a thermo-photometric method.

A number of toxic metals were highly enriched in the breathing zone due to anthropogenic activities. Elemental analyses revealed significant presences of toxic heavy metals (Hg, Cd, Cr, Pb, As) in PM$_{2.5}$ and that, among 25 detected elements, (S, Ca, K, Fe) showed concentrations larger than 1.0 µg·m^{-3}. Enrichment factor analysis revealed that (Cd, Se, S, Mo, Zn, Pb, Cu, As) in PM$_{2.5}$ were dominated by anthropogenic sources. Nine soluble ions detected during the campaign contributed 42 µg·m^{-3} to PM$_{2.5}$ concentration on average, and three secondary inorganic ions (SO$_4^{2-}$, NO$_3^-$, NH$_4^+$) accounted for 85% of total ionic concentration. Nitrate concentration was close to and occasionally exceeded sulfate concentration during the campaign, which reflects rising vehicle activities. Molar ratios of sulfate/(total sulfur) and nitrate/(total oxidized nitrogen) frequently exceeded 10%, which suggests significant effects of photochemical reactions on PM$_{2.5}$ pollution in the breathing zones.

The average total carbon concentration was 32 µg·m^{-3}, and the average ratio of OC/EC was 2. During the campaign, seasonal variation of TC was dominated by OC, and both concentrations peaked in fall and were low in summer. The EC concentration showed little seasonality, and OC was well correlated with EC in winter and spring. Secondary organic carbon was estimated to contribute 1.2 - 9.6 µg·m^{-3} to PM$_{2.5}$ concentration in the breathing zones. Comparison of concurrent measurements of sulfur element and sulfate suggests that the average concentration of organic sulfur (0.3 µg·S·m^{-3}) in PM$_{2.5}$ samples was relatively small compared with inorganic sulfur during the campaign.

5. Declaration

This work has been reviewed by staff at Zhejiang Environmental Protection Agency, China, and has been approved for publication. Approval does not signify the endorsement of commercial products mentioned in this article, and opinions in this article are solely those of authors.

6. Acknowledgements

We thank Dr. Shaocai Yu at US EPA, Dr. Jinyou Liang at Zhejiang University, and anonymous reviewers for having critically reviewed this article. This work was partly supported by Hangzhou Bureau of Science and Technology (Grant 20091133B15), China.

REFERENCES

[1] K. Huang, G. Zhuang, Y. Lin, J. S. Fu, Q. Wang, T. Liu, R. Zhang, Y. Jiang, C. Deng, Q. Fu, N. C. Hsu and B. Cao, "Typical Types and Formation Mechanisms of Haze in an Eastern Asia Megacity, Shanghai," *Atmospheric Chemistry and Physics*, Vol. 12, No. 1, 2012, pp. 105-124.

[2] B. Hou, G. Zhuang, R. Zhang, T. Liu, Z. Guo and Y. Chen, "The Implication of Carbonaceous Aerosol to the Formation of Haze: Revealed from the Characteristics and Sources of OC/EC over a Mega-City in China," *Journal of Hazardous Materials*, Vol. 190, No. 1-3, 2011, pp. 529-536.

[3] X. N. Ye, Z. Ma, J. C. Zhang, H. H. Du, J. M. Chen, H. Chen, X. Yang, W. Gao and F. H. Geng, "Important Role of Ammonia on Haze Formation in Shanghai," *Environmental Research Letters*, Vol. 6, No. 2, 2011, p. 024019.

[4] A. P. K. Tai, L. J. Mickley and D. J. Jacob, "Correlations between Fine Particulate Matter (PM$_{2.5}$) and Meteorological Variables in the United States: Implications for the Sensitivity of PM$_{2.5}$ to Climate Change," *Atmospheric Environment*, Vol. 44, No. 32, 2010, pp. 3976-3984.

[5] J. Lewtas, "Air Pollution Combustion Emissions: Characterization of Causative Agents and Mechanisms Associated with Cancer, Reproductive, and Cardiovascular Effects," *Mutation Research*, Vol. 636, No. 1-3, 2007, pp. 95-133.

[6] D. S. Grass, J. M. Ross, F. Family, J. Barbour, H. James Simpson, D. Coulibaly, J. Hernandez, Y. Chen, V. Slavkovich, Y. Li, J. Graziano, R. M. Santella, P. Brandt-Rauf and S. N. Chillrud, "Airborne Particulate Metals in the New York City Subway: A Pilot Study to Assess the Potential for Health Impacts," *Environmental Research*, Vol. 110, No. 1, 2010, pp. 1-11.

[7] A. Zanobetti and J. Schwartz, "The Effect of Fine and Coarse Particulate Air Pollution on Mortality: A National Analysis," *Environmental Health Perspectives*, Vol. 117, No. 6, 2009, pp. 898-903.

[8] J. Feng and W. Yang, "Effects of Particulate Air Pollution on Cardiovascular Health: A Population Health Risk Assessment," *Plos One*, Vol. 7, No. 3, 2012, p. e33385.

[9] Z. Q. Lin, Z. G. Xi, D. F. Yang, F. H. Chao, H. S. Zhang, W. Zhang, H. L. Liu, Z. M. Yang and R. B. Sun, "Oxidative Damage to Lung Tissue and Peripheral Blood in Endotracheal PM$_{2.5}$-Treated Rats," *Biomedical and Environmental Sciences*, Vol. 22, No. 3, 2009, pp. 223-228.

[10] Y. Wei, I. K. Han, M. Hu, M. Shao, J. J. Zhang and X. Tang, "Personal Exposure to Particulate PAHs and An-

thraquinone and Oxidative DNA Damages in Humans," *Chemosphere*, Vol. 81, No. 10, 2010, pp. 1280-1285.

[11] Y. J. Lee, Y. W. Lim, J. Y. Yang, C. S. Kim, Y. C. Shin and D. C. Shin, "Evaluating the PM Damage Cost Due to Urban Air Pollution and Vehicle Emissions in Seoul, Korea," *Journal of Environmental Management*, Vol. 92, No. 3, 2011, pp. 603-609.

[12] N. Z. Muller and R. Mendelsohn, "Measuring the Damages of Air Pollution in the United States," *Journal of Environmental Economics and Management*, Vol. 54, No. 1, 2007, pp. 1-14.

[13] M. Sillanpaa, A. Frey, R. Hillamo, A. S. Pennanen and R. O. Salonen, "Organic, Elemental and Inorganic Carbon in Particulate Matter of Six Urban Environments in Europe," *Atmospheric Chemistry and Physics*, Vol. 5, No. 11, 2005, pp. 2869-2879.

[14] Y. F. Lam, J. S. Fu, S. Wu and L. J. Mickley, "Impacts of Future Climate Change and Effects of Biogenic Emissions on Surface Ozone and Particulate Matter Concentrations in the United States," *Atmospheric Chemistry and Physics*, Vol. 11, No. 10, 2011, pp. 4789-4806.

[15] X. Y. Zhang, Y. Q. Wang, T. Niu, X. C. Zhang, S. L. Gong, Y. M. Zhang and J. Y. Sun, "Atmospheric Aerosol Compositions in China: Spatial/Temporal Variability, Chemical Signature, Regional Haze Distribution and Comparisons with Global Aerosols," *Atmospheric Chemistry and Physics*, Vol. 12, No. 2, 2012, pp. 779-799.

[16] D. McCubbin, "Health Benefits of Alternative $PM_{2.5}$ Standards," American Lung Association, Clean Air Task Force, Earthjustice, Washington, 2011. http://www.earthjustice.org/sites/default/files/Health-Benefits-Alternative-PM2.5-Standards.pdf.

[17] X. Li, L. Wang, Y. Wang, T. Wen, Y. Yang, Y. Zhao and Y. Wang, "Chemical Composition and Size Distribution of Airborne Particulate Matters in Beijing during the 2008 Olympics," *Atmospheric Environment*, Vol. 50, 2012, pp. 278-286.

[18] S.-C. Lai, S.-C. Zou, J.-J. Cao, S.-C. Lee and K.-F. Ho, "Characterizing Ionic Species in $PM_{2.5}$ and PM_{10} in Four Pearl River Delta Cities, South China," *Journal of Environmental Sciences*, Vol. 19, No. 8, 2007, pp. 939-947.

[19] D. Houthuijs, O. Breugelmans, G. Hoek, E. Vaskovi, E. Mihalikova, J. S. Pastuszka, V. Jirik, S. Sachelarescu, D. Lolova, K. Meliefste, E. Uzunova, C. Marinescu, J. Volf, F. de Leeuw, H. van de Wiel, T. Fletcher, E. Lebret and B. Brunekreef, "PM_{10} and $PM_{2.5}$ Concentrations in Central and Eastern Europe: Results from the Cesar Study," *Atmospheric Environment*, Vol. 35, No. 15, 2001, pp. 2757-2771.

[20] F. Yang, J. Tan, Q. Zhao, Z. Du, K. He, Y. Ma, F. Duan, G. Chen and Q. Zhao, "Characteristics of $PM_{2.5}$ Speciation in Representative Megacities and across China," *Atmospheric Chemistry and Physics*, Vol. 11, No. 11, 2011, pp. 5207-5219.

[21] J. C. Chow, J. G. Watson, L. C. Pritchett, W. R. Pierson,

C. A. Frazier and R. G. Purcell, "The DRI Thermal Optical Reflectance Carbon Analysis System—Description, Evaluation and Applications in United-States Air-Quality Studies," *Atmospheric Environment Part A—General Topics*, Vol. 27, No. 8, 1993, pp. 1185-1201.

[22] J. G. Watson, J. C. Chow, D. H. Lowenthal, L. C. Pritchett, C. A. Frazier, G. R. Neuroth and R. Robbins, "Differences in the Carbon Composition of Source Profiles for Diesel-Powered and Gasoline-Powered Vehicles," *Atmospheric Environment*, Vol. 28, No. 15, 1994, pp. 2493-2505.

[23] K. F. Ho, S. C. Lee, J. C. Chow and J. G. Watson, "Characterization of PM_{10} and $PM_{2.5}$ Source Profiles for Fugitive Dust in Hong Kong," *Atmospheric Environment*, Vol. 37, No. 8, 2003, pp. 1023-1032.

[24] J. C. Chow, J. G. Watson, L. W. A. Chen, W. P. Arnott and H. Moosmuller, "Equivalence of Elemental Carbon by Thermal/Optical Reflectance and Transmittance with Different Temperature Protocols," *Environmental Science & Technology*, Vol. 38, No. 16, 2004, pp. 4414-4422.

[25] J. C. Chow, J. G. Watson, L. W. A. Chen, G. Paredes-Miranda, M. C. O. Chang, D. Trimble, K. K. Fung, H. Zhang and J. Z. Yu, "Refining Temperature Measures in Thermal/Optical Carbon Analysis," *Atmospheric Chemistry and Physics*, Vol. 5, No. 11, 2005, pp. 2961-2972.

[26] H. S. El-Zanan, D. H. Lowenthal, B. Zielinska, J. C. Chow and N. Kumar, "Determination of the Organic Aerosol Mass to Organic Carbon Ratio in IMPROVE Samples," *Chemosphere*, Vol. 60, No. 4, 2005, pp. 485-496.

[27] Y. M. Han, J. J. Cao, J. C. Chow, J. G. Watson, Z. S. An, Z. D. Jin, K. C. Fung and S. X. Liu, "Evaluation of the Thermal/Optical Reflectance Method FOR Discrimination between Char- and Soot-EC," *Chemosphere*, Vol. 69, No. 4, 2007, pp. 569-574.

[28] H. S. El-Zanan, B. Zielinska, L. R. Mazzoleni and D. A. Hansen, "Analytical Determination of the Aerosol Organic Mass-To-Organic Carbon Ratio," *Journal of the Air & Waste Management Association*, Vol. 59, No. 1, 2009, pp. 58-69.

[29] X. D. Feng, Z. Dang and W. L. Huang, "Pollution Level and Chemical Speciation of Heavy Metals in $Pm_{2.5}$ during Autumn in Guangzhou City," *Huan Jing Ke Xue*, Vol. 29, No. 3, 2008, pp. 569-575.

[30] B. J. Turpin and J. J. Huntzicker, "Identification of Secondary Organic Aerosol Episodes and Quantitation of Primary and Secondary Organic Aerosol Concentrations during SCAQS," *Atmospheric Environment*, Vol. 29, No. 23, 1995, pp. 3527-3544.

[31] R. Strader, F. Lurmann and S. N. Pandis, "Evaluation of Secondary Organic Aerosol Formation in Winter," *Atmospheric Environment*, Vol. 33, No. 29, 1999, pp. 4849-4863.

[32] H. A. Gray, G. R. Cass, J. J. Huntzicker, E. K. Heyerdahl and J. A. Rau, "Characteristics of Atmospheric Organic

and Elemental Carbon Particle Concentrations in Los Angeles," *Environmental Science & Technology*, Vol. 20, No. 6, 1986, pp. 580-589.

[33] J. C. Cabada, S. N. Pandis, R. Subramanian, A. L. Robinson, A. Polidori and B. Turpin, "Estimating the Secondary Organic Aerosol Contribution to PM$_{2.5}$ Using the EC Tracer Method," *Aerosol Science and Technology*, Vol. 38, Supplement 1, 2004, pp. 140-155.

[34] S. C. Yu, "Role of Organic Acids (Formic, Acetic, Pyruvic and Oxalic) in the Formation of Cloud Condensation Nuclei (CCN): A Review," *Atmospheric Research*, Vol. 53, No. 4, 2000, pp. 185-217.

[35] S. C. Yu, R. L. Dennis, P. V. Bhave and B. K. Eder, "Primary and Secondary Organic Aerosols over the United States: Estimates on the Basis of Observed Organic Carbon (OC) and Elemental Carbon (EC), and Air Quality Modeled Primary OC/EC ratios," *Atmospheric Environment*, Vol. 38, No. 31, 2004, pp. 5257-5268.

[36] S. C. Yu, R. Mathur, K. Schere, D. W. Kang, J. Pleim, J. Young, D. Tong, G. Pouliot, S. A. McKeen and S. T. Rao, "Evaluation of Real-Time PM$_{2.5}$ Forecasts and Process Analysis for PM$_{2.5}$ Formation over the Eastern United States Using the Eta-CMAQ Forecast Model during the 2004 ICARTT Study," *Journal of Geophysical Research-Atmospheres*, Vol. 113, No. D6, 2008, Article No. D06204.

[37] S. Yu, P. V. Bhave, R. L. Dennis and R. Mathur, "Seasonal and Regional Variations of Primary and Secondary Organic Aerosols over the Continental United States: Semi-Empirical Estimates and Model Evaluation," *Environmental Science & Technology*, Vol. 41, No. 13, 2007, pp. 4690-4697.

Permissions

The contributors of this book come from diverse backgrounds, making this book a truly international effort. This book will bring forth new frontiers with its revolutionizing research information and detailed analysis of the nascent developments around the world.

We would like to thank all the contributing authors for lending their expertise to make the book truly unique. They have played a crucial role in the development of this book. Without their invaluable contributions this book wouldn't have been possible. They have made vital efforts to compile up to date information on the varied aspects of this subject to make this book a valuable addition to the collection of many professionals and students.

This book was conceptualized with the vision of imparting up-to-date information and advanced data in this field. To ensure the same, a matchless editorial board was set up. Every individual on the board went through rigorous rounds of assessment to prove their worth. After which they invested a large part of their time researching and compiling the most relevant data for our readers. Conferences and sessions were held from time to time between the editorial board and the contributing authors to present the data in the most comprehensible form. The editorial team has worked tirelessly to provide valuable and valid information to help people across the globe.

Every chapter published in this book has been scrutinized by our experts. Their significance has been extensively debated. The topics covered herein carry significant findings which will fuel the growth of the discipline. They may even be implemented as practical applications or may be referred to as a beginning point for another development. Chapters in this book were first published by Scientific Research Publishing Inc.; hereby published with permission under the Creative Commons Attribution License or equivalent.

The editorial board has been involved in producing this book since its inception. They have spent rigorous hours researching and exploring the diverse topics which have resulted in the successful publishing of this book. They have passed on their knowledge of decades through this book. To expedite this challenging task, the publisher supported the team at every step. A small team of assistant editors was also appointed to further simplify the editing procedure and attain best results for the readers.

Our editorial team has been hand-picked from every corner of the world. Their multi-ethnicity adds dynamic inputs to the discussions which result in innovative outcomes. These outcomes are then further discussed with the researchers and contributors who give their valuable feedback and opinion regarding the same. The feedback is then collaborated with the researches and they are edited in a comprehensive manner to aid the understanding of the subject.

Apart from the editorial board, the designing team has also invested a significant amount of their time in understanding the subject and creating the most relevant covers. They scrutinized every image to scout for the most suitable representation of the subject and create an appropriate cover for the book.

The publishing team has been involved in this book since its early stages. They were actively engaged in every process, be it collecting the data, connecting with the contributors or procuring relevant information. The team has been an ardent support to the editorial, designing and production team. Their endless efforts to recruit the best for this project, has resulted in the accomplishment of this book. They are a veteran in the field of academics and their pool of knowledge is as vast as their experience in printing. Their expertise and guidance has proved useful at every step. Their uncompromising quality standards have made this book an exceptional effort. Their encouragement from time to time has been an inspiration for everyone.

The publisher and the editorial board hope that this book will prove to be a valuable piece of knowledge for researchers, students, practitioners and scholars across the globe.

List of Contributors

Min Zhao
Provincial Key Laboratory of Agricultural Environmental Engineering, College of Resources and Environment, Sichuan Agricultural University, Chengdu, China

Li Li, Zhilin Liu, Bin Chen, Jianqiu Huang, Jinwang Cai and Shihuai Deng
Meishan Environmental Monitoring Center, Meishan, China

P. R. Jayakrishnan and C. A. Babu
Department of Atmospheric Sciences, Cochin University of Science and Technology, Cochin, India

Hironori Sato and Toshitaka Suzuki
Department of Earth and Environmental Sciences, Faculty of Science, Yamagata University, Yamagata, Japan

Motohiro Hirabayashi, Hideaki Motoyama and Yoshiyuki Fujii
National Institute of Polar Research, Tokyo, Japan

Yoshinori Iizuka
Institute of Low Temperature Science, Hokkaido University, Sapporo, Japan

Wei Li
Graduate School of Environmental Studies, Nagoya University, Nagoya, Japan

Tetsuya Hiyama
Research Institute for Humanity and Nature, Kyoto, Japan

Nakako Kobayashi
Hydrospheric Atmospheric Research Center, Nagoya University, Nagoya, Japan

Antonio de la Casa, Gustavo Ovando, Luciano Bressanini and Jorge Martínez
Facultad de Ciencias Agropecuarias (FCA), Universidad Nacional de Córdoba (UNC), Córdoba, Argentina

Christiana F. Olusegun
Department of Physical and Chemical Sciences, Elizade University, Ilara-Mokin, Nigeria

Zachariah D. Adeyewa
Department of Meteorology, Federal University of Technology, Akure, Nigeria

Wei Li
Department of Marine, Earth, and Atmospheric Sciences, North Carolina State University, Raleigh, USA

Jean-Louis Zerbo
Université Polytechnique de Bobo-Dioulasso, Bobo-Dioulasso, Burkina Faso
LPP-Laboratoire de Physique des Plasmas/UPMC/Polytechnique/CNRS, Saint-Maur-des-Fossés, France
Laboratoire d'Energies Thermiques Renouvelables (L.E.T.RE), Université de Ouagadougou, Ouagadougou, Burkina Faso

Frédéric Ouattara
Ecole Normale Supérieure de l'Université de Koudougou, Koudougou, Burkina Faso

Jean-Pierre Legrand
Institut National des Sciences de l'Univers (INSU), Paris, France

John D. Richardson
Kavli Institute for Astrophysics and Space Research, Massachusetts Institute of Technology, Cambridge, USA

Christine Amory Mazaudier
LPP-Laboratoire de Physique des Plasmas/UPMC/Polytechnique/CNRS, Saint-Maur-des-Fossés, France

Marcelo de Carvalho Alves
Soil and Rural Engineering Department, Federal University of Mato Grosso, Mato Grosso, Brazil

Luciana Sanches
Department of Sanitary and Environmental Engineering, Federal University of Mato Grosso, Cuiabá, Brazil

José de Souza Nogueira and Vanessa Augusto Mattos Silva
Department of Physics, Federal University of Mato Grosso, Mato Grosso, Brazil

Isidoro Orlanski
Atmospheric and Ocean Science Program, Princeton University, Princeton, USA

Khalid M. Malik
National Agromet Center, Pakistan Meteorological Department, Islamabad, Pakistan

Peter A. Taylor
York University, Toronto, Canada

Kit Szeto
Climate Research Division, Environment Canada, Toronto, Canada

Azmat Hayat Khan
National Drought Monitoring Center, Pakistan Meteorological Department, Islamabad, Pakistan

Mojtaba Zoljoodi and Ali Didevarasl
Atmospheric Sciences and Meteorological Research Centre (ASMERC), Tehran, Iran

Igor Mingalev, Galina Mingaleva and Victor Mingalev
Polar Geophysical Institute, Kola Scientific Center of the Russian Academy of Sciences, Apatity, Russia

Jorge F. Carrasco
Dirección Meteorológica de Chile, Estación Central, Chile
Centro de Estudios Científicos, Valdivia, Chile
Universidad de Magallanes, Punta Arenas, Chile

Nikolai Romanov and Vasiliy Erankov
Institute of Experimental Meteorology, FBSI RPA "Typhoon", Obninsk, Russia

Alessandro Attanasio and Antonello Pasini
Institute of Atmospheric Pollution Research, National Research Council (CNR), Monterotondo Stazione, Rome, Italy

Umberto Triacca
Department of Computer Engineering, Computer Science and Mathematics, University of L'Aquila, L'Aquila, Italy

Vincent Courtillot, Jean-Louis Le Mouël, Dominique Gibert and Fernando Lopes
Geomagnetism and Paleomagnetism, Institut de Physique du Globe de Paris, Université Paris Diderot, Sorbonne Paris Cité, 1 rue Jussieu, Paris, France

Vladimir Kossobokov
Geomagnetism and Paleomagnetism, Institut de Physique du Globe de Paris, Université Paris Diderot, Sorbonne Paris Cité, 1 rue Jussieu, Paris, France
Institute of Earthquake Prediction Theory and Mathematical Geophysics, Russian Academy of Sciences, Moscow, Russia

Michael Richter and Sonja Deppisch
Urban Planning and Regional Development, Hafen City University, Hamburg, Germany

Hans Von Storch
Institute for Coastal Cesearch, Helmholtz-Zentrum, Geesthacht, Germany

Franco Belosi, Gianni Santachiara and Franco Prodi
Institute of Atmospheric Sciences and Climate (ISAC), National Research Council, Bologna, Italy

Reginald R. Muskett
Geophysical Institute, University of Alaska Fairbanks, Fairbanks, USA

Guojin Sun
Department of Chemical and Biological Engineering, Zhejiang University, Hangzhou, China
Zhejiang Environmental Protection Agency, Hangzhou, China

Yao Shi, Qingyu Zhang, Mengna Tao, Guorong Shan and Yi He
Department of Chemical and Biological Engineering, Zhejiang University, Hangzhou, China

Lin Yao and Li Jiao
Environmental Monitoring Center, Hangzhou, China